Mike

207-602-257C

U.N.E.

Global Insecurity

Global Insecurity

A Strategy for Energy and Economic Renewal

Edited by
Daniel Yergin and
Martin Hillenbrand

1982
HOUGHTON MIFFLIN COMPANY
BOSTON

Library of Congress Cataloging in Publication Data
Main entry under title:

Global insecurity.

Developed from a research project sponsored by the
Atlantic Institute for International Affairs.
Includes bibliographical references and index.
Contents: Crisis and adjustment / by Daniel Yergin —
World energy to the year 2000 / by Robert Stobaugh —
The bedeviled American economy / by Robert Dohner —
[etc.]
1. Energy policy. 2. Economic policy. 3. World
politics. I. Yergin, Daniel. II. Hillenbrand, Martin
Joseph, date. III. Atlantic Institute for
International Affairs.
HD9502.A2G55   338.9   82-3013
ISBN 0-395-30517-9   AACR2

Printed in the United States of America

D 10 9 8 7 6 5 4 3 2 1

To Ulf Lantzke,
executive director of the
International Energy Agency,
who has recognized the dangers ahead

Contents

Contents

Preface

A decade or so ago, high and growing levels of energy consumption were something to be applauded, an indicator of the advance of civilization. Today, such levels are as much a source of worry and insecurity as of satisfaction. So pervasive is the importance of energy in modern life that the insecurity extends beyond concerns about the price and availability of energy to fundamental questions about the possibilities for sustained economic growth and the stability of society, and about war and peace.

This is not a book about energy itself, but about the economic, social, and political consequences of the energy problem. We look not only to the recent past, but also to the future in order to ask how these consequences might be felt in the rest of the 1980s and the 1990s. Our aim is not to offer specific forecasts, but rather to provide a framework for understanding the challenges and choices that the people of the world will face.

Global Insecurity is the result of a four-year research project. Even this relatively brief period has spanned four energy eras — an oil "glut," a major oil shock, a minor oil shock, and another "glut" — all of which have been accompanied by rapid shifts in public attitudes and a great deal of confusion. Indeed, the one sure glut has been that of confusion.

In this work, we try to clear up some of that confusion with what we hope is a coherent and reasonable interpretation.

After 1973 the easy assumptions about limitless cheap energy to fuel economic growth were suddenly dispelled. This change came as a great jolt to the industrial world. It may well have brought full employment to an end, at least in the United States and Western Europe. It certainly brought to the fore grave new security problems. Still, there has been a great temptation to wish the problem away every time the pressure eases a bit; many forget that the easing is bought at the expense of hundreds of billions of dollars of lost economic growth. This is why, during the complacencies and misconceptions that a period of oil surplus encourages, calling attention to realities takes on even greater importance. But this book also aims to point in the directions in which solutions can be found — or, if not solutions, then at least modes of adjustment likely to be less painful than those resulting from neglect or error.

Two basic premises inform this study. One is the need to maintain economic growth. The other is the reasonable probability of recurrent energy problems, which can exact heavy tolls. The interaction of these two issues in the 1980s and 1990s raises serious concern about the social and political stability of democratic systems, about the limits of stress that such systems can endure, and about the potential for conflict.

This work is deliberately international in scope, involving thirteen experts from the United States, Western Europe, and Japan. It is written not only for specialists, but also for decision-makers in the private and public sector, faced with conflicting priorities and dilemmas, and also for citizens understandably perplexed by the developments of recent years and even more perplexed about what to expect in the future. We seek to answer such critical questions as:

How have energy problems affected the political, economic, and social systems of the world? How are these problems interacting with such basic trends as the evolution of technology, shifts in the balance of power, and other kinds of shifts as well — for instance, in values between generations? How might the likely evolution of energy supply make its impact felt through the rest of this century? Need the consequences be more inflation and higher unemployment and continued low growth, accompanied by rising discontent and increased domestic and international tension? What are the dangers and risks both to national stability and to the international order?

As these questions indicate, we are trying to make clear the underlying connections and relations that are so often obscured or totally ignored. In other words, we are offering an antidote to the fragmentary thinking characteristic of so much of the response to the energy problem over the last ten years. For instance, it has been very difficult for many to recognize the primary role the oil shocks have played in the economic downturn, or to see America's economic discontents in a global context, or to relate the imperatives of foreign and security policy to domestic energy policy.

We also hope that this multidisciplinary study will contribute to a deeper understanding of the respective countries and societies discussed, as well as of the changing international order. The role of energy in our lives goes beyond its role as a factor of production. In numerous and frequently unappreciated ways, energy has helped determine many aspects of the way we live. Energy developments during the years ahead could, depending on the adjustment process, change daily life. Some of the changes could be quite minor. Others, if the adjustment process proves difficult or inadequate, could be major and quite uncomfortable.

Our focus is on those countries known as the "consumers" — the United States, Western Europe, Japan, and the oil-importing developing world. It became obvious early on that to focus only on the industrial countries was to deal only with part of the problem. The developing world also directly and seriously felt the impact of developments in oil prices and supplies. Future energy developments will have a major, even a decisive, impact on these countries as they attempt to struggle, perhaps under increasingly unfavorable circumstances, out of poverty and economic backwardness.

A quick word on the plot. Chapter 1 is an overview that sets out the themes. Chapter 2 presents two realistic scenarios for energy supply in the rest of this century. We have tried to separate what is practical and likely from what people wish for. The results do suggest further constraints ahead. Chapter 3 explores the strong connection between energy difficulties and the stubborn problems of inflation and low economic growth in the United States. Chapter 4 examines how the energy issue has challenged and might continue to challenge America's political system and its society. Chapters 5 and 6 follow in the same sequence for Japan, as do Chapters 7 and 8 for Western Europe. The onerous burdens and lost opportunities imposed on the developing countries by the energy problem provide the subject of Chapter 9. In Chapter 10, we analyze the difficulties that have developed in the international payments and trading systems and the stimulus given to a new protectionism. The strains and demoralization in the Western alliance are discussed in Chapter 11, and Chapter 12 focuses on the dangers for the entire international order.

Not everything can be covered in one book, and so we leave for subsequent research two important subjects: the consequences of the energy question, first, for the Soviet Union and its bloc and, second, for the countries that make up the Organization of Petroleum Exporting Countries. Some of the latter group are going through the most massive and rapid process of modernization in world history. We deal with their problems and choices at some length in Chapters 1 and 10, but the subject requires a considerable research project of its own.

Global Insecurity points to the very real risks and dangers ahead. The costs of two oil shocks have already been great; the cost of future constraints and crises could be even greater. We hope to stimulate an international debate that will help to avoid or at least minimize the risks. This is an optimistic work, for our conclusion is that a reasonable adjustment, while not foreordained by any means, is certainly possible. That, then, is the theme of this work — crisis and adjustment, and the race between the two.

This work has been sponsored by and conducted under the auspices of the Atlantic Institute for International Affairs in Paris. It has been cosponsored by the Harvard University Energy Security Program and conducted in collaboration with the Energy and Environmental Policy Center of Harvard University's Kennedy School of Government in Cambridge, Massachusetts; the Royal Institute of International Affairs in London; the Deutsche Gesellschaft für Auswartige Politik in Bonn; and the Istituto Affari Internazionali in Rome.

It would not be feasible to list everyone who gave advice, attended conferences, or otherwise contributed to the completion of this work, but we do thank them all. Certain acknowledgments must be made, however.

Our gratitude to the other members of the Steering Committee created by the Atlantic Institute for this project: Ulf Lantzke, executive director of the International Energy Agency; Robert Belgrave, formerly trading director of British Petroleum and now director of the British Institute's Joint Energy Policy Program; Karl Kaiser, director of the Deutsche Gesellschaft für Auswartige Politik; Cesare Merlini, president of the Istituto Affari Internazionali; and Ian Smart, formerly deputy director of the Royal Institute of International Affairs and now an independent consultant. The committee met a number of times to discuss the organization of the project and then specific chapters. During the course of this project, the Atlantic Institute was able to take advantage of its proximity to the International Energy Agency to seek advice. Dr. Lantzke, J. Wallace Hopkins, the deputy executive director, William Martin, and other members of the IEA staff were most helpful.

Daniel Yergin acknowledges with great gratitude the support he received as a Fellow of the German Marshall Fund of the United States, which allowed him to devote himself almost full time to this project for the last two years.

Our deep thanks to the Keidanren of Japan, the United States Department of Energy, the German Federation of Industry, the al Dir'iyyah Institute, the Rockefeller Foundation, the Industrial Federation of the Netherlands, the Ford Foundation, the AMAX Foundation, British Petroleum, the Janss Foundation, the Volkswagen Foundation, and the City of Berlin, all of which helped to fund the project and its various meetings, which included a major conference in Berlin. Without such generous support, this book would never have been completed — indeed, never attempted. None of them, however, bears responsibility for the contents of the various chapters. That solely belongs to the respective authors and to the editors.

Our appreciation —

To Gregory Flynn and Fabio Basagni, assistant directors of the Atlantic Institute, who helped structure and manage the project.

To Robert Stobaugh, director of the Energy Research Project at the Harvard Business School, whose continuing advice and analysis helped define and shape the entire study.

To William Hogan and Henry Lee of the Energy and Environmental Policy Center, which provided a home on the American side of the Atlantic for this project, and to Alvin Alm, director of the Harvard Energy Security Program, which provided a context.

To Austin Olney, editor in chief at Houghton Mifflin, for his guidance, support, and careful attention.

To Jane Shorall, who expertly coordinated this project and helped edit the manuscript.

To Barbara Weisel, Benjamin Brown, and James Rosenfield, who assisted with the research. To Jane Kumin, for careful translation. To Helen Brann, for her customary expertise. To Luise Erdmann, who gracefully copyedited the manuscript. To the Salzburg Seminar, directed by John Tuthill; the Lehrman Institute's Foreign Policy Roundtable, chaired by Nicholas X. Rizopoulos; and the Conference on Energy and Social Adaptation, chaired by Dorothy Zinberg—where parts of this book were rehearsed.

Finally, we must thank the authors, who took a long view when many others did not, and who patiently and with good humor put up with editors who demanded much more than they had bargained for or could reasonably have anticipated.

Daniel Yergin   *April 1982*
  Cambridge

Martin Hillenbrand
  Paris

Global Insecurity

1

Crisis and Adjustment: An Overview

by Daniel Yergin

The energy question is really a question about economic growth and security, which in turn means it is a question about the future of Western society. The relation of energy to economics and to security is a most appropriate issue to take up in a time much troubled by inflation and unemployment, when confidence is eroding both in the economic machine itself and in the competence of government — or, indeed, in anybody else — to do much to improve its performance. The central role of economic growth for democratic societies is inescapable. It makes possible rising real incomes for people. It means jobs rather than unemployment. It provides opportunities and a reason for optimism and confidence rather than for disenchantment and despair. It helps to resolve conflicts and tensions within the framework of democratic society. And it allows nations to harmonize their goals, laying the basis for common purpose rather than conflict.

These are fundamentals. Stagnation and unemployment and depression sorely tested democratic systems in the years between World War I and World War II. The disaffected turned to political idols on the right and the left, and in some countries democracy gave way to dictatorship, and dictatorships turned to war. The ultimate toll was measured in tens of millions of deaths and in the endless stretch of devastated landscape.

It was hardly surprising, therefore, that economic growth was at the top of the national and international agenda at the end of World War II. For it was essential to demonstrate that democratic societies, based on market systems, could deliver the goods — jobs and opportunities and rising incomes — without the dislocations of inflation and monetary instability.

That may have been the hope; yet, with the memory of interwar stagnation and unemployment still fresh in people's minds and the tasks of reconstruction so large before their eyes, the hope had to be kept within modest bounds.[1] As it happened, the actual results went far beyond what anybody might have reasonably anticipated. We can look back on a very impressive — some might even say stunning — achievement, and one in which great pride can be taken. Between 1950 and 1973, Japan's gross national product increased tenfold; Western Europe's, almost three and a half times; and the United States', almost two and a half times.

Many factors drove this sustained surge of growth: the momentum of postwar reconstruction and recovery, successive waves of trade liberalization, the stability provided by an international economic regime based on the dollar, European integration, technological innovation and higher productivity, managerial dynamism, various government policies — and energy, in particular, oil.

The Oil Way of Life

For oil, increasingly cheap and available, was a key ingredient, truly the fuel of economic growth. Between 1950 and 1973, petroleum reserves in the free world increased eightfold. Almost 90 percent of that growth was in the Middle East and Africa, where the costs of extraction were very low. This increasing abundance was reflected in the price of oil, which, in real terms, declined by 50 percent during those years. Industrial and individual consumers responded, and some governments encouraged switching to oil as a way both to modernize their industries and to escape the social problem of coal mining. The growth was awesome; oil consumption in the free world increased fivefold between 1950 and 1973, and oil came to play a larger and larger role in the total energy mix (see Table 1.1). To give one example, between 1950 and 1973, the share of coal — the traditional fuel

Table 1.1 The Structure of Energy Consumption in the Industrial World*

	1950	1973	1980
Oil	29%	52%	47%
Natural Gas	12	23	24
Coal	57	22	25
Primary Electricity*	2	3	4

* hydropower, nuclear, geothermal
Source: World Energy Industry Information Services.

of industrial society — in Western Europe's total energy mix declined from 86 to 25 percent, while oil's share rose from 12 to 59 percent.[2]

But there were limits to the available supply of oil. In the late 1960s and early 1970s, the demand for oil had been growing very rapidly because of economic growth, because oil was cheap, and because oil burns more cleanly than coal. This interest had been accentuated by a growing concern about air quality in the industrial world, which provided yet a further reason to replace coal with oil. In 1970, oil production in the United States reached its peak, then began to decline. Meanwhile, consumption continued to grow, and the United States began to draw heavily on the world oil market to meet its requirements. At the same time, market conditions made it possible for the oil-exporting countries to win control of oil production from the international companies.

The Cost of the Oil Shocks

A year of "almost universal boom" is how 1973 has been described. It was also a watershed in modern economic history. The world's production of goods and services grew almost 7 percent in that year, the last year of that kind of growth, with the result that the demand for oil strained the supply system. At the same time, the high concentration of oil production and reserves in the Middle East made the overall supply system highly vulnerable to "accidents," particularly political accidents. In 1973, the accident was the October War, which interacted with the basic market conditions to set off the first oil shock — a fourfold increase in the price of oil. Five years later, in the midst of a widespread complacency about a permanent oil glut, the shah of Iran was ousted from the Peacock Throne, setting off the second oil shock — this time raising the price of oil to two and a half times its former level. Though not widely recognized, in absolute terms, the second shock was actually more significant than the first. In 1973–74, the price of oil went up about $8 a barrel; in 1979–80, the increase was about $21.

With oil so important in the economy, the two oil shocks inevitably had to have a dramatic impact upon the fortunes of an industrial world that had become dependent upon this fuel. And these shocks did have pervasive effects, both in that which can be measured — inflation, recession, and unemployment — and that which cannot so easily be measured — eroding confidence and growing discontent.

These are the consequences we struggle with today. The oil shocks appear to have ended the era of high growth and full employment — what has been called the era of "flamboyant growthmanship." In its place, they have initiated a new and uncertain and uncomfortable era of "stagflation," a dual visitation of high inflation and low growth.[3]

After 1979, national concern was primarily on inflation, which engenders insecurity and bitterness, devalues savings and currencies, destabilizes the social order, and shortens the time horizon against which people spend and invest. But, very recently, attention has shifted to the issue of employment.

While the goal of full employment may always have extended somewhat further than the reality, there was still a rough approximation of full employment throughout the Western world. But now there is increasing evidence that the goal has been forsaken both in the United States and in Western Europe. Only Japan still delivers on the promise, although there are those who ask how long Japan can maintain its commitment. This abandonment is a major development; but its implications have hardly begun to be addressed, so intense has been the attention given to inflation.

Certainly our intention is not to suggest that growing economies must be a thing of the past. Rather, it is to point out the considerable challenge: Industrial economies that advanced so far on the basis of cheap and easy oil in the postwar years have to find ways to change and adjust in order to resume stable economic growth in a world in which oil is expensive and not so secure. To be sure, it would be foolish, a case of tunnel vision, to attribute all the unpleasant economic news of recent years to the oil shocks. World inflation was already gathering momentum in the early 1970s in response to such factors as food prices and a commodity boom, declining productivity, the financing of the Vietnam War, and growing rigidity in the labor markets — perhaps even in response to rapid growth itself.[4] The Bretton Woods international monetary system was under pressure. Even in Japan there was a growing concern about the "refraction" of its remarkable growth trend. Yet perspective is required. The inflation rate that terrified the Nixon administration into wage and price controls in August 1971 was just 3.8 percent.[5]

Indeed, the two surges in oil prices clearly have been a driving force behind today's stagflation. Curiously, though, some analysts have gone into very considerable intellectual acrobatics to deny this obvious reality, with the result that a part of recent economic debate has been strangely irrelevant.

The rapid rise in oil prices has driven inflation in several ways. It has, most obviously, affected the price index directly. So pervasive is the place of oil in the economy that its price has also had a significant impact on the cost of much partly finished and finished production. The prices of substitutes or alternatives have also been bid up, whether they be coal or housing insulation or Hondas. Finally, increases in oil prices can, through compensating wage hikes and shifts in expectations, become embedded in the "home-grown" or underlying inflation rate.

At the same time, the oil shocks have extracted a considerable toll on economic growth through severe recessions and economic slowdown. This has occurred first through what has been called the OPEC tax, the purchasing power that is withdrawn from the domestic economy by the oil exporters. The International Monetary Fund has described the workings of this tax as a "real resource transfer of unprecedented magnitude."[6] In other words, the income of the population of the industrial countries is reduced, as is aggregate domestic demand, with a consequent lowering of economic activity and a rising of unemployment. The oil shocks have also been a major cause of the productivity slowdown, which in turn has contributed to the economic slowdown. The fact that the recessions were worldwide reinforced their impact in each country.[7]

These dual effects — higher inflation and lower growth — have confronted policy-makers in the Western world with a difficult choice: To aim first at reducing inflation — to use restrictive monetary and fiscal policies to try to prevent the oil price increases from becoming embedded in the underlying inflation rate. Or to aim first at reducing unemployment — to pursue stimulative fiscal and monetary policies to balance the depressive effects of the OPEC tax.

The dilemma is obvious. The first option, to fight inflation, reinforces the downturn brought on by the OPEC tax. The second option, to fight unemployment, could provide further fuel for inflation. The fury and smoke of the battle among competing economic models in recent years is testament to the extent — and the pain — of the dilemma.

Nevertheless, the general trend of policy has been to focus first on inflation as the more immediate and dangerous threat. This has resulted in a historic retreat from the postwar commitment to full employment. Meanwhile, the general stagflation — plus high interest rates, uncertainty, and increased energy costs themselves — has retarded the long-term investment required for higher productivity.

The effects of the oil shocks on the "big seven" industrial nations are illustrated in Figure 1.1 (and laid out in the appendix). The oil shocks cost these countries $1.2 trillion in lost economic growth (comparing the 1966–73 period to 1974–81).[8] Cumulative inflation was almost three times as high in the aftermath of the oil shocks as before, and the average unemployment rate doubled. In addition, the growth of world trade registered a striking decline. The record has been getting worse in the last few years. The growth rate for the entire industrial world, which averaged 5 percent in the 1960s, declined to a little over 1 percent in 1980 and 1981 and will probably be no better in 1982. Unemployment in Europe, which averaged 2.8 percent between 1966 and 1973, reached 9 percent by the spring of 1982, the highest level since the reconstruction period after World War II, and was at about the same level in the United States. Full employment — the ambition that had become a near reality in the years of high growth — has, at least for the time being, come to an end.

While the oil shocks have caused considerable difficulties for the indus-

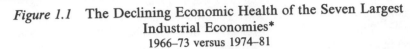

Figure 1.1   The Declining Economic Health of the Seven Largest
Industrial Economies*
1966–73 versus 1974–81

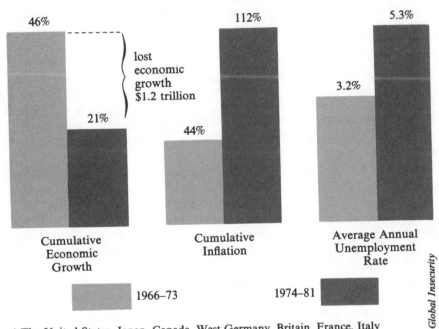

* The United States, Japan, Canada, West Germany, Britain, France, Italy

trial nations, no countries have suffered more than the oil-importing developing countries. There is more than one irony in this. To begin with, they are very modest players in the world oil market. Moreover, the majority of them — "trapped in the solidarity of the decolonized," as one Third World leader has expressed it — have tended publicly to blame the industrial world for their recent economic problems and only in private to acknowledge the heavy weight of the oil price shocks. But this weight has proved devastating. The rise in real oil prices has imposed enormous pressures on the balance of payments of these countries, creating what the World Bank has described as "unsustainable trade deficits." Their oil bill rose from $4.8 billion in 1973 to $47.2 billion in 1981 and now accounts for a quarter of their total imports in value terms. Lower growth in the industrial world has reduced export opportunities, and thus much of their oil deficit has been financed by a stunning rise in indebtedness — their long-term debt increasing from $97 billion in 1973 to $425 billion in 1981.

Taken together, these trends have seriously impaired economic performance in the developing world. Only the availability of foreign financing, and in particular bank lending, has kept a bad situation from getting worse. The burden has fallen unevenly. A handful, known as the newly industrializing countries, have been able to maintain brisk growth rates by substantially expanding their exports, particularly manufactured goods, into the burgeoning OPEC market. Leaving out the new industrializers makes for a grimmer picture; the average annual GNP growth per person declined from 1.8 percent in the 1960s to 0.8 percent in the 1970s in the rest of the oil-importing developing countries. The weight of debt makes further borrowing more difficult, which puts even more pressures on incomes and development. In the words of the World Bank, the burden of energy costs could mean for many developing countries "the difference between rising and stagnating income per person."[9]

But the energy problems of the developing world do not stop with the oil deficit. Wood and other traditional fuels provide a quarter of its total energy consumption, and the high price of imported energy has increased the pressure on these resources, including an acceleration of deforestation, thus adding another dimension to the energy problem.

The unhappy set of economic circumstances set in motion by the oil shocks contains the potential for far-reaching crises. In the industrial nations, high inflation, low growth, and high unemployment can erode the national consensus and undermine the stability and legitimacy of the political system. In the developing world, zero growth leads to misery and upheavals. Protectionism and the accumulation of debt threaten the international trade and payments system. And, of course, there is the tinder of international politics, particularly involving the Middle East, where political and social upheavals can cause major oil disruptions and where fears about and threats to energy supplies can lead to war.

The Theme of Adjustment

But this gloomy picture is incomplete. Impaired economic performance and decay need not be the only response to the oil shocks. For modern economies, unlike traditional subsistence economies, are not static; they are changing all the time in response to new forces. And, with the passing of almost a decade since the first oil shock, we can observe that a profound and promising process of adjustment is taking place. Perhaps it began when the prices first began to tighten, even before the 1973 embargo. But its considerable extent and potential have only recently become visible, partly in response to the second round of price hikes, but also because there is an inevitable lag between germination and harvest. Enough is now clear that we can argue that adjustment is one of the central energy themes of the 1980s. It is certainly one of the main themes of this book.

What do we mean by adjustment? Low economic growth has been described as one form of adjustment, admittedly the least desirable form — "when all else fails."[10] But we can refine the matter. In our view, adjustment should be conceptualized as the *alternative* to reduced growth and as the *requirement* for sustained growth. Indeed, adjustment may be thought of as the "all else." It is a reorientation of key patterns related to energy — of energy production and consumption themselves; of trade and payments; of social, political, and security relationships; and, indeed, of attitudes and values — that encourages more satisfactory and stable economic prospects.

In terms of energy production and consumption, adjustment takes three forms: diversification away from OPEC to other sources of oil; substitution of other energy sources for oil; and greater energy efficiency, otherwise known as conservation. In each category, there have been significant developments.

Three major sources of oil outside OPEC — the North Sea, the North Slope of Alaska, and a new oil province in Mexico — became available in the second half of the 1970s. The first two, at least, might well not have been economic at pre-embargo prices. Together, the three added about 6 million barrels a day to world supplies between 1973 and 1981. This diversification adds an important element of stability to the world oil market. Another important form of diversification was the growth of strategic or national oil stockpiles, which are to be used as buffers in a crisis.

The long-term trend toward increased dependence on oil was reversed after 1973. While oil consumption remained flat in the industrial world in 1979 and declined by about 7 to 8 percent in 1980 and by another 5 to 6 percent in 1981, non-oil energy consumption increased by 5 percent in 1979 and by 2 percent in both 1980 and 1981. In some countries, particularly in France and Japan, nuclear power made an increasing contribution. In the United States, Germany, and other countries, plans for extensive nuclear expansion have been stymied by environmental and safety concerns, economics, and by uncertainty about the growth in the demand for electricity.

The most important substitution has resulted in a reversal in the long-term decline in coal. It has hardly been as smooth or as rapid as some might have believed in the 1970s, but it is definitely taking place. The most dramatic shift has been in Japan, both in electric generation and in industry. Japanese coal imports increased by 25 percent between 1977 and 1980. International trade in coal has been substantially expanding. In 1981, 255 million metric tons of coal moved in international commerce — a 30 percent growth over 1978.[11]

But by far the most important and powerful form of adjustment has been on the demand side, in terms of greater energy efficiency. Indeed, it has proved much more potent than most analysts would have expected even in the middle 1970s. Between 1968 and 1973, total primary energy consumption in the industrial economies had grown at an average annual rate of 5.6 percent. Between 1973 and 1979, this rate slowed to an average of 1.2 percent per year.[12] Some of this decline was in response to the reduction of economic growth, certainly. But some of it represented a much more hopeful process — the partial decoupling of energy consumption and economic growth; in other words, the discovery that there is much greater flexibility in how energy is used than had been conventionally expected.

One can see why conventional wisdom was unprepared. In the 1960s, GNP and energy consumption grew pretty much at the same rate. Only after 1973 did this relationship break down. Indeed, the divergence was such that by 1981, the industrial world consumed about 5 percent more energy than it had in 1973, despite real GNP growth of about 21 percent. In other words, the industrial world used 16 percent less energy in 1981 than would have been the case if the previous one-to-one relationship had been maintained. This achievement has been, in part, at the expense of some industries and some jobs. Such heavy energy users as steel and aluminum have declined. During a period of recession, the less efficient, older units are not operated, leading to a better ratio of energy to production.

Still, the broad trend is clearly toward greater efficiency. The record of the individual industrial countries is quite impressive (see Table 1.2). (Canada is an exception, presumably because of price controls and the buildup of an energy-intensive petrochemical industry in Alberta.) The decoupling spread across all sectors — changes in building practices, behavioral changes of homeowners, more efficient industrial processes and better industrial housekeeping, greater efficiency in transportation. Indeed, since one out of nine barrels of oil used in the world in the late 1970s was burned as gasoline on American highways, the legislation mandating greater efficiency in the American automobile fleet, combined with the change in consumer preference, in itself will have a continuing strong influence on the world energy balance.

The effects of both adjustment and economic downturn were felt on the world oil market in the early 1980s. The OPEC producers saw demand for their product fall precipitously by 40 percent — from 31 million barrels a day (mbd) in 1979 to less than 18 mbd in early 1982. As revenues fell below expenditures, this created a kind of oil crisis for the petroleum exporters, forcing many of them to cut back on development plans and to increase borrowing in international capital markets. One country, Saudi Arabia, had consistently tried to put the brakes on the continuing drive by other producers to keep pushing up the price. With the largest oil reserves in the world, Saudi Arabia had reason to be most concerned about how further adjustment might erode the long-term market for OPEC oil. Higher prices,

Table 1.2 Index of Energy Consumption
(Ratio to real GNP)

	U.S.	Japan	France	West Germany	Italy	Britain	Canada
1973	100	100	100	100	100	100	100
1980	86	82	90	85	88	88	103

Note: This index means that if it took 100 barrels of oil equivalent to produce a given amount of GNP in 1973, that same amount was, for instance, produced in the United States by 86 barrels in 1980.
Source: International Monetary Fund, *World Economic Outlook,* June 1981, p. 147.

the Saudis recognized, would encourage more conservation, more fuel switching, more non-OPEC oil development, and more investment in alternative energy sources.[13] The OPEC countries would pay a price for their own prices.

Most of the other producers, however, ignored such warnings and instead competed among themselves in raising prices and adding premiums — until the warning came true. In fact, the decline in the demand for OPEC oil was far greater than most had anticipated. These new circumstances created great tensions among the producers, and now they competed not in raising prices but in cutting prices and offering discounts. The situation raised questions as to whether the producers could maintain their unity under such pressures. Moreover, another deeply disquieting question awaited them in the future: How could the already saturated market accommodate a significant increase in production by Iran and Iraq that would accompany an end to the war between those two countries?

Still, the future was not as bleak for the OPEC countries as portrayed in the first months of 1982. For it was not only adjustment, but also the downturn in the world economy that was responsible for the reduced demand for OPEC oil, the energy source whose consumption is most sensitive to the level of economic activity. And economic recovery would stimulate demand for OPEC oil again. In addition, high interest rates and sluggish demand had led to a rapid running down of petroleum inventories, which had been rapidly built up by companies, especially during the peak year of 1979. This inventory depletion exaggerated the actual drop in demand for OPEC oil, perhaps by two or three million barrels a day. Finally, while demand for OPEC oil was down, it is important to remember that, on January 1, 1982, OPEC producers were receiving two and a half times as much for each barrel of oil as on January 1, 1979.

Will the momentum of adjustment continue throughout the industrial world? Perhaps — as technical innovations that began in the 1970s come to fruition in the years ahead. Perhaps not — if price signals, public policy, and popular attitudes lose clarity and a resumption of economic growth stimulates energy demand anew without a concurrent investment in energy efficiency.

What this question reflects is that changes in demand and supply do not take place in a vacuum. Adjustment encompasses a larger canvas. Indeed, we can conceptualize adjustment as emerging from an interaction of forces and changes in six realms — in the energy markets themselves; in attitudes, perceptions, expectations, and values; in government policies and programs; in technology; in the availability of capital; and in the vast external political and economic realm. The difficulties and varieties of the economic, social, and political response, which form part of the adjustment process, are explored in some detail in this book.

But there is an important question to address now. What is most striking about Table 1.2 is that the country that has adjusted the most is not the United States or Canada, the countries with the greatest potential, but rather Japan, which began from the most energy-spare base. In 1973, Japan used only 57 percent as much energy for every unit of GNP as did the United States. By 1980, it used only 43 percent as much. How can we explain Japan's relatively better record on adjustment? To begin with, 60 percent of its energy consumption is in the industrial sector, as opposed to a third in Western Europe and the United States, and this sector has generally proved to be the most responsive to the new energy situation. Japanese companies have been willing and able to invest heavily both directly in energy efficiency and in new plants that embody greater efficiency. This attitude reflects the greater propensity among Japanese firms than among their counterparts elsewhere to make longer-term investments. Moreover, a more dynamic economy has made more resources available for investment. Also, companies concluded that increased energy efficiency was necessary to maintain competitiveness in the world market. A system of flexible labor-management relations has given Japan more latitude in making structural changes in the economy. This was aided by the considerable success of its export efforts, which enabled it to both pay its oil bill and still deliver rising real incomes to its citizens. This export drive, in turn, was assisted by the fact that Japanese products, such as more efficient cars, were more suited to the post-1973 world market than those of its competitors. But the overall continued economic growth also marked a substantial shift away from and reduction of those industries, such as petrochemicals and aluminum, that are heavy energy users.

The reasons for Japan's success extend into the character of Japanese society and its basic beliefs. A consensus emerged quickly after October 1973 about the nature of the problem and what to do: Japan was dangerously dependent on imported oil; the rising costs threatened the balance of payments and were a major burden on the nation; efficiency had to be pushed as much as diversification and substitution; the economy had to be reoriented toward "knowledge-intensive" industries, which use relatively little energy. Once a consensus emerges in Japan, society moves quickly, perhaps because the Japanese are more accustomed than other peoples to change and adaptation. Certainly, that happened in this case. The government provided "guidance" and incentives to push conservation. The oil shocks propelled a shift in attitudes away from consumption values and reinvigorated a traditional sense of constraint. This made for a stronger consensus. In this overall context, national policy was able to attack in succession the four disequilibria introduced by the oil shocks — extremely high inflation, large balance of payments deficits, slow growth, and high unemployment. Energy efficiency was seen as a central element in the long-term adjustment process.

This coherent response contrasts sharply with that of the United States, where a new energy policy seemed to be announced every two years and where there was a protracted debate as to whether there was a problem and, if so, what kind of problem it was and what to do about it. In America, the tradition was not one of constraints but of abundance. After all, the United States was also a significant energy producer; as late as 1952, it produced half the world's oil. Within little more than two decades, it became the world's largest importer of oil. Yet, it had had little experience or preparation for being so major a consumer in a world market. Moreover, unlike Japan, the United States has a rich resource base, so higher energy prices resulted in significant income transfers not only out of the country but also within it. Domestic interests and perceptions diverged widely. Instead of consensus, there were bitter battles — over prices, the reality of the problem, the role of the state, and the respective virtues of the marketplace and regulation; between producers and consumers; between producers and environmentalists; and among regions.

In the aftermath of the second oil shock, however, a consensus did appear to emerge in the United States as the public, in response to the events in the Middle East, began to accept the reality of the energy problem. Yet ample supplies in the world oil market in 1981 and the first part of 1982 acted as a sedative, suggesting a powerful drive to deny the challenge. Almost as soon as prices leveled out, a glut psychology took hold once again and, as in 1977 and 1978, people dismissed the problem as a thing of the past.

Is it?

Two Energy Futures

Our aim is not only to understand the recent rapid changes in energy availability and their consequences but also to look at the future — not, however, in order to make specific forecasts but to provide a structure for thinking about the questions and issues that energy considerations may raise for economic, political, and social life in the 1980s and 1990s. We are looking for basic trends in order to identify the themes, issues, and conflicts that could dominate the agenda in the years ahead, and to ask what might be required for a better rather than a worse outcome.

We need a starting point, a framework, for evaluating how the energy situation may evolve in the rest of this decade and into the next. The long lead times associated with major energy problems enable us to see pretty far into the 1980s. For the 1990s, we are on less sure ground. Recognizing the considerable uncertainty, Robert Stobaugh, in Chapter 2, develops two scenarios for future energy supply and demand: the Upper Bound and the Lower Bound. Both are based upon reasonable assumptions. We do not take prices as the only predictor of supply and demand, but attempt to incorporate political, social, institutional, and technical considerations as well. The year 1978 is used as the base, since it is the most recent "normal" year. Both scenarios differ from the conventional wisdom, in that they point to significant constraints on the energy supply, which, on the basis of current relationships between economic activity and energy consumption, suggest either a moderately tight energy balance or a dangerously taut one.

The more optimistic scenario, the Upper Bound, assumes that no major political upheavals interfere with oil production, that there is not another nuclear accident like the Three Mile Island incident, and that energy development in the industrial world goes forward. In the Upper Bound, we see the overall world energy supply increasing by 2.6 percent a year. Because of the growing demand in the developing world, particularly in the OPEC countries, this means that the industrial countries can anticipate a 2 percent annual increase in energy supply.

More worrisome is the Lower Bound, which portends energy stringency, with possible grave economic effects. It assumes a pattern similar to that of 1973–80; that is, that things do not go very well on the supply side. It is a case in which major upheavals or political changes further depress OPEC production or where energy production in the industrial countries is retarded by political opposition, an inability to resolve environmental questions, or by capital constraints. Results prove disappointing in the effort to find the new commercial reserves of oil and gas required to make up for the declines in existing reserves. In the Lower Bound, world energy supplies increase by 0.8 percent a year. Again, because of the growth of demand in the developing world, this means zero growth in the energy supplies available to the industrial countries.

Many factors will affect the supply balance. Two are critical. The first is oil production in the Middle East, particularly in Saudi Arabia. In the Upper Bound, Saudi oil production is projected at 11 mbd in 2000; in the Lower Bound, at 6 mbd. Both are plausible levels. Economic and political considerations will shape the actual outcome. The second is oil production in the United States. It varies between 9.3 mbd in 2000 in the Upper Bound and 6.1 mbd in the Lower Bound. Here, geology will be the prime determinant.

Both of the scenarios would have significant price consequences. In the Upper Bound, oil prices might increase at an average rate of 2 percent per year, reaching $45 a barrel (in 1980 dollars) by the year 2000. This would be half the increase registered in the years 1973–80. In the Lower Bound, they might increase at 4.5 percent per year, reaching $72 a barrel (in 1980 dollars) by 2000. This increase would be one and a half times that of the years 1973–80. Prices will probably not move in regular annual increases, but rather in irregular and large jumps — a most important consideration.

Of course, there are other possible scenarios. Even more foreboding than the Lower Bound is a major shutdown of oil production in the Arabian/Persian Gulf, which could have devasating effects. That possibility requires a wide range of contingency planning, to which we shall return. It is also possible that the energy situation will evolve in a more favorable manner than the Upper Bound. A number of factors might bring about such a happy evolution — good luck in finding and exploiting new reserves, technical advances in current energy production, the successful deployment of essentially new technologies. If such an evolution occurs, many of the more worrying questions we pose in this book can be laid aside. But on the basis of what we know today, it would not be at all prudent to count on such an outcome, and it seems likely that the future will fall somewhere between our two bounds.

If energy supply develops in rough approximation to the Upper Bound, then the situation appears manageable — so long as the industrial nations are able to maintain their current level of economic efficiency. Efficiency improvement of 1.5 to 2 percent a year could well allow 3 to 4 percent growth rates. Thus, the Upper Bound need not exact a penalty on economic growth; reasonable levels of employment could be maintained, and there would be time to develop energy-related innovations. Certainly, there is considerable potential in the United States, as well as in Western Europe, for further improvements in energy efficiency.[14] Japan poses more of a question. Even the Upper Bound would impose further pressure on the troubled economies of the developing world through consistently adverse terms of trade.

Overall, however, the Upper Bound appears manageable — if the momentum of energy efficiency is maintained, if price movements are not too jerky, if investment proceeds steadily, if the national political will is brought to bear, and if a reasonable amount of international coordination and cooperation is maintained.

What happens if there is bad luck on finding new conventional supplies, if there are technical problems in developing alternatives, if there is low investment on both the supply and the demand side, and if obstacles to energy production turn out to be difficult to surmount? In other words, what happens if the world finds itself at the Lower Bound? The simple answer is that a much larger increase in energy efficiency than so far registered would be required to prevent stagnation and severe, even crippling economic problems.

Three issues — the manner in which oil prices move, the rate of investment, and protectionism — hold particular importance for the way in which the adjustment process proceeds as well as for how close to the Upper or Lower Bound the world finds itself.

If prices actually increase in a regular fashion, then both behavior and investment — in energy efficiency, energy supply, and new technologies — would be responsive. Moreover, governments would tend to provide correct signals on energy policy and, at the same time, be better equipped to avoid destabilizing lurches in macroeconomic policy. But prices are unlikely to angle up in a straight line on a graph, but rather to take big jumps at irregular intervals, followed by periods of flat or declining real price. This we describe as the pattern of "jagged peaks and sloping plateaus." One might reasonably anticipate one or two price shocks in the Upper Bound, while several price shocks, with shorter intervals of time, might characterize the Lower Bound. This kind of pattern imposes very high costs in terms of inflation, lost economic growth, and monetary instability. It also tends to work against long-term capital investment. The effect is accentuated by the oscillations in government policies and public attitudes it engenders — from panic and hysteria to the complacency of the glut psychology. The realities of world energy are not necessarily very appealing, and when things appear to be going right, there is a natural enough predilection to want to forget about them, to persuade oneself that the problem is solved.

This certainly is what happened in the United States after the second oil shock. By 1982, many people wanted to pretend that the events of 1979–80 had occurred in the distant past, as though they had no relevance to the present and the future. At a time when it was impossible to sort out the effects of short- and long-term conservation, recession, and oil inventory reductions, many nevertheless hurried to pronounce the problem "solved." There was little recognition of the volatility and uncertainty inherent in energy demand. Rather, the same kind of linear thinking that had assumed in the middle 1970s that energy demand had to grow in lock step with economic growth now assumed that a flattening out or decline in demand was inevitable. In the United States, the arrival of the Reagan administration in 1981 led to a revised government policy that strongly reinforced the glut psychology. The infrastructure — programs, institutions, financing, people — needed to accelerate alternatives and conservation was virtually dismantled. The subject of post-1985 fuel efficiency standards for the automobile industry was laid aside, and even contingency planning for energy emergencies was downgraded. Such moves not only had a significant impact on the adjustment process in itself, but also delivered a confusing message of complacency to the public.

Irregularity in price movements has a most inhibiting effect on investment, a very serious concern, as a considerable part of the adjustment process is really a question of investment. Since the early 1970s, there has been a general tendency in the industrial world either to avoid investment or to concentrate on short-term rather than long-term investment. "Euro-pessimism" has been reflected in a dramatic curtailment of investment in West ern Europe.

Many have pointed to the short-term time horizons that characterize much business investment in the United States. "We're dominated by quarterly results," is how the chairman of a major capital goods company in the United States explained the matter. "If we have a bad quarter, everyone goes around here with a long face. It's as though your wife had died." Although investment horizons are longer in Japan, even there a foreshortening tendency has been noted. An economic environment marked by high uncertainty and pessimism, with oscillating real energy prices, and which alternates between oil-induced inflation, with high interest rates, and oil-induced recession, with shrinking profits — such an environment does not encourage stable economic growth or steady investment. Indeed, the European Community has estimated that Western Europe needs 4 percent economic growth in order to make investment sufficient for energy adjustment — a goal far from being met today.[15]

Many significant energy supply plans announced in 1979 and 1980 were, by 1982, either scaled back or indefinitely postponed, and much conservation investment lost its priority. This was the result not only of poor overall economic conditions. For there is a strong feedback effect on conservation investment, based upon changing expectations. In this case, the feedback arose from the "glut psychology" and an expectation for future prices sharply different from that of two years earlier. The airline industry provided a vivid example. In 1980, a number of airlines, reeling from higher fuel costs, placed orders for much more efficient aircraft. But in 1982, several of the airlines, beset by poor business conditions, concluded that fuel prices were not going to rise as they had expected, so they decided that they could afford to cancel or defer their orders. While this may make sense from the viewpoint of the individual firm, at least in the short term, the cumulative impact of such decisions is to reduce the general level of energy efficiency in the years ahead and so to increase the likelihood of pressure on energy supplies — and thus on prices — in the future. Meanwhile, government policies meant to develop over a longer period tend to get scrapped in this kind of environment. This all means that the capital stock does not turn over as rapidly and that therefore adjustment goes more slowly, leaving the economy more vulnerable to further price shocks — and to yet another cycle of low growth and low investment.

Less adjustment means that once economic recovery begins, rising incomes will exert a stronger pull on energy demand, which could well pick up more rapidly and to a greater extent than might be anticipated. This is especially true for oil, the consumption of which is particularly sensitive to the business cycle and which seems to be the marginal source at times of rising economic activity.

Here we come to a key danger. "Accidents," particularly political events, alone need not ignite an oil shock, as was made clear with the Iran-Iraq war. Energy demand in the last quarter of 1980 was on a downward curve in response to the price rise following the second oil shock, and oil inventories were high. The more likely flash point occurs when accidents interact with a market in which demand is rising — as was discovered in 1973 and 1979. Thus, the world enters the danger zone when economic activity and energy demand are on an upswing.

Future shocks can actually be more debilitating than the past two. If production peaks in some OPEC producers in the late 1980s or early 1990s and if a higher proportion of world oil exports are coming from the Arabian/Persian Gulf, then any interruption there can have strong effects. Moreover, there is good reason to think that the vulnerability of economies to future shocks has actually increased. The share of GNP in Western economies devoted to energy expenditures has tripled, from about 3 percent to 8 to 10 percent. Thus, an oil price rise of any given magnitude will now have a greater effect on the consuming nations than the first or second shock did.

The third issue involves the international trading system. To what extent will nations try to buffer themselves against economic, political, and social distress by pushing the adjustment process off onto the trading system; that is, by trying to buy some form of short-term benefit through protectionism? The oil shocks directly engender or accentuate some problems that stimulate protectionism — balance of payments difficulties, structural weaknesses in some industries, differences in energy availability and efficiency. These are compounded by other factors — for instance, the outward thrust of some exporters (Japan and the newly industrializing countries), the "globalization" of the automotive and other industries, rigidities in some labor markets, and varying paces of technological change.

The trend toward protectionism is already clear in the rise of nontariff barriers and voluntary restraints meant to protect economically and politically sensitive industries and employment. There is also a creeping trend into bilateralism through "government-dependent" or "managed" trade in such ways as subsidized exports and state-to-state deals. Protectionism arises in yet another form, the use of divergent energy prices and policies as a way to subsidize domestic industry. For instance, the United States has, by decontrolling oil prices, become more completely integrated with the OECD energy market. But further shocks could give domestic politics a powerful impetus to seek a buffer through renewed price controls, an approach that might also be followed by such other OECD energy producers as Canada, Britain, and Norway.

The pressures to expand protectionism will be increased by future oil shocks. Constriction and rigidity in world trade will lead to lower production, lower growth, and lower investment and so impede the adjustment process throughout the world. One of the major questions for the 1980s will be whether the international trading system can survive such pressures. The answer will depend very much on the attitudes and strategies adopted by governments and private enterprise under what could be trying circumstances.

The Lower Bound would pose an overriding challenge: Can the adjustment process move faster and in a steady fashion? Here, more starkly, the world would confront the race between crisis and adjustment. For the developing countries, the Lower Bound would be nothing short of a disaster. Their balance of payments problems would become truly enormous; oil prices would be rising at the same time that their export markets were contracting because of economic downturn and greater protectionism in the industrial countries. But also consider what the Lower Bound might mean for Japan. It is not prepared for the very low growth that the Lower Bound would impose. The result would be high unemployment and perhaps the sundering of its basic fabric of labor-management flexibility. Weak governments in Japan would institute austerity programs that respond to the budgetary shortfalls but not to the more fundamental problems. How long could a national consensus survive such pressures?

Without considerable acceleration in energy efficiency, tension and conflicts would develop within the industrial nations over such issues as how to apportion the effects of rising prices, how to reduce oil dependence, how to handle emergencies, how to restructure and reorganize the domestic economy, and how to allocate lost economic growth. Contention on these issues would raise even more basic questions — about the soundness of the social contract, the consensus that is fundamental to the political and social order; about the distribution of power; and about how people live and what they expect. These questions would become increasingly acute as unemployment rose, real incomes declined, and governments appeared increasingly helpless and incompetent.

Efforts to find solutions would not necessarily stop with attempts at protectionism, "special relationships," and energy autarky. The further possible consequences of deep discontent should not be underestimated. One would be a turbulence and viciousness in domestic societies, a search for villains and scapegoats on whom to blame the problem — with the problem itself so redefined as to obscure reality. Citizens might also put pressure on national governments to resort to foreign policy and military force to attempt to "solve" the problem. Finally, democratic governments, appearing unable to deal with a pervasive economic and social crisis, might well be succeeded by more authoritarian regimes that offer order and slogans, if not stable economic growth.

These possibilities are unpleasant to contemplate. But it is important to think about a range of futures and to ask difficult questions. No one can be certain about the stability of consensual government during a protracted period of harsh economic circumstances, very different from that of the great postwar boom. The stakes are high enough, the uncertainties great enough, the effects pervasive enough, and the dangers grave enough that it would be foolish to close our eyes to what might result from a combination of bad luck on the energy supply side and insufficient adjustment on the demand side. For it is the character of national societies and the nature of the international system — not whether motorists pay 70 cents or $1.40 for a gallon of gas — that is truly at stake in what has been called the energy crisis.

Energy and World Politics

So far we have explored how energy problems have interacted with the economic, political, and social life of the consuming nations and how these interactions might unfold in the future. Our theme has been crisis and adjustment.

This theme pertains no less in another dimension, that of international relations. For the escalation of energy questions from being matters attended to by bureaus and ministries of mines to being central issues of national concern has also had a dramatic impact on the international system. In one way or another, international politics is always about the competition of nations for security, power, prestige, position, and wealth. Energy affects each of these objectives and so considerably alters the character of the competition. The change is far-reaching. In the middle of the twentieth century, the advent of nuclear weapons and the East-West rivalry, as well as decolonization, reshaped the international order. In the last quarter of the century, the increased importance of energy has made for no less qualitative a change. By so significantly altering the international economy and the fortunes of nations, it has made a dramatic impact on world politics.

Energy has added a new meaning to the definitions of power and has changed the ranking of nations and, indeed, the international balance of power, and has done so with remarkable swiftness. For some countries, oil was formerly their only internationally significant asset, but in a decade they have succeeded in transferring this asset into other forms, beginning with wealth. The cumulative transfer of income to the OPEC countries made possible by the price increases that began in 1973 exceeds $1 trillion. Countries that only a decade ago were minor players in world politics are now among the arbiters of the world economy. In turn, they have translated oil and wealth into still other assets — political influence, technology, and military prowess. These changes have resulted in a major diffusion of advanced weaponry and have also pushed the proliferation of nuclear weapons.

For the Western countries, energy has become a security issue of no less salience than the East-West balance, but it is more difficult to bring into focus. For the energy question is highly fragmented, with many divergent interests colliding and coinciding even within a single country as well as among nations. It is also one in which the basic interests have a very practical, even immediate, tangibility that is missing from the stark abstraction of the nuclear equation. For these interests can be measured in the number of barrels per day required to keep industrial society open for business. This basic requirement has generated a whole range of vulnerabilities for the Western nations, binding economic and security concerns tightly together. The mutual interests and common problems of the industrial nations are considerable in this context; yet, at the same time, the direct and indirect consequences of energy problems could strain or even fragment the Western alliance, which is fundamental to world stability.

The energy question has placed enormous weight for the entire international order on what happens in one area of the world, the Middle East. It is an area torn by regional and religious rivalries, by social and ideological tensions, as well as by East-West competition. This means that instability there can quickly be transmitted to the international order, and such dependence makes the world order more unstable.

We say energy, but more than anything else, that means petroleum. While exports of natural gas and coal are growing, neither raises at this time sizable security interests, save as they contribute to adjustment. Even the controversial Soviet–Western European natural gas deal, for instance, only raises the total contribution of Soviet gas to Germany's and France's overall energy mix to just 5 percent. Nuclear energy does raise fundamental security questions. It is a key element in some national strategies to reduce dependence on imported oil. But it also carries its own worrisome baggage — nuclear weapons proliferation, safety, terrorism, dependence on foreign suppliers.

But it is oil that is at the center of the energy question and our current predicament. It is the change in the supply of oil that has changed the character of international politics. It is a series of oil shocks, not coal or natural gas shocks, that has sent such reverberations through the world economy. It is oil that provides half the energy used in the world every day. It is oil that has made possible an extraordinary accumulation of wealth by very small nations. It is oil that has made world politics so vulnerable to what happens in the Middle East. It is oil that has turned the matter of energy supply into an overarching security issue.

The security problems arise from a basic asymmetry between who consumes and who produces oil. The problem was laid out in modern terms by Winston Churchill on the eve of the First World War, when the decision was made to shift Britain's Royal Navy from coal to oil:

> If we overcame the difficulties and surmounted the risks [and converted to oil], we should be able to raise the whole power and efficiency of the Navy to a definitely higher level; better ships, better crews, higher economies, more intense forms of war power — in a word, mastery itself was the prize of the venture ... To build any large additional number of oil burning ships meant basing our supremacy on oil. But oil was not found in appreciable quantities in our islands. If we required it, we must carry it by sea in peace or war from distant countries. We had, on the other hand, the finest supply of the best steam coal in the world, safe in our mines, under our own hands.[16]

The problem then was to assure secure supplies of oil sufficient to keep the Royal Navy operating. The problem today is to ensure a sufficient supply to keep the world economy afloat.

The asymmetry is the result of geography and geology — the location of most reserves of cheap and easily accessible oil in the Middle East — which has led to the increasing dependence of the industrial world on a handful of producers in an explosive part of the world. Some 60 percent of the free world's known reserves are in the Middle East; 65 percent of the oil moving in world trade flows from this area. About half of the oil moving in world trade flows through the Strait of Hormuz. Moreover, the dependence on the Arabian/Persian Gulf could conceivably grow later in the 1980s and in the 1990s as producers elsewhere in the world pass their natural peak of production and begin to decline.

At the beginning of the 1970s, this dependence started to introduce major instability into the international system. The problem became clear to the world in 1973 with the application of the "oil weapon." And a sense of false security was destroyed in 1979 with the fall of the shah of Iran.

The Dangers

Three overlapping sets of dangers exist. The first concerns the internal stability of the key oil-producing countries. Of course, strong nations have always depended on weaker and more unstable nations for many vital supplies without great danger or concern. But there is considerable danger when the industrial world is so acutely dependent on states where two centuries of political, economic, and social change are being compressed into a matter of years and the outcome of the interaction of so many forces and ideologies is so uncertain. There is, of course, much speculation about what will happen, but there is no sure way of predicting. Yet, as one recent report warned, "There are at least half a dozen countries in the area whose regimes must be regarded as precarious in a ten-year perspective."[17] Certainly, instability in key oil producers threatens both oil supply and regional equilibrium. Revolutions, terrorism, coups, or social upheavals that give way to additional anti-Western regimes would pose even more of a threat. Such changes would add to the "hostile oil," oil produced by countries fundamentally antagonistic to the West.

The second danger is regional conflict, for the rivalries in the area are many and varied — the Arabs versus Israel, Iran versus Iraq, Syria versus Iraq, Egypt versus Libya, South Yemen versus Saudi Arabia, Christians versus Moslems in Lebanon, radical versus traditional, Sunni versus Shi'ite, and on and on. These rivalries can result in hostilities, which can threaten the oil supply and which can set off the trigger that draws the superpowers into conflict. The vast influx into the region of advanced weaponry, the best that East and West have to offer, has added to the volatility, as will the proliferation of nuclear weapons.

The third danger is an expansion of Soviet influence, directly or indirectly, among the oil producers. East-West competition is hardly new. After all, the first postwar crisis concerned the unwillingness of the Soviet Union to withdraw its troops from Iran after World War II. It took a determined stand in the United Nations in 1946 to bring that about — something noticeably unsuccessful in the case of Afghanistan in 1979.

But since 1979, there has been a new and more direct overlap between the energy problem and the East-West rivalry. The circumstances were set in motion by the diminution of Western power and influence, which began before the oil shock but was much accentuated by it. Iran had moved forward to assume the role of regional stabilizer — a reasonable response to a difficult security problem. The collapse of the shah's regime has left a considerable power vacuum and makes the security problem more acute. At the end of 1979, the Soviet Union invaded Afghanistan, projecting its power in a new way. Some argue that the invasion was part of a larger strategy to win suzerainty over oil supplies and thus over the international economy. Others see it as more of an ill-thought-out way to stabilize a pro-Soviet regime close to collapse. Whatever, the consequence has been potentially to change the balance of power, to enlarge greatly the Soviet presence in the area, and to provide a political as well as a physical staging point for further expansion of Soviet influence. This influence already extends directly and through proxies into Yemen, Ethiopia, and Libya. Moreover, the Soviet Union has not found the international costs of Afghanistan particularly burdensome. Despite the creation of millions of refugees, that entity known as "world public opinion" has generally delivered a verdict of blatant inattention.

Several objectives could motivate a Soviet drive to penetrate or exert increased influence over the oil producers themselves. At the simplest level, some kind of control could assure additional oil for Comecon nations at a time when further growth in Soviet oil production is problematic. It could also provide an excellent source of hard currency; the annual hard currency earnings of the Soviet Union are roughly one tenth of the revenue flows to the Gulf nations. But potential gains go far beyond this. For to control the oil producers of the Arabian/Persian Gulf would be to control the most golden of geese in the history of the world — the regulator and even the banker of the modern industrial economy. The world balance of power would be dramatically changed if Western Europe, Japan, and perhaps even the United States were dependent for a substantial part of their energy on the Soviet Union. The Soviets are aided by their close proximity to the Middle East; they might well encounter little resistance were they to move swiftly in a blitzkrieg to seize some significant piece of real estate.

But they would also have reason to be fearful of the consequences. For Western interests in the region are much greater than their own, and they could not be certain that they would not set off some larger conflict, perhaps involving nuclear weapons.

Corrosion is the more likely method by which Soviet influence would spread, feeding on regional instability, ethnic tensions, and national upheavals, winning invitations and access from one faction or another. The disorder in Iran or some local war could offer tempting opportunities. In other words, the Soviet Union could become the ultimate beneficiary of instability by capitalizing on local troubles. Moreover, the Soviets' ideological framework, however worn in practice, would still persuade them that social developments will move in directions favorable to them, that time is on their side, and therefore that patience will have its rich rewards.

Although a direct Soviet military offensive could have the most apocalyptic results, that specter is in some ways easier to visualize, plan for, and organize against than is the case with the other two dangers. The more likely problems are those emerging from internal and regional conflicts, which can disrupt or threaten oil supplies and may or may not be accompanied by a possible expansion of Soviet power or influence. But it is more difficult to prepare for or respond to these kinds of conflicts.[18]

The range of dangers poses a formidable agenda for the Western industrial countries at a time when the Western system has been taxed by the emergence of the energy problem. This system is the product of a series of overlapping political, economic, and military associations, formal and informal, joined by a common set of interests and values. The energy problem has cut sharply across these relations. It has created great conflicting pressures toward both cooperation and fragmentation in terms of political and economic as well as security issues.

Anxieties about supply and export competition have led to tensions as states try to preserve or restore their own particular position. Tensions have also emerged from the differences in the energy balances — between, for instance, those that have large indigenous energy resources and those that do not. The world is going to look different to a Japan, 85 percent dependent on imports for its energy, than it is to a United States, which is 85 percent self-sufficient in energy. Differences in attitudes and beliefs create strains. An obvious case concerns the Arab-Israeli dispute, but there are strains on other issues as well, such as the conflict between the United States and the Federal Republic of Germany over the Western European natural gas pipeline deal with the Soviet Union. The American administration focused on the possible dangers of German dependence for gas on the Soviet Union and on the transfer of hard currency and technology to the Soviet Union. The Germans saw it as an important element in diversification away from OPEC oil, less of a threat in terms of dependence, and a boon to depressed German industries; they also found more concrete benefits from a policy of détente.[19] To the list of tensions must be added the confusion and suspicion created by the difficulties of dealing with the unexpected, the shocks, the complex problems for which there are no easy answers — especially when there is considerable pressure to respond to the short-term and parochial perspectives of domestic constituencies.

Yet this very complexity also provides the driving force for cooperation and cohesion. No single Western nation can cope with the energy problem by pursuing an isolationist or nationalist strategy, for both it and the countries most important to it would all likely end up worse off. Problems would be inescapably transmitted through the international economy. Neither planning for dealing with a military crisis in the Arabian/Persian Gulf nor meaningful domestic responses during a supply disruption are likely to be effective without coordination with other Western countries. Nor can the consequences — whether they be balance of payments difficulties, Third World debt, economic slump, or protectionism — be effectively countered without cooperation. The failure of cooperation can be costly, as was discovered in 1973–74 and again in 1979.

Still, a good deal of progress has occurred since the confused and panicky days of late 1973. An energy security system for the industrial nations, complementing their other associations, has emerged through the institutional mechanisms of the International Energy Agency, the economic summits, and the European Community. They provide a multilateral framework for creating parallel and mutually reinforcing energy policies, for mediating differences, for assessing and analyzing problems, and for formal and informal crisis management and sharing in an emergency. But progress on a common framework for dealing with the strategic issue has been slower and much more difficult to attain. The Carter Doctrine, enunciating the vital interest in the Arabian/Persian Gulf, was not so happily received by some other consumers who are actually much more dependent on the Gulf than is the United States. In fact, the doctrine simply made more explicit what was already an inescapable fact — the centrality of Western interests in the region. Clarifying this reduces the risk of miscalculation.

The challenge before the consuming nations is formidable. The greatest danger of East-West conflict was in the first five years after World War II, when there were large, unresolved questions about the future orientation of the various countries, especially of Germany. Once those questions were settled, the likelihood of conflict was greatly reduced. Today, the question marks loom increasingly large over the future orientation of the key oil producers. The instability, uncertainty, and high stakes can lead to large-scale military conflict. Upheavals well short of direct Soviet intervention can nevertheless result in a de facto shift in the global balance of power. Any type of conflict in the area can disrupt the flow of oil. A taut supply balance by itself or in interaction with some regional crisis can add further to the already costly economic, political, and social strains that have resulted from the first two oil shocks.

Energy security is a particularly difficult challenge to meet because it is so complex, because the uncertainties are great and the connections ambiguous. There is, therefore, a tendency to postpone and ignore when possible, to argue and quibble, and then in a crunch respond with hysteria or find oneself faced with a fait accompli.

Can Adjustment Outrun Crisis?

Yet the energy question does appear to be manageable in many dimensions, and, in those where it is less tractable, damage can at least be limited. And, without minimizing the extent of the challenge, we are optimistic. There is tremendous flexibility and vitality in modern industrial societies, as evidenced by the adjustment that has already begun. But adjustment needs to keep ahead of potential crises to allow maximum adaptability.

We can point to several basic requirements for the Western world: a definition of common agenda, consistency, a parallelism and complementarity of purpose and program, an ability to tolerate a longer time horizon — and a recognition of the consequences of the energy problem. It is too easy to fall prey to the lure of complacency. To ignore the basic links among energy, inflation, and reduced economic activity is to misconstrue reality and to ignore appropriate responses. Ulf Lantzke, the executive director of the International Energy Agency, has expressed the matter succinctly: "There still is an oil gap; only now it is expressed in the number of unemployed."[20]

A major aspect of adjustment is to deal with the underlying situation, aiming at a reduction of the dependence on oil and, in particular, on Middle Eastern oil. This would make the economies of the Western and developing worlds, as well as international politics, less vulnerable to future shocks and pressures. The key steps are the diversification of oil supplies, the substitution of other energy sources, and energy efficiency. Of the three, the last might have been taken the least seriously half a decade ago. Today energy efficiency is proving the most potent short- and medium-term energy source available. Further success at tapping it will reduce the threat of future shocks, make either an Upper Bound or a Lower Bound scenario more comfortable, reduce pressure on oil producers, make such crises as do occur more manageable by introducing more slack and flexibility into the overall energy supply system and provide time for a stable longer-term transition.

What elements will prove conducive to a successful adjustment? The first need is for sustained attention to the problem, neither hysterical nor complacent, but with some greater consensus about its nature and the appropriate responses. The second need, which follows directly, is for a broad effort to accelerate the adjustment process. A great deal of change in energy availability and price has been crammed into a very few years. A speedy adjustment is required. Price signals that tell the truth are important, for prices are a significant part of the solution even as they are also a major part of the problem. But simple reliance on price does not appear adequate to those conscious of the broad range of dangers and the compressed time horizons. Compensation must be sought for the jerky and confusing signals that the market delivers and to make up for the short time horizons of the private sector. Here there is an important and appropriate role for government in addressing such market imperfections as lack of information and knowledge and in bridging the distance between the shorter horizons of private interest and the longer ones of the national interest. Particular attention needs to be given to the issue of investment. Price changes over the last ten years have been such that, at this juncture, almost anything that stimulates investment will stimulate energy efficiency. But there is still a specific need to encourage investment aimed at energy efficiency.

However, that investment will not necessarily occur in the present economic environment. Huge sums of money are required to increase energy efficiency in Western industrial society, and there are other claimants both for investment funds in general and for energy funds in particular — as well as a propensity to avoid investment. Moreover, at least in the United States, funds may be much more easily channeled into energy supply, even in the quest of extremely marginal barrels, rather than into the much more bountiful terrain of energy efficiency.[21] National governments must also stave off the protectionism that seeks short-term respite from the adjustment process at considerable long-term costs. Finally, the developed nations (as well as the oil producers) must attend to the problem of the oil-importing developing countries, the weakest link in an interconnected system, on simple grounds of self-interest. This involves a range of measures, from foreign aid to trade facilitation to technical assistance, that will help them become less dependent on imported oil.

Adjustment is also required in the traditional security orientation of the industrial nations, for they need to develop a credible ability to respond quickly and in sufficient force to protect the integrity of the oil supply system. There is a particular need to compensate for the Soviet Union's considerable geographic advantage, and the time advantage that goes with it, yet not do so in such a heavy-handed way as to feed nationalistic antagonisms in the gulf area. A cooperative framework needs to be developed, involving both the industrial consumers and the states in the region. This will require a considerable effort, for a Western alliance more than three decades old, born out of a sense of sharp and shared danger, must go through quite a process of redefinition to accommodate the new dangers. This is made all the more difficult by the need to involve the key states in the region, where positions are complicated by the Arab-Israeli dispute and other regional tensions, the dependence on foreign workers, the pressures of nationalism, the influx of wealth and weapons, and the conflicts between modernization and tradition. An Arab-Israeli settlement, following on the one between Israel and Egypt, might also contribute considerably to stability in the region and so reduce the risk to oil supplies.

There are obvious and unresolved gaps in the perspectives of the Europeans, the United States, and states in the gulf that include sharply different calculations of the political and military elements that would enhance regional security. An asymmetry in the Western alliance is clear. The United States, it seems, is to play the preponderant role, while other industrial nations that are actually much more dependent do less. This situation might, in some circumstances, have a corrosive impact on domestic support for that role in the United States. Moreover, the more broadly based and multinational is the effort, the more acceptable and less disturbing it will generally be to both consumers and producers. But the economic problems that have been set in motion by the first two oil shocks have constrained defense spending among allies. Future energy problems would further hamper efforts to establish the credible capabilities. At the same time, it must be clear to America's allies that the United States, which is to oil consumption what Saudi Arabia is to oil production, is taking the domestic steps in terms of energy efficiency to assure a stable world oil market. And the United States must be sensitive to the more precarious energy positions of its partners. Even if all this were to come to pass, the greatest difficulty might well prove to be how to respond effectively and expeditiously to local and perhaps quite obscure crises.

Finally, emergency preparedness measures need development. Fire escape routes in skyscrapers are worked out in advance to allow for an orderly response with minimum losses rather than panic during an actual fire. Responses to oil emergencies need to be demarcated in advance for the same reasons. Currently, such planning is incomplete. The particular subjects that need elaboration include stockpiling, inventory management, emergency-demand restraint, and fuel-switching. The episodes of panic in 1973 and 1979 proved very costly. Merely repeating words like "glut" and "market," as if they were some mantra, will not prove sufficient. Glut can turn to shortfall on little more than an assassin's bullet, and the orderly and reasonable functioning of markets can be brutally overwhelmed by an upheaval and the sudden panic of the unprepared.[22]

What we need, then, is coherent thinking about the energy problem, its manifold consequences, and reasonable responses to it. It hardly makes sense to focus on just one aspect. For instance, the Reagan administration has put great emphasis on developing a military deterrent capability for the Arabian/Persian Gulf, in effect recognizing the extent of the problem. Yet its domestic energy policies appear based on the assumption that all the problems are over. Obviously there will be differences about methods and approaches, for the challenge is hardly simple. But it is inescapable. Much change, some of it very damaging and impoverishing, has been compressed into a few short years. We need to regain that time to permit the flowering of innovation and technology that will allow the twenty-first century to look back on these years as an unpleasant and trying interlude, but one safely navigated. This is a reasonable and prudent goal. But to achieve it, we must cope wisely and realistically with the pervasive global insecurity that has been set in motion by the earthquakes in the prices and in the availability and reliability of world energy supplies.

2

World Energy
to the Year 2000

by Robert Stobaugh

The first thing to make clear in any discussion of the basic energy situation for the world from now to the year 2000 is simply that any such description involves a large amount of uncertainty. This may seem obvious, but it needs to be said. For experience since 1973 has underlined how easily projections can go askew because of the large number of forces at work in the energy situation.

There has certainly been no lack of effort at projecting. Econometricians have sought to predict the future by using models based on economic theory and historical statistical relationships. And technologists have sought to predict the future by using models based primarily on engineering cost estimates.

Since the first oil shock of 1973, however, these models have not proved particularly reliable in predicting either energy supply or energy demand.[1] Understanding their shortcomings will help us arrange our thoughts for discussing future energy developments.

A major reason for the inadequacy of these models has been their entire reliance on price as a predictor of both energy supply and demand. Other important factors — political, social, and institutional — were typically excluded, since they could not be meaningfully quantified. Yet political, social, and institutional considerations also play very important roles in the energy field.

This is obviously the case in energy supply, where often a relatively few production sites are involved. Take, for example, the generation of electricity by nuclear power. Observers predicted that the very large increases in oil prices in 1973–74 would speed the development of nuclear power; indeed, Western leaders made nuclear power the cornerstone of their national energy plans. Yet, political barriers, even well before the Three Mile

Island accident in the United States, slowed nuclear developments to a stalemate in many countries. Clearly, price alone — even if the prices of all competing energy sources were known — is an inadequate predictor of energy supply.

It would seem that price would predict better on the demand side because of the involvement of millions of decision-makers. But even here, systematic barriers prevent accurate estimates from being made. For future prices are likely to be outside the range of observations based on historical experience, and the relationship among income, behavior, and energy consumption is not well understood and is undergoing continuous and significant change.[2] Moreover, there may be major market imperfections — lack of information, inadequate access to capital — that interfere with consumer response. Finally, even if one knew what the equilibrium was likely to be in the year 2000, there would be innumerable paths by which to arrive there. Hence, any one of a number of conditions for the intermediate years would be plausible.

Thus, being all too well aware of the obstacles to accurate estimates, we shall try, with some modesty, to present an overview of the evolution of the world energy situation to the year 2000, with intermediate stops at 1985 and 1990. The year 1978 is used as the base, since it was the last "normal" year; 1979 through 1981 were years of oil shortage and then recessions.

Two scenarios are presented. Except for coal, both are by and large supply-constrained scenarios. True, supply and consumption will come together at some equilibrium, at which different prices will exist for the different energy forms in different markets. And these prices will affect both supply and demand. Geology, of course, will prove to be especially important in determining the availability of oil and gas. And the factors noted earlier — political, social, and institutional — will also play a major role in determining available supply. Because of the difficulty of constructing a formal mathematical model incorporating all these factors in a meaningful way, we base our projections on judgment, which in turn is based on our knowledge of the world energy system and a study of the projections of others, especially those of the International Energy Agency (IEA). It could be said that we are using an "implicit mental model."

The two scenarios provide a frame of reference for our entire project; they are used as a starting point for thinking about the impact of energy on the economic, political, and social systems of the world in the years ahead. Future developments might accord with one or the other, or could go off in quite another direction, but in terms of what we know today, both are based on reasonable assumptions. Most major energy supply technologies — developing an oil province, building a nuclear power plant — require lead times of a decade or more. So we can see in reasonably clear detail pretty far into the 1980s insofar as new energy supplies are concerned. For the 1990s, we are obviously on less sure ground. But, again at this time, it seems likely that the future will fall someplace between our two scenarios.

The rest of this chapter develops the assumptions that lie behind these scenarios. At this point, let us put them out on the table.

One scenario is dubbed the Upper Bound. It calls for an annual growth in energy supply of 2 percent for the industrial nations; that is, for the members of the OECD. The Upper Bound is based primarily on the projections and analyses of the IEA, although it is not identical to any single one of their estimates. This scenario can be thought of as a case in which things go very well on the supply side. There are no major political upheavals to interfere with oil production; there is not another incident like the Three Mile Island accident; the industrial countries adopt sensible policies to encourage production and manage to overcome quite a bit of the political opposition to some of the supply options.

The more worrisome scenario, called the Lower Bound, portends energy stringency. It envisions zero growth in energy supply for the OECD countries through the year 2000. This scenario is based on what might well occur if things do not improve over the pattern of 1973–80 — in other words, if things do not go very well. A major political upheaval, or perhaps just a change in political or economic outlook by the current ruling elite of Saudi Arabia, might further depress OPEC oil production. Energy production in the industrial countries might be retarded by political opposition, an inability to resolve environmental questions, or capital constraints. Geology may deal a bad hand in terms of the new oil and gas resources required to make up for declines in existing reserves.

Of course, there is always an even more foreboding scenario — a shutdown of all or a major part of the oil output of the countries surrounding the Arabian/Persian Gulf. Although mentioned in the other chapters, this possibility is not discussed in detail. It is very difficult to estimate the probability of such an eventuality, but the consequences would be so severe that it should be taken into account in energy planning as well as in other contingency plans.

It is also possible that the energy supply will evolve in a much more favorable manner than that of the Upper Bound case, perhaps because of good luck in finding and exploiting new reserves, or because of technological advances in current energy production, or because of the successful introduction of new energy technologies. In that case, some of the responses needed to cope with the other possibilities might not be needed. Such an outcome would be most welcome, but it would not be prudent to count on it. Moreover, since the societies of the world would be under less pressure and therefore better off if it did happen — a bonus rather than a penalty — we will consider it no further.

Table 2.1   The World Energy Supply to the Year 2000 and
World Oil Price

	Upper Bound	Lower Bound
Available energy supply in 2000	*mbdoe**	*mbdoe*
OECD	119.6	77.4
OPEC	13.2	13.2
Other developing countries	35.2	21.4
TOTAL	168.0	112.0
Annual growth in available energy supply, 1978–2000		
OECD	2%	0%
OPEC	6%	6%
Other developing countries	4.5%	2%
WORLD AVERAGE	2.6%	0.8%
Average annual increase in world price of oil, 1980–2000	2%	4.5%
Price/barrel (1980 $) in 2000	$45	$72

* mbdoe = millions of barrels daily of oil equivalent.
1. The OECD energy consumption in 1980 was 77.1 mbdoe; the price per barrel, $30.
2. "World" includes net exports to or imports from the centrally planned economies.

What happens to prices in these two scenarios? The key price is that for oil. Different patterns of energy consumption can interact with supply to produce very different prices. Still, we can make some reasonable assumptions. In the Upper Bound, with consumers placing only modest price pressure on energy supplies because of their greater availability, we assume an average growth rate of 2 percent per year in the real price of oil between 1980 and 2000. In the Lower Bound, with more pressure brought about by the limited supplies, we assume that oil prices will rise at an average annual rate of 4.5 percent.* In the Upper Bound, starting at about $30 a barrel in 1980, the world price of oil would thus rise in real terms (in 1980 dollars) to $45 a barrel in 2000. In the Lower Bound, it would rise to $72 a barrel in 2000 (see Table 2.1). These two prices are at the upper range of

* Note that in an economic model, these two scenarios represent two different supply curves: an Upper Bound for which at a given price the quantity forthcoming is greater than that for the Lower Bound, at an even higher price.

recent estimates, but seem plausible. After all, during the 1970s, oil prices exhibited an ability to go to unexpectedly high levels. There are many paths to a point of equilibrium. We have put the price changes in terms of average annual rates of real increases, but recent history suggests that price increases are more likely to come in irregular, large jumps, followed by a drop and then gradual decline in real terms. The new price level, however, is substantially higher than the old. In other words, rather than a smooth upward trend in real prices, which fits so well on a graph, much more likely is an unstable pattern of "jagged peaks and sloping plateaus," with each iteration at a substantially higher altitude.

For each of our two supply scenarios, there would be different economic growth and consumption patterns, which will be discussed in detail in other chapters and so are dealt with only briefly here. The Communist nations, the centrally planned economies, are mentioned only as they provide energy to or take energy from the rest of the world.

This chapter thus focuses primarily on world energy supply — and how much of that supply is likely to be available for the OECD countries.

Supply

Oil

OPEC

OPEC's production in 1978 was 30.5 millions of barrels a day (mbd), with Saudi Arabia being the dominant producer at 8.1 mbd. Table 2.2 shows oil supply, with A in mbd and B in metric tons per year. As recently as 1978, authoritative forecasts indicated that OPEC production would climb to 45 mbd by 1990; in 1979 these forecasts tended to be lowered to 36 mbd or so. By early 1982, OPEC production had dropped to less than 20 mbd as a result of a strong worldwide recession, conservation of oil and fuel-switching due to higher oil prices, and a rundown of oil inventories.

Although observers expect an upturn in OPEC oil production once a worldwide economic recovery begins, it is unlikely that OPEC production would climb above 30 mbd, even if the market should demand it. First, some of the earlier estimates were based on incorrect assumptions about the production of some OPEC oilfields. Second, by the beginning of the 1980s, OPEC members had begun to ask whether programs of rapid economic development would damage their long-term economic and social prospects. One estimate indicated that up to half of the $400 billion spent by OPEC countries between 1974 and 1978 had been wasted because of a too hasty development, which stimulated widespread social and political problems, such as inflation, corruption, wild building booms, and a large influx of foreigners.[3] These problems led in turn to a weakening of established social and political values, accompanied by disappointment and resentment. Apprehensions about such tensions, reinforced by the experience of the shah of Iran, have led all the OPEC nations to reevaluate their plans for development and hence their need for revenue. In 1981 and 1982,

Table 2.2 Conventional World Oil Production, Including Net Trade with the Centrally Planned Economies, 1978, 1985, 1990, and 2000

A. Millions of barrels daily of oil[a]

	1978	Upper Bound			Lower Bound		
		1985	1990	2000	1985	1990	2000
OPEC							
Saudi Arabia	8.1	10.5	11.0	11.0	8.0	7.0	6.0
Iran	5.2	3.0	4.0	4.0	2.0	3.0	3.0
Iraq	2.6	4.0	3.5	3.5	4.0	3.2	2.5
Kuwait	1.9	1.5	1.5	1.5	1.5	1.5	1.5
Libya	2.0	1.9	1.9	1.7	1.7	1.6	1.5
Venezuela	2.2	2.0	1.8	1.8	2.0	1.8	1.8
Others	8.5	7.9	7.9	7.5	8.0	7.3	6.7
TOTAL	30.5	30.8	31.6	31.0	27.2	25.4	23.0
Other Developing Countries							
Mexico	1.3	3.4	5.0	6.0	3.3	4.0	5.0
Others	3.9	5.2	6.0	7.0	5.2	6.0	7.0
TOTAL	5.2	8.6	11.0	13.0	8.5	10.0	12.0
OECD Countries							
United States	10.3	10.3	10.3	9.3	9.0	6.8	6.1
Canada	1.6	1.5	1.5	1.5	1.5	1.5	1.5
United Kingdom	1.2	2.6	1.9	1.5	2.6	1.9	1.5
Norway	.4	.6	.6	.6	.6	.6	.6
Australia	.5	.4	.5	.5	.4	.5	.5
Others	.4	.6	.6	.6	.6	.6	.6
TOTAL	14.4	16.0	15.4	14.0	14.7	11.9	10.8
Exports of the Centrally Planned Economies (CPE)	1.3	.4	−1.0	−2.0	.4	−1.5	−3.0
Total available	51.4	55.8	57.0	56.0	50.8	45.8	42.8
OPEC Consumption	−2.5	−4.0	−5.0	−9.0	−4.0	−5.0	−9.0
Net Available, ex-OPEC and CPE	48.9	51.8	52.0	47.0	46.8	40.8	33.8

Table 2.2 (continued)

B. *Millions of tons of oil*[a, b]

	1978	Upper Bound			Lower Bound		
		1985	1990	2000	1985	1990	2000
OPEC							
Saudi Arabia	405	525	550	550	400	350	300
Iran	260	150	200	200	100	150	150
Iraq	130	200	175	175	200	160	125
Kuwait	95	75	75	75	75	75	75
Libya	100	95	95	85	85	80	75
Venezuela	110	100	90	90	100	90	90
Others	425	395	395	375	400	365	335
TOTAL	1525	1540	1580	1550	1360	1270	1150
Other Developing Countries							
Mexico	65	170	250	300	165	200	250
Others	195	260	300	350	260	300	350
TOTAL	260	430	550	650	425	500	600
OECD Countries							
United States	515	515	515	465	450	340	305
Canada	80	75	75	75	75	75	75
United Kingdom	60	130	95	75	130	95	75
Norway	20	30	30	30	30	30	30
Australia	25	20	25	25	20	25	25
Others	20	30	30	30	30	30	30
TOTAL	720	800	770	700	735	595	540
Exports of the Centrally Planned Economies (CPE)	65	20	−50	−100	20	−75	−150
Subtotal available	2570	2790	2850	2800	2540	2290	2140
OPEC Consumption	−125	−200	−250	−450	−200	−250	−450
Net Available, ex-OPEC and CPE	2445	2590	2600	2350	2340	2040	1690

[a] Oil includes natural gas liquids.
[b] Obtained from millions of barrels of oil daily in Table 2.2A by multiplying by 50.
Source: Data for 1978 are from British Petroleum, *BP Statistical Review of the World Oil Industry;* Exxon, *World Energy Outlook,* various editions; and various published and unpublished IEA documents. If sources differed, author's estimate was used. The Upper and Lower Bound estimates are the author's estimates, with Upper Bound figures approximately IEA estimates.

an anticipation of slower growth in the demand for oil because of the sharp price rises in 1979 and 1980 reinforced the lower estimates.

One alternative, of course, would be to maintain high production levels and then invest the revenues in the West. This option, however, is made less attractive by the difficulty of earning a real return on the investment, especially in the face of high rates of inflation. Another alternative is to produce less oil, thereby stretching their oil reserves over a much longer period of time. Moreover, oil in the ground has seemed to offer a sounder return than money invested in rapid domestic development or deposited in Western banks. Even after an economic upturn pushes up demand for OPEC oil again, it is likely that these concerns will continue to have a depressing effect on OPEC oil output.

Furthermore, the lesson of 1979–82 is that OPEC can obtain more revenues with lower production rates. OPEC's annual rate of income in early 1982 was far higher than the pre-1979 price explosion (production of 20 mbd at $30 a barrel versus 30 mbd at $13 a barrel — or $220 billion versus $140 billion). Hence, it is prudent to plan that OPEC countries are likely to continue with cautious investment policies, and they are unlikely to expand their oil production substantially above 30 mbd.

Many observers hope that over the long run, Saudi Arabia will be an exception to this projection because of its ties with the West, its desire for healthy Western economies where it has large investments, and its proven ability to spend large sums of money. The Upper Bound case assumes that Saudi output will rise to 10.5 mbd in 1985 and 11.0 mbd in 1990 and 2000. The Lower Bound case assumes that Saudi production will be 8.0 mbd in 1985, 7.0 mbd in 1990, and 6.0 mbd in 2000, which could be sufficient for the industrial infrastructure under development. There is, of course, great uncertainty about Saudi Arabia's future oil policy. But given the inherent lack of political stability and hence the chance of turmoil in the region, and given the possibility that the rulers who come to power during the next twenty years might well favor a more conservative production policy, it seems at least as likely, if not more likely, that Saudi Arabia will be producing 6.0 mbd as 11.0 mbd in 2000.

The Upper Bound case shows OPEC production rising to 30.8 mbd in 1985, 31.6 mbd in 1990, and easing off to 31.0 mbd in 2000. But if Saudi Arabia adopts the more stringent conservation strategy shown in the Lower Bound, and several other OPEC countries — most likely Iran and Iraq — produce at a somewhat lower level than assumed in the Upper Bound case, then it is possible that OPEC output will approximate 23 mbd by 2000. Furthermore, the oil export potential of a number of OPEC members — Algeria, Indonesia, Libya, Nigeria, and Venezuela — could be reduced because of growth in internal demand and a peaking of oil reserves.[4]

Of course, it is possible that Saudi Arabia's production would be even lower than 6 mbd — if, for example, a radical or religious fundamentalist government should take over the country. Should such an event be accompanied by turmoil in some of the other countries around the Arabian/Persian Gulf, as seems likely, then as much as 8 to 12 mbd could be lost to the world market for an extended period of time. Whether this stoppage would bring forth a Western response, and if so, how effective such a response would be, are matters of much speculation. Although this chapter does not describe this "oil shock" scenario, it would be prudent for the oil-importing nations to make explicit plans to deal with such a loss.

Other Developing Countries

The other developing countries produced 5.2 mbd in 1978. Their production under the Upper Bound case is expected to rise to 8.6 mbd by 1985, 11.0 mbd by 1990, and to 13.0 mbd by 2000. The main provider of additional oil is expected to be Mexico, with 3.4 mbd projected for 1985, 5.0 mbd for 1990, and 6.0 mbd for 2000; in 1980, Mexico's production was 2.1 mbd, up from 1.3 mbd in 1978.

The only difference between the Upper and Lower Bound cases for the "other developing countries" is that the Lower Bound case assumes that Mexico's production will be 4.0 mbd in 1990 and 5.0 mbd in 2000; that is, 1 mbd lower than in the Upper Bound case. True, Mexican prospects have generated quite a bit of optimism, especially in the United States, and estimates of Mexican oil reserves have been pushed upward, with some believing that recoverable reserves of oil and gas are 200 billion barrels. But there is a good deal of confusion about these projections. Others in the oil industry believe that this number refers to oil "in place," only part of which would be recoverable. The current official policy of Mexico is to limit exports to the pace required by domestic economic development rather than adjust it to the size of its oil reserves or to foreign pressure. Mexican leaders are concerned that too much oil money too quickly will result in so much inflation that Mexico's other exports and tourism will be priced out of world markets. If this happened, the country's critical unemployment problem would be made worse. The unemployment issue is especially important in Mexico because its population is expected almost to double between 1980 and the end of the century, rising perhaps to over 125 million. Mexico's leaders have, as have many of OPEC's leaders, publicly rejected the alternatives of piling up dollars in bank accounts or making substantial investments outside Mexico.

OECD Countries

In 1978, the OECD countries produced 14.4 mbd of oil, of which the United States provided 10.3 mbd. The United States, as the major OECD supplier, is the chief uncertainty in future projections.

The Upper Bound case assumes that U.S. production will remain at 10.3 mbd through 1990 and then decline to 9.3 mbd in 2000. The Lower Bound case assumes a drop in U.S. production, to 9.0 mbd in 1985, 6.8 mbd in 1990, and to 6.1 in 2000. Thus, the difference between the Upper and Lower Bound cases for U.S. production is significant: 1.3 mbd in 1985, 3.5 mbd in 1990, and 3.2 mbd in 2000. The Lower Bound projection is about equal in 1990 to a recent Exxon forecast but lower in 2000 than the Exxon forecast.[5] But even any forecast for 2000 is highly speculative, since about 65 percent of that supply would have to come from reserves that have not yet been found.

Why this uncertainty? For some years, the amount of oil produced in the United States has been greater than that replaced through the discovery of new reserves. As a result, U.S. oil reserves dropped 30 percent between 1970 and 1980 despite substantially higher prices and much increased drilling. True, additions to oil reserves in 1980 about equaled oil production, and oil production rose slightly between 1979 and 1980, but this was due to the increased exploitation of high-cost, marginal fields rather than the new giant fields needed to reverse the long-term downward trend. The pattern of U.S. exploration is particularly critical; and wildcat drilling in the high-potential basins in Alaska and on the offshore continental shelf has been modest because of U.S. leasing policies. But there is no strong evidence to support the common assertion that the trend in the reduction of oil reserves could be reversed if more prospective acreage were leased in the United States. Recent explorations in such areas as the Gulf of Alaska, the Baltimore Canyon, and Georges Bank off the East Coast have not been encouraging. Thus, the Lower Bound case appears to be just as likely to occur as does the Upper Bound case. Indeed, the best evidence suggests that although we can hope otherwise, it would not be prudent to base an energy policy on the speculation that U.S. oil production would continue to be close to its 1980 level.

Estimates for the rest of the OECD countries are the same for both Upper and Lower Bound cases. The North Sea production, from the United Kingdom and Norway, is expected to rise from 1.6 mbd in 1978 to 3.2 mbd in 1985 and then decline slightly by 1990 and 2000. The supply of the other OECD countries, principally Canada and Australia, is expected to change little from 1978 through 2000. Other observers project different production levels from the OECD countries; for example, slightly lower levels from the North Sea and slightly higher levels from Canada and

Australia. But any difference in the estimates for the OECD countries (the United States excepted) is likely to be unimportant.

The Centrally Planned Economies: Soviet Union, Eastern Europe, and China

The centrally planned economies, including China, have been net exporters of petroleum — 1.3 mbd in 1978. The central force in this picture is the U.S.S.R., which was the world's largest oil producer in 1980. It was also a sizable exporter, providing about 1.8 mbd to other Communist countries and 1.3 mbd to non-Communist countries. There is considerable uncertainty as to whether the Soviets can maintain oil production at its 1980 levels or whether a substantial drop is likely by 1990, and how much of any drop will be made up by exploiting the Soviet Union's abundant natural gas reserves. Also uncertain is Eastern Europe's appetite, and ability to pay, for oil imports, as dramatized by the upheaval in Poland that began in December 1981. Finally, not much is known about the potential size and timing of the exploitation of offshore Chinese fields. Still, many experts in the world oil industry are now projecting that the centrally planned economies, as a whole, will become a net importer during the latter half of the 1980s and that their imports will increase thereafter.

The Upper Bound case shows net imports of 1.0 mbd in 1990 and 2.0 mbd in 2000; the Lower Bound shows 1.5 mbd in 1990 and 3.0 mbd in 2000.

Summation of Oil

The Upper Bound case indicates that the total world supply of oil available in the period 1985–2000 will be about 10 percent above 1978 levels. But this translates into only a slight increase in the availability of oil for the OECD and the developing countries during the mid-1980s and 1990s and a decrease by 2000. This decline in availability is caused by rising oil consumption in the OPEC nations and the change from a net exporter to a net importer by the centrally planned economies.

In the Lower Bound case, total oil availability will fall below that of the Upper Bound by 5.0 mbd in 1985, 11.2 mbd in 1990, and by 13.2 mbd in 2000. The 33.8 mbd estimated for 2000 in the Lower Bound case, to be available after OPEC consumption and the centrally planned economies' imports, represents only about two thirds of the oil that was available in 1978. Thus, the Lower Bound case, although perhaps just as likely as the Upper Bound case, presents a bleak outlook for oil consumers.

The oil estimates do not include the production of synthetic fuels, which are addressed separately.

Natural Gas

The free world production and consumption of natural gas, net of gas flared or reinjected, was about 17.2 mbdoe in 1978; 80 percent was produced in the OECD countries, especially in the United States. This total includes only net exports of the centrally planned economies and therefore excludes another 7.6 mbdoe of natural gas produced and consumed in the Soviet Union, Eastern Europe, and China — principally the Soviet Union.

The Upper Bound estimates show substantial growth, with gas availability reaching 21.0 mbdoe in 1985, 25.0 mbdoe in 1990, and 29.2 mbdoe in 2000 (Table 2.3). OECD production is expected to grow from 14.0 mbdoe in 1978 to 15.4 mbdoe in 1985, and to 14.2 mbdoe in 1990 before dropping to 12.6 mbdoe in 2000. The United States produced more gas in 1978 than all of the other OECD countries combined, and this status is expected to continue through the rest of this century. Still, there is major uncertainty in projecting U.S. natural gas production. In the United States, natural gas reserves peaked in 1967, and production peaked in 1973; but the increase in prices resulting from the 1978 Natural Gas Policy Act and the prospect of price decontrol of all new gas in 1985 (or earlier) have engendered quite a bit of optimism about U.S. potential. Particular hope exists for large finds of "deep" gas, that is, gas from reservoirs 15,000 feet or more below the earth's surface. But it will be well into the 1980s before the U.S. picture is clarified substantially.

The Soviet Union, with some one third of the world's proven gas reserves, is potentially a much larger producer of natural gas. In fact, the gas industry was the most dynamic sector of the Soviet energy economy during the 1970s. Gas production increased 50 percent between 1975 and 1980, and some observers believe that Soviet gas production could more than double between 1980 and the year 2000. This would require vast quantities of Western high-quality pipe and transmission equipment as well as overcoming hostile geographical conditions. At a minimum, increased U.S.S.R. gas production will help reduce Soviet dependence on oil, thereby making more oil available for export; at a maximum, the Soviets could increase very substantially their gas exports to Europe, where they now provide about a fifth of the natural gas used by the Federal German Republic, France, Italy, and Austria.[6]

The OPEC countries and the non-OPEC developing countries are generating industries that are big users of natural gas. Furthermore, some OPEC countries — Algeria, in particular — plan to deliver gas to major consumers by pipeline. But most of the natural gas from the developing countries, including OPEC, will have to be exported in oceangoing vessels in the form of liquefied natural gas (LNG). Thus, LNG trade will have to expand vastly. A major uncertainty is the connection between

Table 2.3 World Natural Gas Production, Including Net Trade with Centrally Planned Economies, 1978, 1985, 1990, and 2000

A. Millions of barrels daily of oil equivalent

	1978	Upper Bound			Lower Bound		
		1985	1990	2000	1985	1990	2000
United States	9.2	10.2	9.6	8.0	8.4	7.4	6.4
Other OECD	4.8	5.2	4.6	4.6	5.0	4.0	3.6
Non-OPEC developing countries	1.2	2.6	4.6	6.0	2.2	3.6	4.6
OPEC	1.6	2.4	5.0	9.0	1.6	3.0	4.2
Exports of centrally planned economies	.4	.6	1.2	1.6	.4	.8	1.2
TOTAL	17.2	21.0	25.0	29.2	17.6	18.8	20.0

B. Millions of tons of oil equivalent

	1978	Upper Bound			Lower Bound		
		1985	1990	2000	1985	1990	2000
United States	460	510	480	400	420	370	320
Other OECD	240	260	230	230	250	200	180
Non-OPEC developing countries	60	130	230	300	110	180	230
OPEC	80	120	250	450	80	150	210
Exports of centrally planned economies	20	30	60	80	20	40	60
TOTAL	860	1050	1250	1460	880	940	1000

Note: Production is net of flared and reinjected gas.
Source: See Table 2.2.

OPEC oil production and OPEC natural gas production. If the OPEC nations adopt stronger policies to conserve their natural resources, they might force oil customers also to buy some natural gas, thereby avoiding the burning of large volumes of gas as waste.

There are three major obstacles to a vast increase in LNG shipments.[7] Money is the first. It costs as much as $5 billion to build a typical LNG project; most of the money is needed for the liquefaction plant and specialized ships. It costs at least $2.50 per thousand cubic feet (mcf) — about $15 per barrel of oil on a comparable heating value basis — to cover operating expenses and amortizations of the capital investment. To this figure, of course, must be added the price to be paid to the producing nation for the natural gas itself. In earlier years, the OPEC countries were willing to sell natural gas at only a fraction of the price they received for oil. But that situation is changing; in fact, virtually all gas specialists believe that the gap between oil and natural gas prices will narrow in the future. OPEC members are considering ways to raise the world market prices for LNG so that they will receive a price for natural gas more in line with that of crude oil.

The second obstacle to LNG is this. The industrial nations would have to depend heavily on the OPEC nations for LNG supplies, so the use of LNG does little to lessen the risk of supply cutoffs.

The third obstacle is the dispute over safety. Some scientific experts claim that the handling and transportation of LNG pose great hazards to the public; they say that accidents to LNG tankers (either when entering ports or when loading or unloading) and accidents at LNG storage facilities could cause catastrophic explosions and fires. Other equally reputable experts support the gas industry's view that such risks are well under control. They point out that LNG ships and storage facilities have been specially designed and are operated to protect against catastrophe. Currently, there are serious gaps in research and development programs to answer the safety questions.

In spite of the obstacles, Japan has moved ahead aggressively in importing LNG. In 1980, Japan used about two thirds of the world's LNG supplies, which still provided less than 5 percent of Japan's total energy. Japan's LNG imports are mainly from Indonesia and Abu Dhabi, with some from Alaska. In 1981, Japan had under discussion a spate of LNG projects with other OPEC countries as well as with Canada. But even if other countries adopt Japan's enthusiasm for LNG imports, it is still highly possible that the vast increase in natural gas output shown in the Upper Bound case for the OPEC and the non-OPEC developing countries will not be realized. The Lower Bound case assumes an output of 8.4 mbdoe for these two groups of countries instead of 15 mbdoe for the year 2000, still a substantial increase over the 1978 levels of about 2.8 mbdoe. The Lower Bound case also assumes a drop of 2.6 mbdoe in natural gas output in the OECD countries, with the figures for the United States being in line with Exxon's recent projections.[8] Overall, the Lower Bound case shows a world output growing from 17.2 mbdoe in 1978 to 20.0 mbdoe in the year 2000, in spite of a drop of 2.8 mbdoe in the United States.

Synthetic Oil and Gas

Engineers have been dreaming of the large-scale production of synthetic fuels for decades. A report issued in 1918 by the U.S. Geological Survey stated: "The production of oil in this country will continue to grow . . . because of the shale resource . . . No one may be bold enough to foretell what tremendous figure of production may be reached within the next ten years."[9] Of course, the Germans made several hundred thousand barrels of oil a day from coal in order to fuel their war machine during World War II. And by the 1970s, the world was pointing to South Africa as an example of what could be done to produce oil from coal; under the threat of oil cutoffs because of its politics, South Africa had built several commercial facilities. But because of its desperate supply situation, its low-cost labor for coal mining, and its willingness to compromise on pollution standards, South Africa represents a different situation from that in the other OECD countries. Elsewhere, there is some interest in Canada in increasing the modest level of oil being produced from its tar sands; in Australia, in producing oil from shale; and in Venezuela, in producing oil from heavy oil deposits. But no other country has the grandiose plans of the United States.

President Jimmy Carter, in July 1979, attempted to launch an ambitious program to start a synthetic fuels industry in the United States. He proposed an expenditure of $88 billion over ten years, at the end of which 2.5 mbd of oil and gas would be made from coal, shale, and, to a much lesser extent, biomass. Most experts believed that Carter's goal was not realizable, and Congress scaled it down substantially. But a number of private companies have strongly supported the synthetic fuels initiative. Indeed, in June 1980, Exxon stated that a 15 mbd synthetic fuels industry — 8 mbd from shale oil and 7 mbd from coal — was possible in the United States by the early part of the next century and described the required resources, including over $700 billion in 1979 dollars.

But it is still an open question as to whether the combination of the federal government, private industry, and others concerned with energy supplies can mollify the environmentalists and others concerned about the rapid development of a synthetic fuels industry to a sufficient extent to meet the Carter administration's goals and various industry hopes.

Moreover, the critics of the massive effort to develop a synthetic fuels industry on a crash basis point to uncertain but increasing cost estimates, various adverse environmental effects, and technological risks. There is also the fear that high developmental costs, high product costs, the inflationary impact, environmental hazards, premature technological selection, and the creation of a highly visible political undertaking would mean that if the crash program failed after elaborate promises and vast expenditures, the entire synthetic fuels program could be halted or dramatically slowed. In other words, the fear is that a crash program involving the construction of many commercial units based on the very limited experience of a few pilot plants, either in operation now or scheduled to go into operation soon, would do the long-term development of synthetic fuels more harm than good. Each commercial synthetic fuel facility would be one of the largest industrial facilities ever built in the world at one time, and a dozen would be required to meet the Lower Bound case by 1990 (0.6 mbdoe; see Table 2.4) and perhaps 150 would be required to meet the Upper Bound case by the year 2000 (8 mbdoe). Each plant, excluding coal mines and certain infrastructure, would probably cost over $3 billion in 1982 dollars.

The environmental effects of such large-scale operations in the United States are unknown. The production of oil from oil-bearing shale has a special set of environmental problems. Mining and crushing the shale oil results in fine particles that float into the sky. Furthermore, the shale rock that remains after the oil has been extracted expands to a volume greater than it previously occupied, which requires filling canyons with spent rock. Even the so-called in situ process, in which rock is burned underground, involves substantial mining and hence this type of environmental damage. Furthermore, the ranchers and farmers are unhappy with the prospect of their scarce water being used for the shale industry and the prospect that the minerals leeched out of the shale oil waste will cause the water supply to deteriorate. Thus, it is possible that coal — because it is available in the East, where water supply is no problem — will make a bigger contribution to the supply of synthetic fuels than will shale oil.

The overall uncertainty in the United States was heightened further by the election of Ronald Reagan as President. His position on developing synthetic fuels is unclear. But budget-cutting and a drive to restrict federal activity in the energy field have ensured that a lower level of federal funds will be available under the Reagan administration than under the Carter administration. Moreover, in 1981 and 1982, there was a sharp scaling back and slowing down of the synthetic fuels plans in response to inflation, high interest rates, and uncertainty about oil prices and demand in the future. Thus, even the Lower Bound estimates seem somewhat optimistic.

Coal

Many observers argue that the world coal production can double or triple between now and the year 2000. The Upper Bound case assumes that world coal production (including imports from the centrally planned economies) in the year 2000 will be 42 mbdoe, or about 2.5 times that in 1978 (see Table 2.5). This level of production seems to be an achievable goal. There are ample coal reserves, even excluding the enormous ones in the Soviet Union and China, to cater to the coal demand for some centuries.

The OECD countries account for about 80 percent of the world supply, excluding the centrally planned economies. The United States, in turn, accounts for a little over half of OECD production and consumption. By the year 2000, the United States will still be accounting for about half of

Table 2.4 World Production of Synthetic Oil and Gas from Coal and Synthetic Oil from Shale, Excluding Centrally Planned Economies, 1978, 1985, 1990 and 2000

A. Millions of barrels daily of oil equivalent

	1978	Upper Bound			Lower Bound		
		1985	1990	2000	1985	1990	2000
TOTAL	negligible	.2	1.8	8.0	neg.	0.6	3.2

B. Millions of tons of oil equivalent

	1978	Upper Bound			Lower Bound		
		1985	1990	2000	1985	1990	2000
TOTAL	negligible	10	90	400	neg.	30	160

Source: See Table 2.2; Upper Bound estimates are rounded figures from Exxon, *World Energy Outlook,* December 1979.

OECD consumption. But the Upper Bound case assumes that the United States will have become a major exporter, accounting, perhaps, for two thirds of OECD production. The Upper Bound case assumes a substantial increase in world trade in coal. In 1978, about 200 million tons of coal (2.1 mbdoe) were traded internationally; by the year 2000, this trade would have to expand fourfold. Substantial facilities for coal loading and un-loading would have to be built, as would ships to transport coal. And large-scale renovation of railroads, especially in the United States, would be needed to carry the increased quantities of coal. There should be no problem in raising the funds and carrying out the required work to ensure the higher levels of coal production.[10]

But barriers to increased coal production do exist. Perhaps most impor-tant is opposition arising from environmental damage associated with coal.

In mining, the principal areas of public concern are the reclamation of land required after surface mining, the settling of land from underground mining, acid drainage from the refuse of coal mines and coal preparation plants, pneumoconiosis (black lung disease), and underground gas explo-sions and mine flooding. But only relatively small increases in the price of coal would be needed to ensure that the best practices would be observed in mining, and given the great price advantage of coal over oil, such in-creases would not seriously damage coal's competitive position vis-à-vis oil. Today, both in the United States and elsewhere, the accident and ill-ness rates in the best coal mines are much improved and are now compara-ble with those in construction work and many sections of heavy industry.

Table 2.5 World Coal Production, Including Net Trade with
Centrally Planned Economies, 1978, 1985, 1990, and 2000

A. Millions of barrels daily of oil equivalent

	1978	Upper Bound			Lower Bound		
		1985	1990	2000	1985	1990	2000
United States	7.2	10.0	11.6	20.6	9.0	10.4	12.6
Other OECD	6.0	7.2	7.8	10.8	6.6	7.4	7.8
Others*	3.4	5.4	6.6	10.6	4.4	6.2	7.6
TOTAL	16.6	22.6	26.0	42.0	20.0	24.0	28.0

B. Millions of tons of oil equivalent

	1978	Upper Bound			Lower Bound		
		1985	1990	2000	1985	1990	2000
United States	360	500	580	1030	450	520	630
Other OECD	300	360	390	540	330	370	390
Others*	170	270	330	530	220	310	380
TOTAL	830	1130	1300	2100	1000	1200	1400

* Includes a small amount of imports from the centrally planned economies.
Source: See Table 2.2 and note 10.
Note: Production includes both steam and coking coal.

But even with these production barriers, demand rather than supply limitations will determine the future consumption of coal. Demand, in turn, will depend on environmental issues and growth in electricity demand. Perhaps the most serious environmental issue involves the burning of coal.

The major emissions that countries regulate include sulfur dioxide (SO_2), particulate matter, total suspended particulates (TSP), and nitrogen oxides (NO_2). Cleaning up some of the emissions, especially sulfur, will create new problems with waste disposal because large quantities of limestone sludge will result when sulfur is removed from the flue gas.

Joint action among nations to control pollution appears necessary in at least two cases. The first is acid rain, believed to result from the long-range airborne transport of emissions. In other words, it is believed that burning coal in some areas causes acid rain to fall in other areas, thereby damaging lakes and forests. Already, acid rain is falling in part of Canada, the United States, and in the Scandinavian countries. The second is the case of the carbon dioxide emissions produced when any fossil fuel is burned, but the problem is accentuated with coal because burning coal produces more carbon dioxide than burning oil. Furthermore, coal is expected to be a fuel long after the oil supplies have been depleted. It is generally agreed that the carbon dioxide concentration in the atmosphere has been rising and that a continuation of that trend could have a warming effect on the atmosphere. There is still great uncertainty about the overall impact of such changes, but some effects may be detectable before the end of this century, and that, if such proves to be the case, could have a substantial negative impact on coal consumption.

With oil and gas prices several times higher than that of coal on a BTU basis, industry is expected to increase its coal consumption. Still, in the United States and most other industrial countries, coal is and will continue to be used primarily as a fuel to produce heat to boil water to make steam to generate electricity. And in the future, as a fuel for power plants not yet built, coal will compete primarily with nuclear power rather than with oil. Coal's growth will still depend mainly on the growth of electricity, which in turn is difficult to project with certainty because of the lack of historical data in the price ranges in which the future price of electricity will be.

The Lower Bound case assumes that the combination of pollution problems that will arise from burning and the slower growth in electricity consumption will result in a coal output of 28 mbdoe in the year 2000, still about 70 percent more than the 1978 figure. Thus, there will still be substantial growth in coal consumption in either the Upper or the Lower Bound scenario.

Nuclear

The 1973 oil shock led the political establishments of the Western industrial countries to agree on at least one common response to OPEC: a rapid buildup in nuclear power.[11] Today, France is the only Western nation that has proceeded more or less in accordance with the plans developed at that time. But the advent of the Socialist government of François Mitterrand in France raises questions about the speed with which France will continue to develop nuclear power. The Soviet Union has also proceeded with the plans set after the 1973 oil crisis, but the Soviets are behind schedule. Nearly every other country has made basic changes in its plans, with their nuclear plans bogged down by a combination of political quagmires, high costs, and lower forecasts of electricity growth.

In America, as elsewhere, the combination of the low growth in electricity demand after the 1973 oil shock and the greater certainty of being able to complete a coal-fired power plant compared with a nuclear one caused

Table 2.6 World Production of Electricity from Nuclear Power,
Excluding the Centrally Planned Economies,
1978, 1985, 1990, and 2000

A. Millions of barrels daily of oil equivalent

	1978	Upper Bound			Lower Bound		
		1985	1990	2000	1985	1990	2000
United States	1.3	2.2	3.0	5.8	1.6	2.2	2.4
France	.1	.8	1.8	2.6	.6	1.4	1.8
Other OECD	1.0	2.4	3.8	8.0	1.6	2.4	3.0
Others	.1	.4	.8	2.4	.2	.4	.8
TOTAL	2.5	5.8	9.4	18.8	4.0	6.4	8.0

B. Millions of tons of oil equivalent

	1978	Upper Bound			Lower Bound		
		1985	1990	2000	1985	1990	2000
United States	66	110	150	290	80	110	120
France	7	40	90	130	30	70	190
Other OECD	49	120	190	400	80	120	150
Others	3	20	40	120	10	20	40
TOTAL	125	290	470	940	200	320	400

Source: See Table 2.2; 1978 figures from U.S. Department of Energy, *Monthly Energy Review.*
Note: The oil equivalents listed above are those equal to the amount of oil that would be consumed in producing the electrical output equivalent to that produced by nuclear power.

the electric companies to reduce their expectations for nuclear power. By 1975, American electric utilities had all but stopped ordering nuclear power plants; between 1975 and 1980, they purchased fewer than half a dozen reactors while canceling perhaps fifty previous orders and deferring a hundred others for periods ranging from five to ten years. There were similar moratoriums on new orders in the Federal Republic of Germany and in Sweden. And by 1980, ambitious plans to develop nuclear power had been indefinitely postponed in Norway, Denmark, Austria, Australia, and in New Zealand. Meanwhile, the situation in Canada, Japan, and in the United Kingdom lay between the French and the American extremes.

The outlook for nuclear power has also changed drastically in the developing world. The number of nuclear power plants to be built in these countries during the 1980s and 1990s is likely to be half or less than the number expected in the mid-1970s. In fact, significant additions of nuclear-generating capacity in the developing countries appear likely only in Argentina, South Korea, and Taiwan, although the governments of Brazil and India still hold to their public intent to move ahead with major programs.

Political opposition to nuclear power is not a homogeneous political movement either across countries or within a given country. Furthermore, the opposition is at different levels of intensities. Some critics, for example, will accept plants that are now operating or under construction but oppose new commitments. Others, however, oppose the technology in absolute terms and demand that existing facilities be dismantled as soon as possible. Many opponents of nuclear power worry about specific issues, such as reactor safety, nuclear weapons proliferation, and waste disposal. Others, however, especially outside the United States, oppose nuclear power as a symbol of an entire social structure that they wish to change.

There are three political situations that enhance the ability of critics to stop a nuclear program: first, if reactor siting decisions are subject to the approval of local or regional governments; second, if licensing decisions can be challenged in court; and third, if the national government does not have clear and total authority to certify the need for a new reactor. It is no coincidence that in the two countries in which all of these political obstacles favor nuclear critics — the United States and the Federal Republic of Germany — the development of nuclear power is at a virtual standstill. In France, on the other hand, none of these three conditions exists. Neither do they exist in the United Kingdom, but even there the program is proceeding more slowly than had previously been planned, primarily because of concern over the high costs and Britain's relative abundance of hydrocarbons.

During 1978, nuclear power provided the energy equivalent of about 2.5 mbdoe, plus another 0.4 mbdoe or so in the centrally planned economies. Slightly over half the total output outside the centrally planned economies was in the United States; the rest was widely dispersed among seventeen other countries, none of which accounted for more than 10 percent. By the beginning of 1980, an additional 116 gigawatts of operational nuclear capacity had been installed that was equivalent to about 3.2 mbd (at the ratio of 1.4 mtoe or .028 mbd per 1 GW of installed capacity). This installed capacity represented a little less than one third of the capacity that had been ordered since the beginning of the industry. The remaining two thirds —

some 270 reactors — were still under construction, and all were scheduled
to enter operation before 1990. By 1990, the total number of countries
using nuclear power reactors to generate electricity was expected to double, but only France and the United States would account for more than 10
percent of the world nuclear-generating capacity.

Compared with these numbers, the Upper Bound estimates of
9.4 mbdoe (equivalent to 336 gigawatts of installed capacity) for 1990
and 18.8 mbdoe for the year 2000 seem to be easily realizable goals (see
Table 2.6).

But there is a major question as to whether all the plants that are now on
order will be completed, either on schedule or at all. The Lower Bound
case is built on the assumption that less than half the growth projected between 1978 and 2000 by the Upper Bound case will actually occur. This
Lower Bound case is consistent with the view of some experts, who believe
that unless there is unassailable validation of the operational safety of light
water reactors and of the environmental safety of waste disposal during the
1980s, the nuclear option will not be available at the turn of the century.
Much of the capacity on order would be canceled; some of the existing
reactors—especially those near major metropolitan areas—would be
shut down; and the moratorium on new orders that has been in existence
in the United States for half a decade would continue there and elsewhere.

*Solar**

Solar energies, principally hydroelectric power, provided about 6.8 mbdoe
in 1978. The Upper Bound case shows this growing to 14 mbdoe by
the year 2000, or a growth rate of 3.3 percent per year (see Table 2.7). Such
a growth rate for a new technology would seem quite low and quite easily
achieved. On the other hand, much of the existing production is from hydroelectric power, and so many hydroelectric sites have already been exploited that electricity from this source is likely to grow more slowly than
will the energy supply from other renewable sources. Thus, the growth rate
of the newer energy sources will have to be quite large in order to achieve
the Upper Bound projections.

With proper government support, the Upper Bound projections could be
met, but there are so many variables and uncertainties regarding technological development and diffusion that no predictions can be made with
confidence. The Lower Bound scenario uses a somewhat arbitrary figure of
10 mbdoe by the year 2000.

* "Solar" encompasses a wide range of renewable sources, such as windpower, hydroelectric, wood, rooftop collectors, and photovoltaics.

Table 2.7 World Energy Production of Solar, Excluding the Centrally Planned Economies, 1978, 1985, 1990, and 2000

A. Millions of barrels daily of oil equivalent

	1978	Upper Bound			Lower Bound		
		1985	1990	2000	1985	1990	2000
OECD	5.6	5.8	6.0	10.0	5.0	5.6	7.0
Others	1.2	1.6	2.4	4.0	1.6	2.0	3.0
TOTAL	6.8	7.4	8.4	14.0	6.6	7.6	10.0

B. Millions of tons of oil equivalent

	1978	Upper Bound			Lower Bound		
		1985	1990	2000	1985	1990	2000
OECD	280	290	300	500	250	280	350
Others	60	80	120	200	80	100	150
TOTAL	340	370	420	700	330	380	500

Source: See Table 2.2.

Supply Summary

The Upper Bound case indicates that the total world supply of energy, including the net trade of the centrally planned economies, will increase from 94.6 mbdoe in 1978 to 168 mbdoe in 2000, an annual growth rate of 2.6 percent (see Table 2.8). In contrast, the Lower Bound scenario shows substantially lower growth for all types of energy supplies, with the result that the total energy supply in the year 2000 is shown as 112 mbdoe, equivalent to an annual growth rate of 0.8 percent.

World Oil Prices and the Overall Balance

World oil prices are of particular concern for several important reasons: their relation to inflation, their effects on domestic demand, the heavy foreign exchange flows associated with the oil trade, and the possibility that the international financial system will fail because of the large amount of funds to be recycled from surplus nations. Furthermore, the price of natural gas that is traded internationally will be linked to the price of oil, although exactly how was not clear as of 1982. The prices of oil and gas, including synthetic fuels, produced and sold within one country will

Table 2.8 World Energy Consumption and Supply,
Including Net Trade with the Centrally Planned Economies,
1978, 1985, 1990, and 2000

A. Millions of barrels daily of oil equivalent

	1978	Upper Bound			Lower Bound		
		1985	1990	2000	1985	1990	2000
Supply							
Oil	51.4	55.8	57.0	56.0	50.8	45.8	42.8
Natural Gas	17.2	21.0	25.0	29.2	17.6	18.8	20.0
Synthetic, from coal or shale	neg.	.2	1.8	8.0	neg.	.6	3.2
Coal	16.6	22.6	26.0	42.0	20.0	24.0	28.0
Nuclear	2.6	5.8	9.4	18.8	4.0	6.4	8.0
Solar (including hydro-electric), geothermal, other new sources	6.8	7.4	8.4	14.0	6.6	7.6	10.0
TOTAL	94.6	112.8	127.6	168.0	99.0	103.2	112.0
Demand							
OECD	77.4	88.8	97.0	119.6	77.4	77.4	77.4
OPEC	3.6	5.4	7.6	13.2	5.4	7.6	13.2
Other developing countries	13.6	18.0	23.0	35.2	16.2	18.2	21.4
TOTAL	94.6	112.2	127.6	168.0	99.0	103.2	112.0

B. Millions of tons of oil equivalent

	1978	Upper Bound			Lower Bound		
		1985	1990	2000	1985	1990	2000
Supply							
Oil	2570	2790	2850	2800	2540	2290	2140
Natural Gas	860	1050	1250	1460	880	940	1000
Synthetic, from coal or shale	neg.	10	90	400	neg.	30	160
Coal	830	1130	1300	2100	1000	1200	1400
Nuclear[a]	130	290	470	940	200	320	400
Solar (including hydroelectric), geothermal, other new sources	340	370	420	700	330	380	500
TOTAL	4730	5640	6380	8400	4950	5160	5600

Table 2.8   (*Continued*)

	1978	Upper Bound			Lower Bound		
		1985	1990	2000	1985	1990	2000
Demand							
OECD	3870	4440	4850	5980	3870	3870	3870
OPEC	180	270	380	660	270	380	660
Other developing countries	680	900	1150	1760	810	910	1070
TOTAL	4730	5610	6380	8400	4950	5160	5600

ªRounded to nearest .2 mbd in 1978.
Source: See Tables 2.2–2.7.

probably be heavily influenced by world oil prices, while the prices of coal and nuclear energy will probably not be directly linked to oil prices except as they affect inflation. Competition in the supply of coal is such that from 1970 to 1980, the price of coal, corrected for inflation, rose 50 percent compared to the 1000 percent increase for oil. And nuclear plants must compete with coal plants in supplying electricity.

One could calculate the different patterns of future oil prices that would result in energy demands that would align with the two supply scenarios just sketched. But one would need perfect knowledge about the effect of world oil prices on other energy prices and the effect of the price of any one energy source on the consumption of that energy and competing energies within various time periods.

Alas, in spite of considerable effort by econometricians, our knowledge of these price effects is currently too elementary to permit an accurate forecast; in any event, energy consumption is affected by numerous factors other than price and indirect price mechanisms — such as rationing, regulations on automobile speed and efficiency, and incentives for conservation investments — as well as by the level of general economic activity and the structure of industry. The repetition in 1979–82 of the pattern of "jagged peaks and sloping plateaus" has created a good deal of confusion and uncertainty about oil prices and energy demand. The president of Exxon summed up the situation when he observed in February 1982: "If you should gain the impression that we feel somewhat less certain than usual about forecasting near-term economic and energy trends, you are absolutely correct. Forecasting oil demand these days is a bit like solving Rubik's Cube — except that there are no pamphlets available to let one peek at the right answer." Still, the dominant 1980 assumption of regular real increases in the price of oil had, by 1982, given way to a dominant view that oil prices were likely to remain flat or decline for many years to come. Both views, but especially the latter one, were based upon extrapolations of short-term trends without sufficient reference to the overall context.[12] The oil price scenarios given below are based partly on the judgments of informed analysts, partly on the econometric models of others, and partly on plausible extrapolations from the last ten years.

The Upper Bound case assumes real price increases for oil *equivalent* to an average annual growth rate of 2 percent between 1980 and 2000. Thus, starting from about $30 per barrel in 1980, the price would rise to about $45 (in 1980 dollars) in 2000. On the theory that prices will rise faster if the supply is more limited, the Lower Bound case assumes an average annual increase of 4.5 percent, or a price of $72 (in 1980 dollars) by 2000. These two assumptions are on the upper end of the range of current speculation about the cost of producing synthetic oil. But, given the uncertain production costs, refining costs, and side effects involved in synthetic oil and the possibility that the price of natural oil might overshoot the cost of synthetic for an extended period of time, this range of $45 to $72 seems realistic.

The consumption of energy by some OPEC nations is likely to be fairly insensitive to the price of oil because the lower prices coinciding with high volumes produce — at least approximately — about the same amount of revenue as the higher prices and lower volumes. Thus, under either the Upper or Lower Bound scenario, the income of these countries from oil and gas exports in 2000 will probably be in the neighborhood of $450 billion (in 1980 dollars), or about 40 percent more than in 1980. True, one can hypothesize that the OPEC countries might well have more industry, and hence use more energy, in the Upper Bound case because the faster growth in the industrial nations under this scenario would create for OPEC more export opportunities in such products as petrochemicals. And the OPEC industry could be expected to be less energy efficient at lower oil prices than at higher oil prices. But any difference in the amount of industry might well be minor and, in any event, is likely to be strongly affected by developments, such as internal politics, that are independent of the scenarios. Thus, the same annual rate of growth in energy consumption, about 6 percent, is used for the OPEC countries in each scenario. This relatively high rate reflects both an expected high rate of economic growth and the development of some energy-intensive industries.

The energy growth rate for the developing countries outside OPEC obviously will be lower than that of the OPEC nations. On the other hand, it should be higher than that of the industrial nations for at least two reasons. First, the rate of population growth of the developing countries will be perhaps 2 percent higher than that in the industrial nations. Second, the earlier stages of industrial development, including electrification programs, typically have resulted in higher rates of energy growth than during the more advanced stages of development.[13] The different prices projected for the two scenarios, of course, will affect the growth rate of energy consumption of the developing countries outside OPEC. Although this effect might be alleviated somewhat by grants and low-interest loans from some of the OPEC nations, these relatively poor countries will not have an easy time paying for higher fuel bills.

An annual rate of growth of energy consumption for the developing countries outside OPEC of about 4.5 percent is assumed for the Upper Bound case and of about 2 percent for the Lower Bound case. The outlook for these countries is explored later in some detail, but a quick analysis reveals the possibly bleak situation facing them. It is often assumed that these countries will require at least 1 percent of annual energy growth per 1 percent of annual GNP growth; that is, a one-to-one relationship between energy and economic growth, or an energy coefficient of 1.0. This relationship, of course, would indicate an economic growth rate of not over 4.5 percent annually for the Upper Bound case. But if these developing countries could achieve the goal through this century that the seven major industrial nations set for themselves for the 1980s at the 1980 Venice summit — a coefficient of 0.6 — then an unlikely high economic growth rate of as much as 7.5 percent could result without being limited by the energy supply. For the Lower Bound case, the two different energy coefficients result in a much more dismal picture, with estimates of annual economic growth of 2 to 3.3 percent. With their populations growing by 2 to 3 percent yearly, it is clear that these countries would require an unexpectedly low energy coefficient — in other words, vast improvements in energy efficiency by some as yet unknown route — in order to achieve real annual increases in GNP per capita.

The OECD nations are expected to continue to be the world's major energy consumers, using about two thirds of the non-Communist world's energy by the turn of the century (see Table 2.8).

If the Upper Bound scenario should develop, the conventional analysis suggests that the Venice summit's goal of an energy coefficient of 0.6 could result in a satisfactory economic growth rate for the industrialized countries. For with an energy growth rate of 2.0 percent, a coefficient of 0.6 would allow an annual economic growth rate of 3.3 percent.

Another way to view the issue, however, is to estimate the annual improvement that will occur in the efficiency with which the total energy supply of a nation (or group of nations) is used. In this view, economic growth is a function of annual increases in energy efficiency — defined as

Table 2.9 World Consumption of Oil and Synthetic Liquids,
Excluding the Centrally Planned Economies,
1978, 1985, 1990, and 2000

A. Millions of barrels daily of oil equivalent

	1978	Upper Bound			Lower Bound		
		1985	1990	2000	1985	1990	2000
OECD	11.6	13.1	13.3	10.2	39.0	33.7	28.3
OPEC	2.5	4.0	5.0	9.0	4.0	5.0	9.0
Other developing countries	7.3	8.5	9.6	10.8	7.8	7.4	7.5
TOTAL	51.4	55.9	57.9	60.0	50.8	46.1	44.0

B. Millions of tons of oil equivalent

	1978	Upper Bound			Lower Bound		
		1985	1990	2000	1985	1990	2000
OECD	2080	2170	2165	2010	1950	1685	1415
OPEC	125	200	250	450	200	250	450
Other developing countries	365	425	480	540	390	370	375
TOTAL	2570	2795	2895	3000	2540	2305	2200

Note: The OECD and other developing countries' shares of oil consumption were assumed to be the same as their respective shares of total energy. Oil consumption includes synthetic liquids, which are assumed to be equal to half of the total synthetics shown in Table 2.4 in 1990 and slightly less than half in 2000.
Source: See Table 2.2.

BTUs or any other measure of energy usage per unit of GNP — and any annual increase in energy availability. From 1973 to 1978, for example, the IEA countries increased their GNP per unit of energy consumed at an average annual rate of 1.7 percent, and the IEA expects that this figure will be 1.6 percent between 1978 and 1990. This means that in the Upper Bound case, the rate of annual economic growth could be a satisfactory 3.6 percent — 1.6 percent increase as the result of improved efficiency plus the 2.0 percent increase in energy consumption.

For the Lower Bound case, it is not meaningful to think in terms of an energy coefficient. The reason is that zero growth for the OECD nations is expected in energy supplies through the year 2000, and an energy coefficient of any size would still result in zero economic growth. *In that case, economic growth would be limited by the ability to use the available energy more efficiently.* For the Lower Bound case, the annual improvements in energy efficiency expected by the IEA would limit economic growth to the dismal rate of only 1.6 percent annually. But for 1977 through 1980, the United States achieved substantially better results in improving its ratio of GNP to energy use and hence improving its implied energy efficiency: 2.6 percent in 1977, 2.0 percent in 1978, 2.6 percent in 1979, and 3.3 percent in 1980. These annual improvements in energy efficiency were accompanied by the annual growth in GNP of 5.3, 4.3, 2.4, and −0.2 percent. In 1979, for example, the 2.6 percent efficiency improvement combined with the 2.4 percent GNP growth resulted in a drop in energy consumption of about 0.2 percent. These U.S. results clearly look promising, but even greater annual increases in efficiency would be required with zero energy growth to realize the economic growth rate of 3 to 3.5 percent that many feel is necessary. Another imponderable is the extent to which political opposition to energy projects would be overcome if the alternative were slow economic growth, periodic energy shortages, and possible international conflict.

But either the Upper or Lower Bound case will call for a considerable shift away from oil to other energy forms. For the OECD oil supply, instead of growing along with other energy sources, will remain about level in the Upper Bound scenario (see Table 2.9). And in the Lower Bound scenario, the supply of oil available to the OECD, instead of remaining at the 1978 level (as does total energy supply), will fall by 32 percent between 1978 and the year 2000. The OECD countries are therefore facing not only a possible stringent supply of total energy but also the adjustment away from a heavy dependence on oil — with oil providing only 35 percent of total OECD energy in the year 2000 under either the Upper or Lower Bound scenario, compared with 54 percent in 1978.

* * *

The chapters that follow examine in depth the problems facing the United States, Japan, and Western Europe, respectively, in reaching adequate levels of economic growth with limited energy supplies. They also consider the consequences that might follow from failing to do so.

3

The Bedeviled
American Economy

by Robert Dohner

What the 1960s were for the United States politically, the 1970s were eco-nomically — a decade of turmoil and disappointment. Economic growth was harder to achieve and inflation harder to subdue than had been imag-ined before. As inflation rates climbed, even during periods of substantial unemployment, attention focused on what came to be known as supply shocks, events that sharply raised the prices of basic commodities, begin-ning with foodstuffs and raw materials. But after 1973, these concerns were overwhelmed by successive increases in OPEC oil prices.

The approximate quadrupling of world oil prices between October 1973 and January 1974 was followed in the United States by the worst recession since World War II, and a second price rise in 1979 brought to an end what was in many respects a hesitant recovery. In the early 1980s, Americans faced imported oil prices six times higher in real terms than those in 1970. And, although adjustments to higher energy prices have already begun, the U.S. economy remains vulnerable to future oil price increases. In 1982, the American economy was still plagued by high unemployment, and a solid recovery had yet to emerge. Indeed, the reestablishment of stable economic growth stands as a dominant challenge for the United States in the 1980s.

By any economic measure, the 1970s was a frustrating decade (see Table 3.1). Economic growth was lower and unemployment higher than in the 1960s. Inflation was much higher than in the previous decade and, despite ten years of anti-inflation policy, higher at the end of the seventies than at the beginning. In addition, the 1970s were much more unstable than the 1960s, making planning and profitability less certain. While oil prices are not solely to blame for the problems of the 1970s and the early 1980s, they bear considerable responsibility.

Prelude to the Embargo

The events of the early seventies drastically changed the character of the world oil market. Not only did the price of OPEC oil skyrocket, but the ownership of production facilities and the control of production decisions passed from the oil companies to the producer governments. The Tripoli Agreements in 1971, the Arab-Israeli war in 1973, and other political events leading up to the January 1974 meeting of the oil ministers all made headlines. Less attention has been paid to the concurrent changes in the economic environment that greatly facilitated the revolution in the world oil market. Chief among these were the exhaustion of surplus producing capacity in the United States in the early 1970s and the coordinated economic expansion in 1973 of the Western countries that compose the Organization for Economic Cooperation and Development (OECD).

During the 1960s, the oil import program of the United States limited petroleum imports (with the exception of imports of residual fuel oil) to 12 percent of domestic production. Thus the United States drew on world markets for only a fraction of its additional oil demand. The United States also maintained surplus oil-producing capacity, which could be called upon to dampen fluctuations in world supplies, but this margin diminished during the sixties.[1] U.S. oil production peaked in 1970, and spare production capacity was exhausted in 1972. Henceforth, the United States would draw on world markets for all of its additional oil requirements as well as enough to make up for falling domestic production, and so add signifi-

---

Table 3.1   Economic Performance in the United States, 1960–1980

	1960–70	1970–80
Average Real GNP Growth Rate	3.9%	3.2%
Variability of Growth Rate[a]	1.9	2.6
Average CPI Inflation Rate	2.7	7.8
Variability of Inflation Rate[a]	1.8	3.1
Average Unemployment Rate[b]	5.3	6.4

[a] Standard deviation of annual rates, in percentage points.
[b] The higher average unemployment rate in the 1970s reflects in part a change in the demographic composition of the labor force toward groups with higher average unemployment. Individual demographic groups still showed higher unemployment in the 1970s. For example, the average unemployment rate of married men was 2.7% in the 1960s and 3.2% in the 1970s.
Source: U.S. Council of Economic Advisers, *Economic Report of the President, 1981.*

---

cantly to the demand for OPEC production. The effect was dramatic. U.S. imports of crude oil rose by 145 percent from 1970 to 1973, by 46 percent between 1972 and 1973 alone.[2]

The simultaneous expansion of the OECD economies in 1973 also exerted pressure on world oil markets. After a relatively mild but widespread slowing of economic activity in 1970, a majority of the OECD countries adopted stimulative policies the next year. The resulting upward turn was notable for both its breadth and speed; between the first half of 1972 and the first half of 1973, real output rose by 7.5 percent and industrial production by 10 percent in the OECD countries. This rise in activity, coupled with the turn by the United States toward imported oil, resulted in a vast increase in oil import demand. Between 1971 and 1973, net oil imports into the OECD countries rose by 22 percent, straining world oil production capacity.[3]

A variety of other events combined to make this a period of great uncertainty, even upheaval. The Russian crop failure of 1972, coming at a time when world grain stocks were dangerously low, had greatly increased food prices. The rise in OECD economic activity along with shortages in basic materials industries put pressure on raw materials prices beginning in 1972. Additional pressure came from the breakdown of the pegged exchange rate system and accelerating rates of inflation, which fueled speculative demand for materials. Raw materials prices jumped alarmingly; the London *Economist* index of twenty-eight raw materials prices more than doubled between 1972 and early 1974, an extraordinary rise, unparalleled in its one-hundred-and-fifteen-year history. The rapid rise in oil demand, coming in a period of underlying uncertainty, set the stage for the scramble in spot markets that followed the oil embargo and the ensuing rise in OPEC prices.

The Gross National Product and the Oil Shocks

The oil embargo, with first a 10 percent, then a 25 percent, fall in production by the Arab members of OPEC, had a shattering effect in the consuming countries and sent policy-makers rushing to prepare allocation plans and to shift the burden of any shortfall to less essential sectors. The period between the time at which supply reductions first began to be felt and the announcement of the end of the embargo was quite short, and the limited availability of oil itself seems to have had little impact on the OECD economies. The real effect followed the January rise in prices. The simultaneous expansion of the OECD economies in 1972 and 1973 gave way to a simultaneous contraction in 1974 and 1975, the worst recession in the industrial world since the 1930s. In the United States, GNP fell by 6 percent from its peak at the end of 1973 to the beginning of 1975, while the unemployment rate jumped from 4.7 to 9 percent.

The way in which an oil price rise affects the level of GNP in the consuming countries has been described as the OPEC tax, since the rise in price transfers income from the oil-purchasing to the oil-producing nations. In the consuming countries, an oil price hike raises the costs of direct consumption of energy and indirectly raises the prices of all goods that require energy for their production, all without raising the incomes of residents. Since, over short periods of time, the quantity of energy purchased is not very sensitive to price, higher energy prices reduce the income left over to purchase domestic goods, and so demand and production fall.[4]

The amount of the OPEC "oil tax" may be estimated by calculating how much more had to be spent for the 1974 level of oil imports after the price increase. For the United States, measuring the price change from the third quarter of 1973 to the fourth quarter of 1974, the oil tax was $16.4 billion (1974 dollars), or 1.2 percent of GNP.[5] Higher prices for oil meant lower real incomes for American consumers, and the real personal income of Americans fell by 1 percent from 1973 to 1974. The character of the tax system also contributed to the fall in consumer incomes. The inflation of 1973 and 1974 pushed taxpayers into higher tax brackets, taking a larger proportion of their incomes in taxes. The remainder, personal disposable income, fell by 1.5 percent in real terms from 1973 to 1974.[6]

The weakness in consumer demand was a major cause of the recession in the United States, especially in the last quarter of 1974, when demand plummeted. The automobile industry was particularly hard hit, with domestic sales declining by 20 percent. Three additional features made this recession more severe. Businesses overbuilt inventories in 1974, and massive sales from inventories drastically reduced production at the end of that year and at the beginning of 1975. Second, all of the OECD countries were affected by the oil shock, which made their recessions mutually reinforcing, and the demand for U.S. exports fell. (Overall, the volume of OECD exports fell by 4 percent from 1974 to 1975, primarily owing to a decline in trade among themselves. OECD exports had only fallen once before in postwar history.)[7]

Finally, fiscal and monetary policy tightened during this period, worsening the recession. Fiscal policy automatically tightened as inflation forced taxpayers into higher brackets, as noted above, draining consumer incomes. No compensating change in fiscal policy was taken until the March 1975 tax cuts; by that time the economy had hit bottom. Monetary policy was also restrictive. The Federal Reserve maintained the previous year's rate of money growth in 1974, and the real money supply fell by 4 percent during the year.[8] Interest rates reached record heights in midyear, and housing starts plummeted. Although the economy received limited help from the tax cuts, monetary growth remained very cautious in 1975 and 1976.[9]

A revival of consumer demand, including residential construction demand, and an increase in imports by OPEC countries led the recovery from the recession. Although the growth of output in the United States was reasonably firm, and much stronger than the recovery in Europe or Japan, business investment remained weak long after the recession had passed, and the path of productivity was particularly disturbing. By the end of 1978, shortly before the second oil shock, the U.S. unemployment rate had been reduced to 5.8 percent, but productivity was only 5 percent above its 1973 level.

The reduction and then the suspension of oil exports during the revolutionary turmoil in Iran prompted a second round of speculative buying and soaring spot market prices, leading to the second oil shock in 1979. In percentage terms, this price rise was not as great as that of 1974; the price of imported oil rose 118 percent from the fourth quarter of 1978 to the first quarter of 1980.[10] However, a variety of factors made this as serious as the previous oil shock. The United States was more dependent on imported oil in 1979 than in 1973, and the approximate doubling was of a much higher initial oil price. So the absolute increase in price was much higher than in 1973–74. The OPEC oil tax, the increased cost of maintaining the 1979 volume of imported oil in 1980, jumped by $44 billion (in 1979 dollars), or 1.8 percent of GNP.

There were also significant income shifts within the United States. During the first oil shock, the oil price control program limited the redistribution of income within the United States. In 1979, a much greater fraction of U.S. oil production was in uncontrolled categories, and the phased decontrol of oil prices increased this fraction through the year. The extent of the income shift may be measured by the rise in domestic oil prices. By the first quarter of 1980, refiners were paying an average of $8 (in 1979 dollars) more for each barrel of domestic oil than they had at the end of 1978, or $30 billion more at 1979 volumes, with prices continuing to rise through 1980. If income losers cut back their expenditures faster than gainers increase theirs, then a major internal income shift such as that caused by an oil shock would depress aggregate demand.

Finally — in comparison to the rapid rises in primary energy prices — processing costs, distribution costs, and energy taxes were much more stable. These margins softened the increase in prices to energy users during 1974 but accounted for a much smaller share of the delivered price of energy in 1979. This gave the second oil price increase more leverage on delivered energy prices, and prices paid by energy users actually rose by more in percentage terms than they had with the quadrupling of OPEC prices in

1974 (see Table 3.2). The President's Council of Economic Advisers esti-mated that the net drag on aggregate demand from all of these sources would amount to 3 percent of GNP by the end of 1980.[11] No attempt was made to support production; the Carter administration had taken actions to slow the growth of the economy in 1979, and these were maintained after the OPEC price rise.[12] Thus, the second oil shock had every indica-tion of being as severe as the first.

With the rapid rise in energy prices and the move of economic policy to-ward restraint, it is not surprising that economic growth slowed considera-bly in 1979. What was surprising was the economy's resilience during this period, as production continued to advance through the first quarter of 1980. A rapid rise of interest rates and the virtual collapse of the bond market early in 1980 led to a fear that control over inflation had been lost.[13] In March the administration proposed expenditure cuts, the Federal Reserve instituted credit controls, and the sharpest and briefest postwar re-cession occurred in the second quarter. But the recovery from the recession was also short-lived. Strict monetary policy and high interest rates brought on a further recession in mid-1981. By March of 1982, the unemployment rate hovered around 9 percent.

Inflation

No phenomenon bedeviled forecasters more than the movement of prices in the 1970s. In large part this was due to events that were perhaps unfore-seeable, as inflation rates fluctuated in response to energy and food price trends. Even the very definition of "the inflation rate" became difficult, since the general trend of price increase was often masked by large relative price changes. To illustrate the effects of oil shocks on inflation, Table 3.2 shows various measures of price change. Imported oil prices and prices to energy consumers led all rates of price increase, but their rise varied con-siderably over the decade. Two measures of the general price level follow in the table. One, the consumption deflator, measures the prices of the goods people buy (including imported oil). The second, the GNP deflator, charts the prices of the things Americans produce and earn their incomes from. The real income effect of the rise in imported oil prices can be seen by comparing these two indexes in 1974 and in 1979–80, when the prices of consumption goods jumped ahead of output prices.

The surge in both deflators after the two oil shocks, evident from the table, occurred as energy prices affected the indexes directly and as they worked their way through into costs of finished goods.[14] Once the adjust-ments to higher external oil prices took place, inflation would of course not vanish, but would recede to some "underlying" inflation rate deter-

Table 3.2 Rates of Price Increase
(fourth quarter to fourth quarter)

	1972	1973	1974	1975	1976	1977	1978	1979	1980	1981
Imported oil	2.0%	38.1%	150.2%	17.8%	-8.4%	7.7%	0.9%	82.7%	30.1%	2.8%
User price of energy	4.3	20.4	36.1	12.2	7.6	9.7	7.5	42.2	22.7	13.2
Consumption deflator	3.5	7.5	11.9	6.2	4.7	5.7	7.6	9.5	10.1	7.8
GNP deflator	4.2	7.5	11.0	7.5	4.8	6.2	8.2	8.1	9.8	8.8
Hourly earnings	6.7	6.7	9.3	7.8	7.0	6.8	8.4	8.1	9.7	8.4

Source: Imported Oil Price, 1972–74: U.S. Customs data (unit value of crude oil imports and $0.95 transportation charge), 1974–1980. U.S. Department of Energy, *Monthly Energy Review*, refiner's acquisition price of imported oil. User Price of Energy: Constructed index based on energy component of consumer price index and wholesale price index for fuels and power. Consumption and GNP deflators: U.S. Department of Commerce *Business Statistics, 1977*, and U.S. Council of Economic Advisers, *Economic Report of the President, 1982.* Hourly earnings: Index of hourly earnings adjusted for overtime and interindustry shifts, U.S. Department of Labor, *Employment and Earnings.*

mined by increases in domestic prices and costs. Labor costs are by far the largest component of total production costs, which makes the labor market and the wage process of particular importance. The way in which wages react to an oil price change is crucial for determining the effects of the shock on the underlying rate of inflation. If wages accelerate after an oil price increase, as workers try to make up for their loss of real income, then an oil shock may leave a legacy of a higher underlying inflation rate.

The wage process in the United States differs from that of most other countries in ways that make adjustment to oil price changes both easier and more difficult. Wage contracts in the United States are longer, with two- to three-year durations the norm, and bargaining is decentralized. Furthermore, contracts are not synchronized but overlap substantially. There is no annual wage round in the United States, no spring struggle, as in Japan, to determine labor costs for the next year. This pattern of overlapping employment contracts imparts a considerable degree of inertia to labor costs, and it is the momentum of wages that complicates adjustment in the United States.

Although wage inflation accelerated in 1974 in the United States, it took nowhere near the jump that characterized many OECD countries. As a result, real wages fell in the United States while they were maintained in other OECD countries, squeezing profits. This was an important factor in the relative strength of the U.S. recovery, as expansion of GNP was profitable when spending increased.[15]

Several of the European countries and Japan were able to overcome this obstacle after the second oil shock. Wage negotiators agreed not to attempt compensation for the external price rise, and profits remained at a level that would allow growth to resume.[16] With its decentralized negotiation and overlapping wage contracts, such coordination is much more difficult to achieve in the United States. A persistent fear was that energy-induced inflation would spill over into the wage-setting process if negotiators attempted to maintain previous real income levels, and thus increase the momentum of the underlying inflation rate. This fear conditioned American economic policy after the two oil shocks.

Economic Policy

As explained above, U.S. monetary and fiscal policies were restrictive in the wake of both oil shocks, further depressing economic activity. Although the failure of economic policy to support GNP and employment has been criticized, it probably was the better of two unappealing choices.[17] A supply shock, such as an oil price rise, raises both inflation rates and unemployment, and monetary or fiscal policy that responds to one worsens the other. To have maintained the level of GNP after the oil shocks risked making the higher inflation rates of the oil price adjustment a permanent part of the underlying, or "home-grown," inflation rate. With the inertia of the wage-setting process in the United States, the incorporation of oil inflation into wage agreements would be extremely hard to reverse. This fear of exacerbating domestic inflation is expressed in several of the Presidents' annual *Economic Reports:* "There is still a danger that [1979's] sharp rise in energy and housing prices may spill over into [1980's] wages and costs and thus become built into the underlying inflation rate. Fighting inflation must therefore remain the top priority of economic policy."[18]

The rise in inflation and unemployment after an oil price rise poses a cruel policy choice, and both the Ford and Carter administrations restrained demand at times when GNP and employment were weakening. The response had the character of a holding action, and the success of the policy could only be measured by the higher inflation rates that were avoided. Yet by 1981, labor costs were rising about 4 percentage points faster than they had in the early seventies, the combination of a two-percentage-point acceleration of wages and an equal decline in productivity growth.

Energy, Investment, and the Productivity Problem

Two additional features of the American economy in the last decade stand out: the weak recovery of investment from the 1974–75 recession and the disappointingly poor performance of productivity. Both have roots in the sharply higher price of energy. Since energy is an important factor in the production process, a steep rise in relative energy prices will change production decisions involving the mix of inputs, may make certain facilities obsolete, and may also affect the profitability of additional investment or hiring. Indeed, the effects on supply decisions and the future growth of productive capacity may outweigh any other effects of higher energy prices on the industrial economies.

As energy prices rise, producers will substitute, if possible, other factors of production for now more expensive energy. The greater use of these other factors lowers their productivity (output per unit of input). For example, oil tankers now travel at slower speeds to save fuel, thereby substituting labor (crew days) and capital services (ship days) for energy. At the same time, this reduces measured labor and capital productivity in shipping (ton miles per crew day, etc.). To the extent that producers generally could substitute labor for energy, they would lower measured labor productivity.

The role of higher energy prices in the drop of productivity in 1974 and its slow recovery since then is the subject of considerable controversy. If the rise in energy prices led producers to substitute labor for energy, it must have been because they could lower costs. This casts some doubt on the importance of energy in explaining the productivity lapse. Energy's share of production costs is quite small, about 5 percent for the economy as a whole, while the share of labor is over 60 percent.[19] A 1 percent increase in labor would have to be matched by a sizable reduction in energy use to result in lower production costs. No study to date has been successful in explaining the 1974 fall in productivity, and measuring the contribution of energy prices has been hampered by lack of adequate energy data.[20]

Data on the use of energy by producers are difficult to assemble and subject to error. Later in this chapter, we present data for energy use in manufacturing: As Table 3.6 reports, energy use per unit of output in manufacturing (labeled "efficiency" in the table) dropped by 14 percent between 1971 and 1974.[21] A drop of this size is large enough to account for much of the fall in labor productivity (increase in labor use), but a careful study would have to determine how much of this reduction was due to a trend reduction and how much it represented "housekeeping" savings, with little effect either on output or on other factors. One additional piece of evidence deserves mention. The drop in labor productivity in 1974 was sharp and unprecedented, and a second drop took place in 1979. Furthermore, the break in the productivity growth trend that occurred in the United States also occurred in other OECD countries.[22] Both the breadth and the timing of the productivity break are evidence of the importance of higher oil prices.

A related concern is the effect of higher energy prices on investment. Higher energy bills reduce the income available to distribute to labor and capital, and cutting back on the use of energy makes capital and labor less productive. The future growth of the labor force will be determined largely by population growth and its age structure, but prospective returns to new equipment and structures will determine how much capital investment will take place. By lowering the returns to capital, higher energy prices could slow investment and therefore economic growth. The extent to which investment is discouraged depends on how closely capital and energy are associated in production. If energy and capital are easily interchangeable, then higher energy prices are a spur to capital growth. If, instead, the important choice in production is the degree of mechanization (how much of a capital-energy composite to give labor), then higher energy prices discourage the growth of capital.[23] Slower growth of energy and capital of course mean slower economic growth.

The behavior of fixed investment by business since the 1974–75 recession has indeed been troublesome. Investment, the weakest component of the recovery, only regained its previous peak in the second quarter of 1978. Over the 1974–79 period, the net capital stock grew by only 3.0 percent per year, compared to a 1960–74 growth rate of over 4 percent. The labor force grew more quickly after 1974 than before, so that the capital stock per employed worker actually declined slightly between 1974 and 1979.[24] Since an increasing share of investment went to the energy industries in this period, investment in other areas was somewhat weaker than portrayed above.[25] Although other factors slowed investment, such as the uncertainty of the seventies and higher rates of inflation, a forceful case for the effect of higher energy prices has been made by Dale Jorgenson of Harvard University.[26] In projections of economic growth, the effect of higher energy prices in lowering the future capital stock is one of the most important factors.

The London *Economist* has aptly referred to the 1970s as the "oil decade." As we have indicated, the effects of higher oil prices were pervasive. Most economic problems of the decade — high inflation, high unemployment, and slower economic growth — while by no means totally due to higher energy prices, have strong connections to the two oil shocks. Furthermore, the effects of higher energy prices were not solely macroeconomic. At the level of the individual, radically higher prices for energy prompted shifts and adjustments that in turn affected the demand for energy, the demand for various products and services, and the vitality of individual industries.

Adjustments in Energy Use

Despite early pessimism about the ability to separate energy demand from economic growth, the United States has made adjustments in its use of energy. As Table 3.3 shows, the United States reduced the energy use associated with each dollar of GNP by about 17 percent between 1973 and 1981. Had energy demand grown in step with the level of GNP, the United States would have consumed another 7.3 million barrels a day of oil or its equivalent in 1981. In this section we will investigate several reasons for the reduced growth in energy demand. However, while the United States has been successful in trimming the growth of demand, the share of energy consumption derived from petroleum rose steadily until 1978. As is apparent from the last column of the table, petroleum plays the role of the marginal, or "swing," fuel, with its share dropping during economic contractions (1973–75, 1979–80) and rising in expansions (1975–77).

The factors responsible for the decline in energy per unit of GNP fall into two categories, those that involve reduced energy use in individual applications and those that reflect changes in the structure of economic activity.[27]

Table 3.3 United States Energy Consumption, 1960–1981

	Energy Consumption (*mbdoe*)	Energy/GNP*	Petroleum as % of Total Energy Consumption
1960	20.8	59.8	45.2
1965	25.0	57.0	43.9
1970	31.5	61.6	44.2
1971	32.2	60.9	44.7
1972	33.8	60.4	46.0
1973	35.1	59.5	46.7
1974	34.3	58.3	46.0
1975	33.3	57.3	46.3
1976	35.1	57.3	47.2
1977	36.0	55.6	48.6
1978	36.9	54.4	48.6
1979	37.2	53.3	47.0
1980	36.5	51.5	44.9
1981	35.1	49.3	43.2

*1000 BTUs per 1972 $ of GNP.

Source: Department of Energy, Energy Information Administration, *Annual Report to Congress, 1980,* Vol. 2; *Monthly Energy Review,* February 1982; and U.S. Council of Economic Advisers, *Economic Report of the President, 1982.*

The postwar period saw a shift in energy consumption toward an increasing use of electricity because of its convenience and flexibility. However, because of the energy losses in generation and distribution, electricity involves more primary energy per unit of energy delivered. As a result, energy losses as a proportion of total energy consumption increased substantially.

Between 1960 and 1973, the total consumption of electricity grew by 7.2 percent per year, or 3 percent per year faster than the growth of GNP. This shift toward electricity continued after the oil embargo, but at a much slower pace. Between 1973 and 1978, electricity consumption grew by 3.2 percent per year, or only 0.7 percent faster than the growth of production.[28] In addition, the average conversion efficiency of electric generation improved slightly after 1973. While conversion losses per dollar of GNP still rose after the embargo, they did so at a much slower rate, moderating the growth in primary energy consumption.

Changes in residential and commercial energy uses contributed only modestly to changes in the energy-GNP ratio, but modifications in the transportation sector, particularly in private automobiles, were important in slowing energy growth.[29] The large and heavily driven American automobile has come almost to symbolize the use of energy in the United States, and because of the limited possibilities for substituting other fuels, the automobile is a crucial part of the U.S. dependence on petroleum. American automobiles became larger and heavier during the 1960s and their fuel efficiency steadily fell, reaching a low point of 13.1 miles per gallon in 1973. The end of low prices and of the unlimited availability of gasoline marked a turning point. At first hesitantly and then quite rapidly, consumers shifted toward lighter and more efficient vehicles. The average car on the road is about six years old, and so the characteristics of the operating fleet change more slowly, but by 1980, average fuel consumption was 15.2 miles per gallon, the highest level since the 1950s.[30]

It is in the industrial sector that the use of energy relative to total GNP has declined most sharply. Between 1973 and 1980, while GNP rose by 18 percent, the industrial use of energy (excluding electricity conversion losses) actually fell slightly. Both increasing energy efficiency and changes in the product mix played important roles in the reduction of energy use. To understand the changing industrial use of energy we proceed somewhat indirectly, looking first at changes in consumer demand.

The effect of rising energy prices on the price of goods would vary depending upon the energy contents. Prices for direct energy consumption would rise by the greatest amount, and prices of goods would rise vis-à-vis those of services, which require less energy. In the face of the dramatic increases in energy prices in the 1970s, one would expect demand to shift away from direct energy consumption and consumption of goods toward services.[31] This is confirmed in Table 3.4, which separates consumption expenditures into quantities of goods, nontransportation services, energy, and purchased transportation.[32] A gradual shift in demand from goods to services is evident throughout the period covered in the table, but it accelerated after 1973, as energy and materials price increases pushed up the prices of goods.[33] The effect of higher energy prices on direct energy consumption is also noticeable in the table, and the volume share of energy expenditures fell back to its level in the mid-1950s.

The last two columns of Table 3.4 reflect the growing dominance of the automobile in personal transportation in the United States, as it replaced other forms of transportation. Energy price increases in the 1970s not only raised the cost of operating a car, but also its purchase price due to its high material content. The rise in the relative price of operation is reflected in the table in a slowing of expenditure growth after 1973. Automobile sales are highly sensitive to the business cycle and to interest rates, which makes

Table 3.4 Personal Consumption Expenditures at Constant (1972) Prices
Volume Shares (%)

	Total PCE (billion $)	Goods[a]	Services[b]	Energy[c]	Purchased Transport[d]	Addendum: Motor Vehicle Sales/ Operations[e]
1978	900.8	50.7	42.2	6.3	0.8	12.9
1973	767.7	52.0	40.4	6.7	0.8	13.2
1969	655.4	52.1	40.3	6.6	1.0	12.1
1965	558.1	52.9	39.7	6.4	1.1	11.9
1956	406.3	55.2	37.1	6.3	1.5	11.1
1951	342.3	55.9	36.3	5.7	2.1	10.4

[a] Durable plus nondurable goods less gasoline, fuel oil, and coal.
[b] Services less gas, electricity, and purchased transportation.
[c] Gasoline, fuel oil, gas, and electricity.
[d] Airline, bus, railway and transit systems.
[e] Motor vehicle sales, parts, repair, gasoline, insurance, parking, and tolls.
Source: U.S. Department of Commerce, Bureau of Economic Analysis, *National Income and Product Accounts, 1929–74*, p. 92, and *Survey of Current Business*, July 1976, July 1979.

precise comparisons difficult, but 1973 may prove to have been a turning point for the share of expenditure devoted to the automobile.

The shift in demand toward services was of course reflected in the pattern of GNP, which accounts for part of the reduced industrial use of energy. Here we focus on manufacturing, which accounts for over 90 percent of industrial energy use. The shift in demand away from goods may be seen in the difference in growth between total GNP and manufacturing. Between 1956 and 1973, manufacturing grew at a slightly faster rate than GNP as a whole. Between 1973 and 1978, GNP grew by 14.5 percent, while manufacturing output grew by only 10 percent.[34] The slower growth rate of manufacturing accounted for some of the reduction in industrial energy use; had manufacturing grown in step with total GNP, its use of energy would have been 4 percent higher in 1978.

More important changes were taking place within manufacturing, both in the composition of goods and in the efficiency with which energy was used in individual industries. The use of energy in manufacturing is concentrated in a few industries: metals, paper, chemicals, and stone, clay, and glass.[35] Together, these industries account for 23 percent of the value of

Table 3.5   Energy Use in Manufacturing

	Energy Intensity (1000 BTUs per 1972 $ Output)			% Share of Total Manufacturing Output		
	1971	1974	1977	1971	1974	1977
Primary Metals	199.1	169.9	187.8	7.8	8.5	6.1
Stone, Clay, & Glass	141.8	130.5	123.3	3.6	3.4	3.1
Chemicals	138.3	124.8	108.2	7.8	7.8	8.3
Paper	128.7	102.9	100.7	3.9	4.3	3.9
All Others	18.5	15.6	13.7	77.0	76.2	78.6
Total Manufacturing	50.6	44.7	38.9	100	100	100
Addendum: Petroleum Refining	376.9	379.6	340.5	2.9*	2.7*	2.4*

* Output in petroleum refining divided by total output of all manufacturing less petroleum refining, expressed as a percentage.

Source: Energy data: Bureau of the Census, *Census of Manufactures, 1972, 1977,* and *Annual Survey of Manufactures,* energy data for petroleum refining from Department of Energy, *Monthly Petroleum Statement* (annual summaries). Energy consumption in primary metals is adjusted for captive coal consumption. Output data: Bureau of Economic Analysis, unpublished data. For petroleum refining, *changes* in output are calculated using a shipments measure, with 1971 value added as a base. For further details, see Robert Dohner, "Energy Consumption in Manufacturing" (1981, mimeo). There are significant data lags.

manufacturing production but 73 percent of manufacturing energy use. For this reason, manufacturing energy use is sensitive to any changes in the shares of these four. Table 3.5 gives the output shares and energy intensities of the big four energy users and those of all other industries. As the table points out, the seventies were marked by a general reduction in the energy required per unit of output, despite an uneven performance by the primary metals industry. There were also important proportional changes, with the share of primary metals and stone, clay, and glass falling, while those of chemicals and other industries rose.

It is possible to break down the reduction of the total energy-output ratio in manufacturing into changes in energy efficiency within individual industries and changes due to shifts in demand toward less energy-intensive products. This is done in Table 3.6. In the 1971–74 period, changes in output shares actually added to energy consumption per unit of output in manufacturing, while increases in efficiency reduced the total energy re-

Table 3.6 Factors Responsible for Changing Manufacturing Energy
Intensity*
(BTUs per 1972 $ Output)

1971 energy intensity	50,600	
Change due to shifting demand mix 1971–74	1,000	
Change due to increased efficiency 1971–74	−7,000	
EQUALS: 1974 energy intensity	44,700	
Percent change 1971–74		−11.7%
1974 energy intensity	44,700	
Change due to shifting demand mix 1974–77	−3,700	
Change due to increased efficiency 1974–77	−2,000	
EQUALS: 1977 energy intensity	38,900	
Percent change 1974–77		−12.9%
Percent change 1971–77		−23.1%

* Two-digit SIC level, less SIC 29.

quirement of a dollar's worth of manufacturing output by 12 percent. In
the 1974–77 period, changing shares and increasing efficiency both contributed to the reduction in manufacturing energy requirements. Over the
entire 1971–77 period, shifts away from energy-intensive industries and
increasing efficiency within industries together resulted in a 23 percent decline in the energy-output ratio in manufacturing.[36]

Energy Prices and Economic Structure

The fact that higher energy prices lead to a shift in demand toward products with lower energy content, or toward products that reduce energy
consumption, has additional implications — for the profitability of industries, for the employment and incomes of workers, and for the location of
economic growth. The adjustments to shifting product demand and activity may involve painful relocations of workers and business as well as
writing off assets made unprofitable by the shift. In fact, such adjustments
are always taking place in the market economics as demand moves to new
products or as competitors emerge. But the huge rise in energy prices in
1974 and 1979, because of its size and its differing impacts across industries, has created severe dislocations in the industrial economies. The

United States, with its indigenous energy resources, energy-using technology, and low rate of turnover of capital stock, was particularly sensitive to the shifting currents of profitability caused by the oil price change. The years since the first price increase have seen rapid expansion in some industries and near-stagnation and painful adjustment in others.

The most spectacular growth has come in the energy extraction industries and the equipment industries that serve them. Employment in the oil and gas industry rose from 274,000 in 1973 to 547,000 in 1980, with much of the increase coming after the announcement of price decontrol in 1979. Coal mining employment rose by 62 percent, to 262,000, in the same period. The equipment industries serving the oil, gas, and coal industries have also shared in the boom. Employment in the oil and gas field equipment industry rose from 45,000 to 96,000, while employment in the coal mining equipment industry rose by over one third.[37]

These gains in employment have been highly concentrated. Texas, Louisiana, and Oklahoma alone account for two thirds of the employment in oil and gas extraction, and West Virginia, Pennsylvania, and Kentucky for the bulk of coal mining employment. The mining machinery industries are also clustered in the producing states, contributing to their employment gains. In percentage terms, the largest increases have come in the sparsely populated western states of Montana, Wyoming, and Colorado, which lack the infrastructure to support rapid growth.[38]

Other industries have not been so fortunate. Two American industries that have suffered under the new energy regime are automobiles and steel, both very large industries, both concentrated in a single region of the country. While, as we noted, the automobile industry is by nature sensitive to the business cycle, the fluctuations in response to the oil crises have been particularly great as consumer demand oscillated between small and large cars. Employment in the industry dropped by 19 percent between 1973 and 1975, rose to a record of almost a million workers in 1979, then dropped by 21 percent in 1980 and by a further 9 percent in 1981.[39] These fluctuations, coming at a time of increasing international competition, forced one of the three large American firms close to bankruptcy and impaired the financial health of a second.

The steel industry has been in a depressed state worldwide since the peak year of 1974, the result of declining ship orders and investment and expanding steel capacity in the newly industrialized and less developed world. The U.S. steel industry has fared better than Europe's, in part because of its ownership of coal resources. The U.S. industry has, however, been adversely affected by the weakness in investment and motor vehicle sales, the reduction in automobile size, and the shift to lighter materials in automobile manufacture. The situation became acute in 1977, when domestic production fell slightly but imports rose by 35 percent. A spate of steel plant closings resulted in the loss of 20,000 jobs in the industry, and more than 50,000 steelworkers were declared eligible for Trade Adjustment Assistance in that year.[40] The Carter administration was able to forestall antidumping suits by domestic steel producers by instituting a trigger (floor) price mechanism for steel imports. Steel trade issues have remained extremely sensitive, however, and the Reagan administration filed dumping suits against European producers on behalf of American steel firms.

Much of American steel capacity is very old, and it uses considerably more energy per ton of produced steel than its competitors.[41] The rise in energy prices accelerated the obsolescence of U.S. facilities and increased the vulnerability of the industry to import competition.[42] However, a uniform characterization does not give an adequate description of the industry. Recent years have seen the growth of small-scale, highly efficient plants using electric furnaces and steel scrap. These "mini-mills" operate in a restricted segment of the industry, primarily producing steel bars and smaller shapes.

Both the steel industry and its critics agree that the 1980s will be a critical period for the industry. The prospects for the growth of the smaller mills seem almost assured. However, substantial investments will be required if the industry is to remain competitive in its present form, supplying a broad range of products, even though current earnings, and the industry climate, are poor.

The American Economy at the Beginning of the 1980s

The eight years since the first oil shock have seen important adaptations in the U.S. economy. Higher energy prices have had their effect on producer and consumer behavior, leading to more energy-efficient production techniques and to a pattern of production that requires less energy. Although it has not always been easy, resources have been reallocated, from energy-using to energy-saving and energy-producing industries. Yet, in spite of these adaptations, the United States remains vulnerable to further oil shocks. In fact, there are reasons for thinking that U.S. vulnerability has increased.

A good indicator of the sensitivity of an economy to an energy price hike is the share of total expenditure (both direct and indirect) that goes toward the purchase of energy. This is a measure of the leverage of an energy price rise, the extent to which total costs in an economy can be raised by a given percentage rise in energy prices. In the early seventies this fraction was quite small; energy expenses were about 3 percent of total expenditure.[43] After the succession of oil shocks this fraction has risen, in spite of the trimming of energy use. An estimate of energy's share of total expenditure would be about 8 to 9 percent in 1980. An oil crisis of any given magnitude will now have a greater effect on the United States and other consuming nations.

In addition, we may have exhausted much of our ability to curtail energy use in the very short run. In a society not previously conscious of energy use, a variety of economy measures exist: lowering lighting and temperature levels, reducing driving speeds, eliminating nonessential automobile trips, and plugging thermal leaks. With higher energy prices, it may be more difficult to reduce consumption without changing the existing stock of equipment or without lowering GNP and changing life-styles. Over time, higher prices will reduce energy use, but over short periods we may have become vulnerable to higher energy prices.[44]

The share of petroleum in OECD energy consumption varies over the business cycle, the percentage rising during expansions and falling during recessions, just as it does in the United States.[45] This feature of oil demand and the observed nature of OPEC pricing suggest that considerable volatility will characterize the real price of oil. The past two oil shocks have led to recessions in the OECD countries, producing a slackening of oil demand and a temporary excess of oil. Continuing OECD inflation has lowered real oil prices while economic recovery has led to increased oil demand, eliminating the slack and eventually raising the use of production capacity. The past two shocks, while triggered by political events, have also come at times when OPEC production was near capacity.[46] The sharp rise in oil prices more than recovers the previous fall in real prices and starts the beginning of a new cycle. This pattern, described in the preceding chapter as one of "jagged peaks and sloping plateaus," appears likely to be repeated in 1982 and 1983, once economic growth resumes in the OECD and as continuing inflation and stable or falling nominal oil prices lower the real price of oil.

Oil consumption and real oil prices are therefore likely to move in fits and starts rather than in the smooth progression of all variables that characterizes most forecasts. Even if one is assured that real energy prices will rise over ten- to twenty-year periods, forecasts of the real price of energy over two- to four-year horizons may be highly uncertain. In such an environment, energy is likely to pose a short-term macroeconomic management problem as well as one of long-term energy supply and demand. The turbulence that is likely to characterize the world oil market should be kept in mind as we trace the pattern of future energy use.

Prospects for Future Energy Use

The years since the oil embargo have seen something of a revolution in American thinking about energy use. Forecasts made early in the 1970s projected energy demand growth rates that were almost identical to the growth of GNP, with very little scope for adaptation. Those projected energy demand levels now seem extraordinarily high. Estimates of energy supply were often optimistic, particularly when they incorporated a backstop technology (e.g., shale oil or coal in solid, liquid, or gaseous form) that would, at some price, make available very large quantities of energy.

Slower economic growth rates in the 1970s and the record of diminishing energy use with higher prices have produced much lower estimates of future energy consumption. But, in addition, the evidence of the adaptability of economies to higher energy prices has opened up a variety of energy use futures, depending upon the level of future prices and the policies taken to influence demand. This is particularly evident in the work of the National Academy of Sciences' Committee on Nuclear and Alternative Energy Systems (CONAES), which details assumptions under which U.S. energy consumption could grow by 4 percent or by 125 percent in thirty-five years.[47] In all of the cases, projected energy consumption lies substantially below an extrapolation of the pre-1973 trend.

Recent experience has also underscored the difficulty of increasing energy supplies. The rapid expansion of a variety of domestic energy sources — shale oil, coal, and nuclear power — faces considerable opposition both nationally and in the specific areas that are affected. In addition, confidence in the continued expansion of oil production from the Middle East has waned and alternative sources of energy have remained tantalizingly uneconomic.

In contrast to the environment in 1960, or even that of the early seventies, energy forecasts now seem much more tenuous. As James Just and Lester Lave have rightly emphasized in their survey of energy projections, the value of an energy forecast is not so much in providing an accurate description of what will occur as in exploring a set of assumptions.[48] The Atlantic Institute study makes two sets of assumptions about energy supplies. The first projects moderate growth in world and OECD energy supplies as well as a shift away from oil toward coal and other energy sources. The second assumes that for a variety of reasons, actions are not taken to increase energy supplies, and that total energy availability to the OECD is no higher at the end of the forecast than in 1978. By putting bounds on energy supplies, one can focus on the adjustments required in the consuming economies. Such an approach is useful in highlighting the value of additional energy supplies or the costs of policies that fail to expand energy supplies. The Atlantic Institute study has the virtue of treating the OECD countries as a system. So long as fuels are internationally traded and market prices are maintained for energy, the supply-price equilibrium facing one country depends upon actions taken outside its borders.

The two cases outlined by Robert Stobaugh in Chapter 2 postulate a

Table 3.7   Shares of OECD Primary Energy Consumption

	1960	1965	1970	1973	1978	1979
United States	54.6%	52.5%	50.7%	49.8%	49.6%	49.2%
Japan	5.1	6.5	9.2	9.5	9.5	9.7
OECD Europe	33.4	34.0	33.3	33.4	32.8	34.0
Other	6.9	7.0	6.8	7.3	8.0	7.2

Source: OECD, *Energy Balances of OECD Countries, 1960 74, 1975-79.*

growth of 2.0 percent per year and zero, respectively, in primary energy supplies available to the OECD countries. The implications of these two cases for the United States in part depend upon how much the United States contributes to the growth of OECD primary energy demand.

The United States is the largest single consumer of energy in the OECD and, to many, the most profligate. Yet the U.S. share of OECD energy consumption has been diminishing over time, primarily due to its generally lower rate of economic growth. As Table 3.7 points out, the U.S. share has dropped by 5 percentage points since 1960; in fact, the U.S. share has fallen throughout the postwar period.[49]

The future growth of U.S. energy consumption compared to that of the OECD as a whole will depend on numerous things, but three factors of particular importance stand out. The first is the growth of the U.S. economy compared to that of the rest of the OECD.

During the 1960–73 period, the United States ranked near the bottom of the OECD countries in its economic growth rate, with an average GNP growth of 4.0 percent per year compared to 5.0 percent for the European OECD countries and 10.4 percent for Japan.[50] If these relative positions were to continue, the growth of energy demand in the United States would certainly be less than that of the OECD as a whole. There are indications, however, that OECD growth rates will tend to converge in the future. The technological adaptation that fueled productivity growth outside the United States may subside as Europe and Japan approach U.S. productivity levels. The forecasted growth rate for Japan has been scaled back considerably, as Chapter 5, by Teruyasu Murakami, points out, and G. F. Ray in Chapter 7 sees reduced European growth rates. However, unless the United States can substantially increase its investment share of GNP, its economic growth is still likely to remain below the OECD average.

The second factor determining the relative growth of U.S. energy demand will be its possibilities for reducing the energy intensity of production and consumption compared to those possibilities in the other OECD countries. The United States currently uses significantly more energy per unit of output than do the European countries or Japan. In part, this can be explained by structural differences; the United States is a large country with low population density and many population centers. As a result, the volume of transport per unit of production is much higher. However, a recent and careful study found that some 60 percent of the difference in energy-GNP ratios was attributable to the difference in the intensity of energy use in individual applications.[51] This in turn reflects the comparatively low taxes on energy in the United States and user prices that are therefore much lower than in Europe or Japan. What is important here is that there are possibilities for reducing American energy use through existing technologies, which were developed or first used in countries where energy prices were historically high. On this score, at least over the first part of the forecast period, U.S. energy demand growth is likely to lag behind the OECD average.

The third factor affecting the future growth of U.S. energy consumption is the allocation within the OECD of production activities that are extremely energy intensive. The rapid expansion of trade relations within the OECD has created the possibility of consuming energy indirectly, by importing finished products with a high energy content. Since energy use in manufacturing is concentrated in semi-manufactures (steel, chemicals, etc.), which are extensively traded, the pattern of specialization in production may account for differences in energy use. Certain OECD countries, in particular Germany, Sweden, and Japan, have large, export-oriented industrial sectors and are net exporters of energy embodied in finished goods. The volumes are quite significant, amounting to over 10 percent of total energy consumption in these three cases. The United States, ironically, is a net importer of energy embodied in other goods, but the amount is small, slightly over 1 percent of total energy consumption.[52]

In the current world energy market, the development of energy-intensive export sectors in countries with very few domestic energy resources is a risky undertaking. Those countries that are highly dependent on imported energy — again, Japan, Germany, and Sweden come to mind — may take steps to diversify out of energy-intensive production. This might occur either through policy, as Japan appears to have done, or by taxes that reflect a vulnerability premium.[53]

Although individual countries can so diversify, world production of such goods will depend upon total demand. Some of the increased production of energy-intensive goods (e.g., steel) will be taken up by the newly industrializing countries; this is already taking place to some extent. The production of petrochemicals may move closer to the sources of oil and form the basis for the industrialization of oil exporters. It is also possible that some of this production would be taken up by OECD countries with indigenous energy resources, such as the United States. It is possible, but not inevitable. Energy costs are not the sole determinant of production costs, even for very energy-intensive goods. Substantial investments would be required, and there may be opposition to reducing the scale of industries in the energy-importing economies. The extent and fluidity of international trade relationships should, however, caution us that energy use in one country may to some degree substitute for energy use in another.

In summary, the growth in energy demand in the United States is likely to continue to be less rapid than that of the OECD as a whole. Rather than attempt precision, we venture a rough guess that 2.0 percent annual energy growth for the OECD would mean 1.4 to 1.8 percent annual growth for the United States, while zero energy growth for the OECD would probably mean a reduction in U.S. energy consumption, perhaps by as much as 0.5 percent per year. This of course assumes that an integrated market for energy is maintained in the OECD, a point to which we return later.

Economic and Energy Growth under Favorable Circumstances

Any discussion of the future energy situation, or the adjustments to stringency in the energy supply, must include some discussion of the variables that drive energy demand. Of these, a principal determinant is the level of economic activity, both because energy is required to produce goods and services and because the income generated by the economy influences consumer demand for energy. If the projected demand for energy outruns the likely supply, other variables, such as the price of energy or a decline in GNP, must do the adjusting.

An important difference between energy forecasts made before the embargo and those made recently is the presumed economic growth rate, which is now expected to be much lower than in the 1960–73 period. One reason is that the labor force, which grew at an average annual rate of 2.1 percent between 1960 and 1980, is now projected to grow by only about 1.3 percent per year. This, and a gradual decline in the average work week, would limit the growth of labor hours to about 1 percent per year.[54]

The second factor lowering the projected economic growth rate is a more modest estimate of productivity growth. The postwar agricultural transition, which moved workers from low- to higher-productivity jobs, boosted productivity growth rates in the 1950s and early 1960s. This transition has been completed and a slowing of the productivity trend after 1965 is now evident. Its poor performance after 1973 has also led to lower expectations of future growth. A current estimate of the productivity trend is 1.5 percent annual growth, which, with the growth in hours, would mean a GNP growth rate of 2.5 percent.[55] But because the causes of the slowdown in productivity are not yet understood, productivity estimates are highly tentative, and a GNP growth rate of 3 percent or more might not be ruled out. A 2.5 percent growth rate beginning in 1978 would mean a level of real GNP 72 percent higher in the year 2000.

The initial energy supply case outlined in Chapter 2 forecasts an annual growth of OECD energy supplies of 2.0 percent. If we assume a conservative annual growth rate of 1.5 percent in U.S. energy consumption, by the year 2000 the United States would be consuming 39 percent more energy than in 1978. This figure, along with the growth in GNP, would imply a reduction in energy use per unit of GNP of 19 percent by the end of the century.

A reduction in energy use of this order is well within current forecasts and should be achieved without straining the American economy. The moderation in energy use is likely to be broadly based, but a few sectors are especially promising. Of these the most important savings are likely to come in the transportation sector, particularly in motor vehicles. Here the growth of transportation demand is likely to be limited, both because car ownership is approaching one vehicle per licensed driver and because the amount of time spent driving per person (currently, about fifty minutes per day) is unlikely to increase greatly.[56] Automotive stock also offers considerable scope for efficiency improvements. Current legislation requires sales-weighted fuel economy averages of 27.5 miles per gallon by 1985, and while actual road performance is somewhat less than in Environmental Protection Agency tests, a doubling of the average fuel efficiency is likely by the year 2000. Current forecasts show transportation energy consumption remaining roughly constant through 1995, and that alone could reduce U.S. energy consumption per unit of GNP by 11 percent.[57]

The residential sector also offers possibilities for energy conservation, even though the rate of turnover of the housing stock is low. Increasing insulation, some redesign of heating systems, and the replacement of electric resistance heating with heat pumps could materially reduce energy consumption. In addition, the rate of household formation will decline in the 1980s and 1990s, and this, combined with a gradual shift toward multifamily units, will decrease energy requirements. An interesting possibility is afforded by the shifting growth patterns in the United States toward the milder climates of the South and West. If this continues, perhaps supported by higher energy prices, it would reduce space heating requirements, although region does not seem to be a major factor as long as a building is well designed.[58]

The industrial sector plays a central role, not only as the largest single consumer of energy, but also as an area whose importance is expected to grow in the future. The industrial sector was particularly successful in reducing its energy use in the 1970s, and the extent to which this reduction can be continued over the next twenty years will be crucial to how the United States copes with reduced energy supplies. There is no doubt that it could reduce the energy intensity of its industrial production, based on international comparisons of industrial processes and analysis of currently available technology.[59] The Industry Resource Group of the CONAES study estimated that energy use per unit of output could be reduced by as much as 40 percent by 2010.[60] The demand mix will also play a role. One key question is whether manufacturing production will continue to grow more slowly than GNP, as it did between 1973 and 1978, or whether it will resume a pattern of growing somewhat faster than total GNP, as it did before 1973. The prices of all energy-intensive goods will rise, but some products, such as aluminum and products of the chemical industry, will face increased demand because of their low weight and energy-saving characteristics. The demand for chemicals and the extent to which the industry remains in the oil-consuming countries is perhaps the most important, because of its large size and rapid growth before 1973.

Economic Adjustment to Declining Energy Supplies

The second energy supply path outlined in Chapter 2, the Lower Bound, describes a very slow growth in world energy supplies and an essentially constant level of energy use by the OECD countries. This outcome is the result of a substantial reduction in OPEC oil production and the failure of other countries to do more than just compensate for that decline. In accounting for the difference between Stobaugh's Upper and Lower Bounds, the United States plays a major role, with rapidly declining oil and gas production and only a minor increase in coal exports. In such an environment, real oil and energy prices increase greatly by the end of the century.

How would this supply path — where "things do not go very well" — come about? To what extent would periodic shortages, rapidly increasing prices, and sluggish growth remove the obstacles to increasing energy production outside of OPEC? The precedents do seem to go in this direction; for example, the concern with energy supplies after 1973 was instrumental in winning approval for the Alaskan pipeline. Yet the possibility that things will not go well on the supply front cannot be ruled out. A low success rate in finding new supplies could lead to a reduction in U.S. oil and gas production, and accidents could mobilize opposition to a number of other energy supplies. One possibility that might result in the Lower Bound case is a continuation of the sluggish economic growth that characterized the last decade. The temporary slack in world oil markets that each recession causes could dilute the sense of urgency surrounding new energy supplies, perhaps leading to their successive postponement. The Lower Bound case could come about because the consuming countries consistently overestimate the level of future energy supplies, and thus the level of energy prices at the end of the century comes as a surprise.

As long as a market in fuels is maintained within the OECD, the prices of energy will be determined within the OECD (or the world) as a whole. If prices move in tandem in OECD countries, then the allocation of a constant level of energy supplies will again depend upon relative growth rates and degrees of adaptability. In the case of very stringent energy supplies, the level of investment in a country would influence both the growth rate and the degree of adaptation, and so OECD energy consumption shares might change less. Still, it is quite probable that some OECD countries would consume less energy in 2000 than in 1978, and the United States may well be among this group.

If a 2.5 percent economic growth rate is maintained, and there is no increase in U.S. energy consumption, then by the year 2000, energy use per unit of GNP would have to fall by at least 42 percent. Alternatively, if economic growth is to continue in a situation of constant or slowly declining energy supplies, then one must identify places where energy use can be reduced. The leading candidates in the United States would be the transportation sector and in residential and commercial buildings. Because of the longevity of structures and the high cost of retrofitting them, reductions of energy use in this area are likely to require a much longer lead time.

Transportation is likely to bear the brunt of stringent energy supplies, both because of its low-priority status and because its energy intensity will greatly increase the cost of transportation services. Already the cost of fuel makes up 30 percent of the operating expenses of airlines, and, although fuel is still only a fraction of the total cost of automobile ownership, it is extremely important in the decision to take a particular trip.[61] In cases where the supply of energy is a strong binding constraint, some restrictions on personal physical mobility are also likely.

A sharp escalation in transportation costs would cause tangible changes in the structure and operation of the economy. They would act like a general increase in tariff barriers, reducing international and regional trade and braking the trend toward increasing economic integration. This in turn would reduce the economies of scale that come from wider geographic markets. Remote areas, such as Appalachia, Alaska, and parts of the American West would be particularly affected. Tourism would be especially hard hit, with obvious effects on areas, such as Hawaii, that depend on tourists. Heavy emphasis would be placed on communications as a substitute for physical mobility, accelerating a development that is already taking place due to technological progress in the communications industry and the economy's steadily increasing premium on time.

The steep rise in energy prices envisaged along a path of constant energy supplies would have a distinct effect upon the composition of GNP. The shift toward services would accelerate, and industrial production would shift away from energy-intensive goods. Energy use in manufacturing is heavily concentrated in the materials industries — that is, primary metals, chemicals, paper, and stone, clay, and glass. The fabrication industries, which use these products, require much less energy. The emphasis in fabrication would be on minimizing the use of materials (making things smaller and lighter), substituting less energy-intensive materials, and recapturing the energy embodied in existing materials through recycling.

Although substitutions among products and increases in efficiency will reduce the energy requirements for any level of output, they alone will probably not be enough.[62] To the extent that price and policy measures fail to reduce energy demand to the available supply, the balancing will be achieved by a lowered rate of economic growth. Restricted supplies and higher energy prices will certainly reduce the growth of GNP in the consuming countries. The question is, How much and how important is the reduction likely to be? Higher prices will reduce the level of GNP because some portion of the productive labor and capital in the economy is used simply to substitute for more expensive energy. Higher prices also lower the return to capital and therefore discourage investment in new plant and equipment. GNP in the year 2000 would be lower because there would be a smaller stock of capital.

The extent to which GNP is affected depends primarily, not surprisingly, on the flexibility of the economy in finding substitutes for energy. Highly flexible economies have considerable resilience, while in inflexible economies the loss in production can be huge.[63] Most current estimates find the degree of substitutability large enough so that the reduction of GNP is small, and much of the reduced growth comes from slowing investment.[64] One study, by Edward Hudson and Dale Jorgenson, finds GNP in the year 2000 reduced by 12 percent, with prices high enough to actually reduce energy use, but these estimates are at best uncertain and describe a level of energy substitution that has never been observed.[65]

In addition, forecasts tend to impose a smoothness upon economic and energy variables and may therefore neglect other influences of the energy environment on economic performance. Given the volatility that the world oil market has displayed in the last several years, it is unlikely that real oil prices will follow a smooth course. If the world oil market and the industrial economies continue to lurch along as they have done in the last decade, GNP could be lost both because economies experience higher unemployment and lower use of capacity and because the uncertainty engendered by the instability reduces investment and economic growth. In countries where wage determination takes place in annual deliberations, it is possible to moderate increases in the face of energy price rises or other price changes that make the country poorer. By limiting the spillover of energy prices into domestic, or "home-grown," inflation, such countries stand a better chance of stabilizing output.

In the United States, with its decentralized wage-bargaining process and longer, overlapping contracts, there may be no practical way to ensure this kind of restraint, since inflation is effectively out of the hands of any single contract negotiation. Policy responses in the United States to oil price rises are likely to continue to be restrictive to prevent an acceleration of domestic inflation, with resulting costs in GNP and unemployment. Finally, the bursts of oil price inflation, oil-induced recessions, and cycles of rising and falling real energy prices will, through the uncertainty they create, reduce investment and therefore economic growth. The economic environment of the 1980s will be inherently more risky than that of the 1960s, and required returns on investment will reflect that risk.

Maintaining the OECD Energy Market

Throughout this discussion, we have assumed that an integrated energy market allocates supplies within the OECD in much the same way that the market allocates energy use within individual countries. The amount of energy that any one OECD country would consume would depend upon its level of GNP, its flexibility in energy use, its share in the production of particularly energy-intensive goods, and the level of prices that balances OECD energy demand with OECD energy supply. Would such a market be maintained within the OECD if energy supplies were progressively tightened? Those countries that are almost totally dependent upon imports for their energy supplies have no choice but to pay world market prices; only subsidies paid by the treasury could lower energy costs to users. But OECD countries differ greatly in their degree of import dependency, and those countries that produce a substantial fraction of their own requirements can limit the rise in domestic prices by divorcing the domestic market from the OECD market.

Both the United States and Canada adopted this strategy after the first oil price hike. The United States maintained controls on the wellhead price of domestic oil and prohibited exports of Alaskan oil. Canada, a net exporter, levied an export tax for the difference between domestic and world market prices. Both countries consumed more oil and produced less than they would have at world market prices. The United States imported more, Canada exported less, and both reduced the volume and raised the price of oil available on world markets.

The development of an energy mercantilism (energy taking the place of gold) is a distinct possibility if world supplies are restricted. Those countries that produce an appreciable fraction of their own energy would limit the rise in domestic energy prices by cutting their energy market off from the world market, either reserving domestic energy for domestic use or subsidizing imports by averaging their price in with that of price-controlled, domestically produced energy. The maintenance of lower prices in countries with indigenous production would limit supplies in the world market, raising the cost of energy to import-dependent countries and increasing the international market's instability.

In several instances this has already taken place. The price controls in the United States and Canada have already been mentioned. Previously, the United States has limited exports of wheat, soy beans, and scrap metal to keep the domestic prices of these commodities from rising.[66] The legislative prohibition of exports of Alaskan oil was counterproductive, but it clearly indicates the "energy as a national consideration" mentality, as, arguably, do the cautious depletion policies of Norway and the United Kingdom. (There is, of course, evidence in the opposite direction: the gradual decontrol of oil and gas prices in the United States and Canada and the refusal of the Thatcher government in the United Kingdom to bow to industry demands for subsidized fuel prices.)

In the United States, another sharp increase in the price of imported oil could lead to renewed controls on domestic oil prices.[67] It is perhaps an unfortunate coincidence that two of the OECD countries most richly endowed with energy resources, the United States and the United Kingdom, have also had the lowest rates of GNP growth and the most difficulty in controlling inflation. Progressively tightened energy supplies and the worsened economic performance that they entail may well make domestic economic considerations overrule energy commitments in the OECD.

An energy mercantilism policy in the OECD would lead to large differences in energy prices across countries. The energy contents of various products differ so widely that patterns of world trade would be disrupted by artificially maintained prices for energy. Several of the most energy-intensive industries, such as steel and chemicals, are industries in which trade issues are most sensitive. The dumping suits brought by European synthetic fiber manufacturers against American exporters, which charged that American textiles were being subsidized by price-controlled oil and gas, are a portent of what could occur. The international trading community has flourished in spite of segmented agricultural markets, but energy is too basic for the trading system to be unaffected. A general rise in protectionism and a contraction of trade is perhaps as likely to arise from a myopic energy policy as it is from poor economic performance and high unemployment.

Factors in the Adjustment

Stobaugh's Upper and Lower Bounds imply quite different outlooks for the American economy. A 2 percent annual growth in OECD energy supplies, even if it means somewhat slower growth in U.S. energy use, should easily support the currently forecast rates of economic growth. On the other hand, zero growth in OECD energy supplies would require a very large reduction in energy intensity if economic growth is to be maintained. The Lower Bound supply path is likely to be played out in recurring oil crises, with sharp increases in oil prices, and economic contractions worsened by anti-inflation policy. The attempts of the consuming countries to deal with energy stringency may divide those countries that have energy resources from those that do not, and in so doing threaten the multilateral trading system that has developed in the last thirty years.

Somewhere between the Upper and Lower Bounds, the energy constraint will begin to be felt, complicating the problems of adjustment and impinging on economic performance. What characteristics of the economy will raise or lower the minimum rate of energy supply growth with no serious economic side effects? And, to the extent that we can control economic variables, can we manage them so that adjustment to a given supply of energy is facilitated and the economic consequences are less severe?

The first and most important influence on energy demand is the rate of economic growth. Although the past few years have demonstrated that the level of GNP is not the only variable affecting the demand for energy, they have done nothing to diminish its importance as a determinant of demand. In situations where "other things are equal," energy demand appears to increase proportionately with an increase in economic activity.[68] Of course, other things — for instance, energy prices — do not remain equal, but adjust to maintain a supply-demand balance. What this analysis does suggest is that variations in the rate of economic growth will greatly influence the level of future energy prices and the extent of required adjustments in use. This is not to say that industrial economies should reduce their economic growth rates in order to moderate the growth of energy demand. That would be choosing as policy what we hope to avoid as consequence. It is important to realize, however, that supply-oriented, high-growth policies, which have recently been advocated as a cure for inflation and unemployment, may founder if steps are not taken to increase energy supplies or improve energy efficiency. By the same token, if the 1970s prove to have been a temporary interlude, and rates of labor productivity growth recover their earlier levels, then a more rapid rate of energy supply growth will be required to avoid adjustment difficulties.

As mentioned above, the flexibility of the industrial economies in reducing energy use will be crucial to how they adapt to restricted energy supplies. The past few years have supplied impressive evidence of the ability to reduce energy use per unit of GNP. Will this flexibility persist as energy supplies are reduced, or will successive substitutions become more difficult? For countries such as the United States, which has had historically low energy prices, possibilities for further reducing consumption may be identified in countries with historically high energy costs. For countries with relatively efficient energy use patterns, or for situations with greatly restricted energy supplies, we begin to venture beyond known technology and practice. The direction that technological development takes will be critical to the adaptability of economies in the longer run. Will higher energy prices, and the prospect of their continued increase, direct technological change toward saving energy, as some have argued, and so relieve some of the pressure on energy supplies?[69] Or was the availability of relatively cheap and abundant energy instrumental to productivity growth, as others have argued, presaging slower rates of future technical advance?[70] Technological forecasting has never been very precise or very accurate, which limits what may be confidently said about longer-term adjustments. But before technological changes can be designed and applied, energy savings will come from the adoption of currently available techniques.

The largest obstacle to the further reduction of energy use is the existing stock of structures and equipment, which embody specific processes and characteristics that reflect outdated investment decisions. Although a certain amount of retrofitting may be done, at some point the existing stock must be replaced by new, energy-efficient structures and equipment. The capital stock itself changes, as new investments are made, to outfit a growing economy and to replace equipment that wears out or can no longer be profitably operated. Changes in economic circumstances, such as energy prices, will affect the kind of investment that is most profitable, and this investment will, over time, change the characteristics of the capital stock. The speed with which the stock changes depends upon the volume of investment that an economy generates.

The United States, as a large consumer and a large (and potentially larger) producer of energy, will require huge investments in the next two decades to increase energy supplies and to use energy more efficiently. Investment in energy-producing industries has already been expanding rapidly, jumping from 25 percent to almost 30 percent of business investment from 1973 to 1979.[71] Over the next decade, capital requirements for oil and gas alone have been estimated by the Department of Energy to be between $25 and $30 billion per year, with an additional $5 to $6 billion for coal.[72] The Bankers Trust Company estimates capital requirements for all energy-producing industries as averaging over $100 billion (in 1979 dollars) per year in the next decade, or about 3½ percent of GNP.[73] Additional investment will be required to modify the energy-using characteristics of the capital stock and to replace the capital made obsolete by higher energy prices.

These are not the only claimants of investable funds; investment will also be required to meet environmental and safety regulations, to provide the capital to accommodate shifts in industrial structure, and to facilitate continued growth in productivity. By one estimate, the investment share of GNP would have to rise by as much as 2½ to 3 percentage points in the years immediately ahead to provide the necessary increments to the capital stock.[74] These requirements are so large that there is concern that investment might be inadequate, for the share of U.S. GNP devoted to investment is lower than in most advanced countries, and personal savings rates in the United States have recently taken a turn for the worse.

This concern reflects a view that investment constitutes a fixed share of total GNP that must be allocated among competing projects. However, in the past, the investment share of GNP has varied considerably, and since 1946 it has moved between 8½ and 11 percent of GNP. Turnarounds have also been abrupt, as in 1963–65, when the investment share rose from just under 9 percent to 10.8 percent of total output.[75] In the longer history of the U.S. economy, huge investment programs, such as the building of the canals and the railroads, have been accommodated and have provided the impetus for sustained economic growth.

The other side of the macroeconomic equation for investment is savings; if the United States is going to divert more of its current GNP to investment, it must generate more savings out of current income. The recent fall in the personal savings rate has caused anxiety. But that rate has varied much during the postwar period and actually hit its highest levels between 1973 and 1975, and new laws encouraging saving for retirement should increase the personal savings rate.[76] Nor does personal savings make up a majority of total savings; a much larger fraction comes from business savings (retained earnings and depreciation allowances), which as a percentage of GNP held up well in the 1970s. Recent tax legislation incorporating faster writeoffs for assets promises to increase the internal funds available for business investment, once the recession ends.

The fiscal stance of the government is another important determinant of total savings. A government budget surplus adds to available funds in the capital market, while a deficit has the government competing with other borrowers for capital market funds. Since tax and expenditure decisions are at least partly policy decisions, a tight fiscal policy can be pursued to accommodate expanded investment.

Finally, with international capital markets, a country's investment need not be limited to its own savings. Countries in which investments exceed savings run current account deficits, financing them through borrowing abroad. This explains the somewhat surprising fact that deficits after the first oil shock bore little or no relation to the degree of countries' oil import dependence. Mexico and Norway, both oil-independent countries with substantial investment opportunities, had current account deficits after 1974, borrowing abroad on the strength of their oil-related investments. Japan and Germany, both oil-dependent high savers, found domestic investment possibilities diminished by the oil price rise and ran surpluses, lending abroad.[77] If profitable investment opportunities exceed available savings in the United States, then a current account deficit is one way to obtain the required materiel.

Apart from the availability of funds or the ability of the nation to devote a sufficient fraction of GNP to investment is the question of whether an adequate amount of investment will, in fact, be undertaken. This has become a matter of recent concern, for, in spite of the rise in energy-related investments and required investments in pollution control, the growth of total investment since the 1974–75 recession has been meager. Particularly troubling is the fact that investments have been concentrated in assets with much shorter lifetimes. Thus, investment in equipment — and within equipment, motor vehicles — has been much stronger than recent investment in industrial and commercial structures.[78] While there are a variety of plausible explanations for this fact, including the difficulty in obtaining approval for the siting of new plants, an extremely important factor has been the greater uncertainty surrounding the profitability of capital assets with long service lives. One reflection of this uncertainty is the low market valuation of equities compared to the replacement value of the assets they represent.[79] Businesses have found it extremely difficult to raise funds through additional equity and have increasingly resorted to debt issue for investment funding. Chronic inflation, and the expectation that the government will periodically adopt restrictive policies to try to slow inflation, is in part responsible for this uncertainty. But to a degree it is also uncertainty about the energy situation: the future level of real energy prices, future demands for energy-using products, and the assurance of continued energy supplies.

Although there is now discussion about policy measures to encourage investment, what is probably most important is a steady growth in GNP and a sense of confidence and optimism about the future. This is especially true in industries such as steel, chemicals, and automobiles, where substantial energy-related investments are required, yet where current earnings are depressed. A continuation of the high and varying rates of inflation, gyrating real energy prices, and the sluggish economic growth that characterized the 1970s would imply rates of investment well below those needed to improve the efficiency of the capital stock. The recurring energy crises and poor economic performance of the zero energy growth path would make adjustments difficult if not impossible. Not only would the supply of energy fail to expand, but the long-term investment required to reduce energy demand would fail to materialize.

Adjustments to higher energy prices require not only additional investment, but the expansion of some industries, the decline of others, and the writing off of assets made unprofitable by the change in relative prices. There are many indications that adjustments of this kind have become more difficult in the industrial economies. As incomes have risen, increasing value has been placed on economic security and the elimination of risk. Among the economic responsibilities acquired by Western governments since World War II has been that for assuring this security. Increasingly, economic security has come to mean the security of particular individuals in particular jobs in particular locations, something that is threatened by almost every change in economic circumstance. In the industrial economies it has become more difficult to dismiss or lay off workers, to write off assets, and to relocate operations, and as a result the ability to adjust to economic dislocations has been impaired.[80]

The rise in protectionist sentiment is a symptom of this change. Often the first visible sign of the need for adjustment in an industry is a rapid increase in imports. Thus recent trade discussions have emphasized "orderly marketing," "managed free trade," and "nondisruption" as principles for avoiding injury to domestic industry. Programs of trade adjustment assistance and "industrial revitalization" have, as often as not, worked to bolster the profitability and continued operation of existing assets rather than transferring them to more profitable activities.[81] Increasingly, the remedy for endangered industries or firms has been government intervention, involving protection from imports and support ranging from loan guarantees to government takeover.

To a greater extent in Europe than in the United States or Japan, governments have intervened to manage the affected industry. One example, with major implications for investment and energy use, is the European steel industry. A European Community–sponsored steel cartel maintains minimum prices for steel products within the EEC and allocates production quotas. This industry is invariably the largest industrial user of energy; furthermore, the energy intensity of its production is related to what share of capacity is actually operated. The EEC steel arrangement increases the European steel industry's consumption of energy in several ways. It enforces low-capacity use among all producers, raising average energy consumption. It also penalizes the more efficient steel producers, who were reluctant participants in the cartel. Finally, it discourages investment in more efficient operations because they cannot be fully used and because investment in additional capacity is a threat to the cartel's stability.

Although the United States has been less willing to ensure the solvency of individual firms, the federal government has provided financial guarantees for firms in the aerospace and automobile industries and has provided import protection for the steel, shipbuilding, automobile, and other industries. The Carter administration's flirtation with industrial revitalization, and the planning and coordination that was discussed, has not continued, but the Reagan administration has already taken steps to protect automobiles and steel from imports. The American steel industry, in particular, is likely to pose continuing problems for policy in the 1980s. Whether the United States is able to restructure it into an internationally competitive industry or whether an import protection policy is pursued to maintain its current profitability will have important implications for the efficiency of American energy use.

The problem is perhaps poorly phrased as one involving an either-or decision. Governments will find it difficult or impossible not to respond to a crisis affecting a major industry or firm. The more relevant question is to what extent the governments of industrial countries can respond in a way that is not open-ended and that helps adjustment to changed economic circumstance rather than impeding it. The ability to effect adjustment to higher energy prices and the structural dislocations that they introduce will be crucial to the transition to less energy-intensive economies. Because such changes in relative prices will create substantial winners and losers and spur calls for government intervention, they will perhaps pose the most difficult challenge for the United States and the other OECD countries.

*   *   *

The economic stability of the United States, as well as of the other industrial countries, will depend on the ability to adjust to volatile but generally rising real oil prices. Adjustments in energy supply and demand are the most obvious of those required. As we stress throughout this study, the ability to adapt to the rising prices and insecurity in world oil markets by increasing energy supplies outside OPEC is particularly important, for if increased energy supplies are not forthcoming, much of the balancing could come from lower real incomes and reduced economic growth.

But the energy problem is not simply one of assuring sufficient supplies; the ability to restructure demand patterns to those consistent with higher real energy prices is also critical. Reducing the demand for energy would limit the rise in prices necessary to clear world oil markets, which in turn would reduce the income transferred to foreign oil producers as well as reduce the expenditure of real resources in producing more expensive alternate energy supplies. Furthermore, adjustments in demand, as well as increases in supplies, will reduce the leverage of the OPEC producers on world energy markets and the vulnerability of the consuming nations.

The past eight years have provided impressive evidence of the adaptability of the OECD countries and the effectiveness of market forces in bringing about those adaptations. While an appreciation of the role of prices and individual decision-making was sorely lacking in earlier analyses of the energy problem, the efficacy of the market in prompting adjustments should not lead policy-makers to sit back, content to let prices do the work. For the price of energy, in addition to being an instrument of adaptation, should also be the object of our attention. The price necessary to balance supply and demand will determine the distribution of income between energy producers and consumers, and the trajectory of energy prices will affect the ability of the consuming countries to continue to provide growth in real per capita incomes. Government policy actions will be instrumental in determining the level of prices that balances supply and demand, particularly in the case of energy supplies, where the leasing, licensing, transportation, and conversion of energy are all areas of major public and policy concern. But actions of policy are also important for demand adjustments. This chapter has stressed the role of industrial policy in influencing energy demand; national policy regarding production and import competition in several industries will greatly influence the efficiency of energy use.

The required adjustments go beyond energy supply and demand. The consuming countries, and particularly the United States, must avoid the unemployment and investment costs of sharp increases in the price of oil. Unless the United States can divorce domestic inflation rates from external oil shocks, the nation will continue to suffer policy-induced contractions as a price for avoiding accelerating inflation. The adjustment will involve a consensus on wage increases based on productivity gains and the international terms of trade (of which oil prices play an important part). The recent profit-sharing schemes in such ailing industries as automobiles are a step in this direction.

This emphasis on adjustment contrasts somewhat with the stated goal of much of American energy policy: achieving independence. While reducing the share of world energy needs met by OPEC members is essential to reducing the instability in world energy markets, the extent to which any single nation is a net importer of energy is less important. Even if individual nations are self-sufficient, they are not immune to external events unless they explicitly divorce domestic from world energy markets. In this case, achieving energy independence means passing vulnerability on to others and, because of the difference in energy prices it would produce, endangering the trading system upon which all OECD countries depend.

4

America in the Strait of Stringency

by Daniel Yergin

The political discovery of economic growth took place in the United States in the years after World War II, and growth became a central focus of national policy and debate. Levels of GNP were achieved in the decades after the war that would have seemed inconceivable in 1945. But economic growth did much more than make possible the ever-increasing production and distribution of goods. It also provided a way to resolve social conflicts and tensions within American society. It became in itself a "social goal . . . the source of individual motivation, the basis of political solidarity, the ground for the mobilization of society for a common purpose."[1] So important were these concerns that they led to a considerable expansion of the responsibilities and activities of the national government in the economy. But the political role of growth extended beyond the nation's borders, for it also helped to harmonize the objectives of industrial nations, to expand the world economy, to avoid the "beggar thy neighbor" policies that would have reduced welfare and increased tensions.

The unfolding of growth opened up an optimistic prospect, a new affirmation of progress, an expectation of continued improvement. All this proved quite different from the expectations at the beginning of the postwar years, when visions of the future were haunted, not only by the harsh vistas of economic devastation and dislocation brought by the war, but also by memories of the interwar years, by fears of renewed high rates of unemployment. Just before his death in 1945, Franklin Roosevelt had predicted that the United States "would presently be in the worst economic jam it had ever seen."[2] Events proved otherwise; the sustained surge of economic growth kept the clock from turning back to the 1930s.

Abundant and cheap energy played an important role in all of this. It was a factor of production, interacting with the other factors, enhancing them, and substituting for them. Increasingly, first in the United States and then in other industrial economies, energy meant cheap oil, and it is only with some exaggeration that we say oil became almost a free good.

Certainly American society developed on something akin to this assumption. The dependence on oil was most clear in transportation and residential patterns, but it was also evident in the commercial and industrial sectors. Moreover, it was generally taken for granted (although not, for the most part, closely examined) that the growth of energy consumption at a vigorous rate was necessary to maintain satisfactory rates of economic growth.

Developments in the last decade have severely shaken these assumptions. Low-cost reliable supplies of energy, and especially of oil, do not appear available. The assumption about the inevitable correlation of energy consumption to economic growth now requires close examination. Perhaps growing energy consumption can no longer be considered a necessary concomitant of economic growth. Perhaps there has even been an actual reversal: the growing demand for energy under the present circumstances may pose a threat to sustained economic growth. Such might well be the case if the Lower Bound, the stringency scenario, turns out to be even moderately close to the mark.

Here we come to the two basic premises of this study. The first is the need to maintain economic growth on a per capita basis in the Western world. The second is the reasonable probability of recurrent energy stringency and tensions. Their connection raises basic questions for the United States in the 1980s and 1990s — about possible modes of adjustment, about the viability of continued economic growth, about the possibility of significant declines in real incomes and employment, about the conflicts that develop when redistribution replaces distribution as the name of the game, and about the tolerance, responsiveness, and adaptability of the political system to the stresses so engendered. These are persistent questions for a democratic system, a system based upon consent, and they have been posed with a starkness and urgency by the pressures on energy supplies. To state the basic issue simply: Energy stringency could pose a serious challenge to the fundamental stability and functioning of America's economic, political, and social system.

We need not look only to future crunches and pressures, for the era of energy stringency has already begun. The evidence has been accumulating since 1973. Chapter 3 outlined the consequences of energy problems in terms of inflation and reduced growth in the United States. But more has occurred than the raising of one set of percentages and the lowering of another. There has also been a shock to the structure of the American economy as well as to the psychological underpinnings of the market system.

The requisite confidence, the compact by which a market economy functions, has been eroded by uncertainty, anxiety, and fear. Coping with energy constraints has been complicated by the deep-seated American belief, even conviction, of perpetual abundance. Moreover, the particular nature of energy dependence has entangled the United States in serious and potentially severe threats to strategic and military security.

Our purpose here is to analyze the social and political responses to the era of energy stringency that clearly manifested itself in 1973. We begin with questions: What kind of system developed in the era of cheap and easy oil? How has the system responded to stringency so far, and what has shaped those responses? How have costs and benefits been distributed? We will observe that a process of adjustment, a quiet but impressive change, has been taking place.

Yet the pressures in terms of supply and price seem likely to grow. And that takes us into a far more difficult realm, looking into the future. What happens in a situation of continued stringency? It would be far too ambitious and dangerous to offer flat predictions. Rather, in order to clarify thinking, we shall highlight two alternative paths of development. In Scenario I, the failure to make an adequate adjustment in the face of accelerating stringency will place very severe strains on the economic and political order, and it could lead to a test of the entire American system on a scale matched in this century only by that of the Great Depression. Scenario II portrays an effective mode of adaptation. Building upon the "quiet change," it proceeds with surprisingly little alteration in the way Americans live. In this case, ingenuity and technology substitute for energy. What it suggests is that the most difficult period for American society will be this decade. Beyond, in the 1990s, innovation may well offer a broadening latitude, a technological abundance, in which economic growth can flourish again.

The problem that emerges for the United States, as well as other Western societies, is how to accelerate adjustment. Some have suggested that this longer-run process of adaptation will be painful, nasty, and traumatic, ending up with a society in North America "very unlikely to be the one which we currently would recognize as a democratic-republican social system."[3] But to us, the contrary appears to be the case, at least in the rest of the century. American society would not look too different a decade or two from now. The major changes would result, not from the adjustment to energy problems alone, but rather from the interaction with other trends, such as those in electronics and population growth. The greater threat to the social and political order arises, not from adapting, but from failing to adapt at an adequate rate. Tardiness will impose its penalty on economic growth — and on political and social stability.

Here we come to the problem of time, which imposes its own logic on developments. For the United States, with its rich and variegated market economy, Scenario I might hardly be worth consideration were there a generation or a generation and a half to wait. Millions and millions of people would make the individual decisions that would add up to a reasonable adjustment. Machines and buildings and factories would come to the end of their useful lifetimes and would be replaced by new stock, responsive to the new circumstances, embodying new techniques. New technologies would diffuse throughout the system. Investment and behavior decisions would be made from fresh perspectives, taking into account relative costs and opportunities and a widening menu of technological alternatives. Perhaps there will be sufficient time for all this. But the reasonable supply scenarios for the 1980s indicate otherwise. The change in the cost, availability, and security of energy supplies in one decade has been swift, and there is a reasonable possibility that such changes could continue to outrun possibilities for a gradual adjustment. That, in a nub, is the problem for the United States and the entire industrial world, and is why we have undertaken this study.

The Gusher: 1950–1973

Land and resources, abundance and self-sufficiency — these are powerful themes in American history and social development. "The Western frontier, from the Alleghenies to the Pacific, constituted the richest free gift that has ever spread out before civilized man," wrote the historian Frederick Jackson Turner, who outlined his famous frontier thesis in 1893. "Not the Constitution, but free land and an abundance of natural resources open to a fit people, made the democratic type of society in America for three centuries." This vision, as David Potter, another historian, argued, shaped the American character, the "people of plenty." But, as Potter points out, the mere existence of abundance did not mean a life of abundance. "A vital distinction separates mere potential abundance — the copious supply of natural resources — and actual abundance — the availability to society of a generous quota of goods ready for use." What makes possible the conversion of potential into service is technology and economic organization.[4]

Certainly, Americans have been people of talent when it comes to applying technology to convert natural resources into useful energy. Between 1850 and 1973, energy consumption increased some thirty-five times over. This was achieved "not merely by accepting nature's proffered gifts but by the application of science, by the elaboration of a complex economic organization, by planning, and by toil."[5] The beginnings of America's industrial economy were firmly laid on the foundation of the ancient energy sources — wood, wind, and water. Wood was by far the most important; as late as 1870, America obtained three quarters of its energy from wood. The abundance was obvious: "All cabin dwellers gloried in the warmth of their fireplaces, their world of surplus trees, where a poor man, even a plantation slave, could burn bigger fires than most noblemen in Europe."[6] And, as the United States shifted from wood to coal to oil, the abundance remained unmistakable. The modern oil industry was born in the United States in 1859. From the beginning, the United States supplied not only its own needs but, as the leading exporter, those of much of the rest of the world as well. As late as 1929, a third of the total demand for oil outside the United States was met by U.S. exports. Out of seven million barrels of oil used by the Allies in World War II, six were provided by the United States.[7]

A fear of oil shortages developed during and after the two world wars, in both cases accentuated by the obvious importance of oil in the war. But the more general problem of the American oil industry was chronic overproduction. The Texas Railroad Commission was empowered in 1932 to regulate the vast overflow of oil in the new fields of Texas, and various state and federal officials tried to stem the tide of surplus "hot" oil from being smuggled illicitly across state borders.[8] Motion pictures and legends stamped the image of the gusher into the popular culture and the public's mind. Natural gas was literally and figuratively less visible but no less abundant; indeed, it was for many years a waste product to be flared away. Coal, too, was abundant, though its image was tarnished by the poverty, conflict, and tragedy that went with mining.

American society operated on this premise of abundance in the years of great economic growth. The initial concern after World War II about a shortage of oil had been eased by the development of offshore production in the Gulf of Mexico and then by the real opening up of the vast oil province of the Middle East, which reinforced the image of a world surplus and masked possible constraints on American supply.

How did energy consumption change between 1950 and 1973? Total consumption more than doubled (see Table 4.1), though there was not the same drastic shift from coal to oil that occurred in Japan and Western Europe. There were no declarations of a "declining industry." Rather, before World War II, oil was already playing proportionally a much larger role in the American economy than in the Japanese or European. Still, the shift away from coal was substantial — from a third to a sixth of total energy consumption. In absolute terms, coal use did increase, but by a mere 3 percent — quite small when measured against the almost tripling in absolute terms of oil consumption and the quadrupling of natural gas consumption.

American society was conditioned to the easy availability of energy,

Table 4.1 U.S. Energy Consumption, 1950–1973

	1950 mbdoe*	1950 % of total	1973 mbdoe	1973 % of total	Absolute growth
Domestic oil	5.7	36%	10.4	29%	82%
Imported oil	.6	4	6.1	17	917
Coal	6.1	38	6.3	18	3
Natural gas	2.8	18	10.7	30	282
Hydro	.7	4	1.4	4	100
Other**	—		.4	1	—
TOTAL	15.9		35.3		122%

* mbdoe = millions of barrels daily of oil equivalent
** includes nuclear

Source: Energy Information Agency, *Report to Congress, 1979*, Vol. II, pp. 7, 13.

especially of oil. That was the message the public received in many forms, beginning in the marketplace. Between 1950 and 1973, the price of domestically produced crude oil declined in real terms by 21 percent. The composite price of all domestically produced fossil fuels fell by 19 percent. Energy consumption and production were directly and indirectly subsidized — through depletion allowances, federal highway programs, nuclear reactor development, and the like. Nonmarket sources said the same thing. "The U.S. oil industry faces a real challenge — a challenge of ideas and indubitable oversupply," warned *Fortune* in 1959. "The glut is certain to last a long time. The reason: too much oil underground too easy to get at, ready to flow at very little additional expense."[9] And the same message was implicitly conveyed in millions of advertising messages.

The effect of all this was most obvious in transportation. America was more and more a society in motion. The number of registered passenger cars increased from 40 million in 1950 to 102 million in 1973. Gas mileage actually dropped; but both consumers and auto manufacturers could easily neglect this factor because gasoline was so cheap that it hardly seemed relevant. The problem for oil companies was to find adequate markets for the oil they had. Alternative modes of urban transportation were allowed to deteriorate. Mass transit was caught up in what is by now a familiar vicious circle — declining ridership, met by higher fares and declining service, precipitating a further decline in ridership. Mass transit was certainly not perceived as a social asset to be preserved.[10] A clear and pungent symbol of

Table 4.2　Principal Means of Transportation to Work, 1976

Total Car	88.9%
Drives self	72.9
Carpools	16.0
Mass transport	5.6
Bike or motorcycle	0.8
Taxicab	0.2
Walks only	4.0
Other means	0.6

Source: Oak Ridge National Laboratory, *Transportation Energy Conservation Book,* Edition 3, I–118.

society's rejection of mass transit was the tearing up of the Red Line Trolley System in Los Angeles, whose tracks had provided the spine on which the city had grown up. The rights of way were converted to freeways. As Table 4.2 indicates, the private car, with only one occupant — its driver — became, for three quarters of Americans, their principal means of getting to work.

Residential living patterns interacted with the private car. Suburbs, composed primarily of single-family structures, spread out from the cities, linked to the larger metropolis by cars and roads. Suburban growth was rapid. In 1960, one tenth of all workers lived in the suburbs of metropolitan areas of 100,000 or more. By 1970, one third of all workers lived in suburbs. Inattention to the energy element characterized new commercial buildings. The average office building constructed in New York City between 1960 and 1965 used twice as much energy per square foot as those constructed between 1945 and 1950.[11]

Industry also operated on the basis of cheap, available energy. True, the energy use per unit of GNP declined slowly in this period. But that appears primarily to have been the result of a switch in fuel mix (from coal to oil to gas) and a rather minor by-product of technological change.[12] A dramatic example of the assumption of cheap and abundant energy at work was the rapid rise of the petrochemical industry in Texas in the 1950s and 1960s on the basis of inexpensive natural gas. Indeed, immediately after World War II, the chemical companies were able to obtain the natural gas supply simply by paying the gathering costs.

Underlying this whole development was what might be called a "first-cost mentality." That is, the selling price of a good was the cost that the purchaser paid attention to, whether for a house or a car or a refrigerator.

The energy operating costs were so minor that they were hardly considered; indeed, in many cases there was no way to find out what the "life cycle" costs were — and even if one could, there was hardly any way to determine their significance.

"It Can't Happen Here"

This pattern of consumption, and the social development that went along with it, had been conditioned by the experience of abundance. But in the late 1960s and early 1970s, constraints on the energy base began to emerge.

One reason was a process of social change that rearranged national priorities. A new political force, environmentalism, started to make its weight felt. Although its roots went back at least to the publication of Rachel Carson's *Silent Spring* in 1962, it became more visible after the blowout of an oil well in the Santa Barbara Channel off California in 1969, which paved scenic beaches with tar. In 1970, 100,000 people paraded down Fifth Avenue in New York City to mark Earth Day. The nation, it was argued, was now rich enough to internalize what had been the externalities of industrial society. Environmentalism made its influence felt in a large number of ways — in such legislation as the National Environmental Policy Act, the Clean Air Act, the Clean Water Act, the Endangered Species Act; in the establishment of environmental impact statements; in the creation of the federal Environmental Protection Agency; and in the development of the significant new industry of pollution control. Environmentalism had its major impact, as far as energy was concerned, on the burning of coal. Concern about air pollution led to fuel switching, especially by electric utilities, away from domestically produced coal to low-sulfur oil, which had to be imported. Although not particularly noticeable at the time, this change led to a significant increase in the demand for oil. Between 1968 and 1973, oil consumption by electric utilities more than tripled, much of this being low-sulfur imported oil.

The peculiar regulatory system on natural gas also imposed constraints. This became clear in the late 1960s, when prices began to diverge sharply between the interstate and intrastate natural gas markets, reflecting the high demand for gas. Natural gas consumption increased at an average rate of over 6 percent between 1966 and 1970 — compared to 5 percent for oil and 1 percent for coal.[13] Obviously, producers preferred to keep incremental gas in the intrastate market, where higher demand could be reflected in higher prices, rather than in the interstate market, where the Federal Power Commission controlled prices. In consequence, spot shortages appeared in the interstate natural gas market.

One constraint was basic: The United States was an aging producer. It was outrunning its geological base. But this highly relevant fact was not represented in the consumption pattern or, really, in prices. Indeed, controls in the early 1970s resulted in prices that increasingly gave the "wrong" information to consumers. The United States was continuing to consume as though it were still essentially self-sufficient, but reserve additions for oil and gas were not congruent with this trend (see Figures 4.1 and 4.2). In other words, the reserve base for oil and gas could not sustain the kind of growth that it had in the past. The marginal source of energy was one that had previously been very small, oil imports. The turning point came in 1970, when U.S. oil production reached its peak and then began to decline. Self-sufficiency — Turner's vision of an "abundance of natural resources open to a fit people" — had, at least in terms of energy, come to an end. The differences in oil were more than just a matter of sulfur content and specific gravity. Imported oil could not be regarded in the same light as domestically produced oil. It was a separate fuel source, and it became the one that balanced supply and demand.

But so deeply ingrained was the assumption of abundance that it was difficult for many people to accept the notion of constraint. At the end of the 1960s the headquarters staff in Michigan for Dow Chemical, one of the nation's largest industrial energy users, became convinced that pressure would mount on natural gas prices. But the company's managers in the Texas division, home of the great postwar petrochemicals development, were reluctant to accept such projections. "People had gone down there when gas was cheap," recalled a senior corporate official. "The people in Texas said, 'It can't happen here.' " It could, and it did.

Yet the assumption of unlimited abundance is so deep-seated that even a decade later, in 1980, the successful Republican presidential candidate was predicting that "with decontrol, we could be producing enough oil to be self-sufficient in five years."[14]

It would be a mistake, however, to think that the United States entered an era of scarcity. While oil was more difficult and more expensive to find, and it was more likely to be found in smaller fields, the United States was still a significant oil producer in the early 1980s, producing a sixth of all the oil used in the world. Unfortunately, there was a catch — it was consuming a third.

This divergence was already becoming marked a decade before, in the early 1970s. Moreover, America was becoming more and more integrated into the world market just as the market was undergoing a number of other significant changes — including the rapidly growing demand for oil throughout the industrial world, the growing nationalistic assertion by oil-producing nations, the rise of the independent oil companies, the erosion of the position of the major oil companies, the running down of coal, and

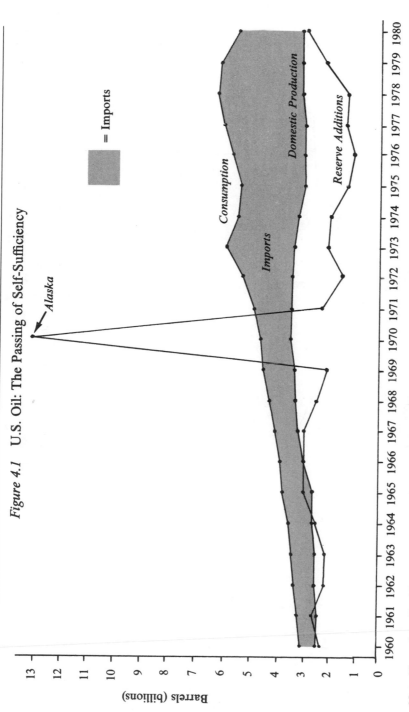

Figure 4.1 U.S. Oil: The Passing of Self-Sufficiency

Note: Natural gas liquid not included in production and consumption.

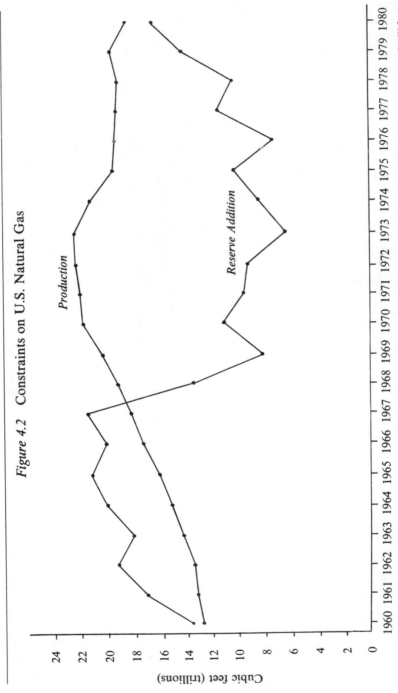

Figure 4.2 Constraints on U.S. Natural Gas

Note: Does not include 26 trillion cubic feet at Prudhoe Bay because not yet available to market, owing to lack of transportation facilities.

Global Insecurity

that growing environmental consciousness, which tended to favor oil consumption. All of these combined to put pressure on the world market, drawing it taut.[15]

In this setting, a political-military crisis erupted, leading to the embargo and the first oil shock, 1973–74. The interruption, the ensuing panic, the new prices, international crises — they all delivered some very important messages to American society: The assumption of abundance was no longer plausible; oil and energy in general could no longer be considered a free good; imported oil was not a neutral energy source; its role should be reduced. If this were not done, there would be a continuing series of crises and perhaps a catastrophe. But how could the role of oil be reduced in ways that did not reduce growth or promote inefficiency?

Politics and Energy: New Transition or Old Wine?

How did the American political system respond to the emerging constraints in the 1970s and early 1980s? As might be expected, the oil crisis drew the national government deeply into the business of determining energy policy. This is not to say that the government was not much involved before, for it had been engaged in everything from building dams to developing nuclear power to regulating natural gas and setting oil import quotas. Nevertheless, all this was done in a highly fragmentary way, with policies for specific fuels and with little effort at coherence. Thus, an official of the Johnson administration could tell an OECD meeting in 1967 that U.S. energy policy meant that "industry is responsible for production, distribution, marketing, and pricing, except in markets where fair prices cannot be guaranteed by competition, such as, for example, gas and electricity in interstate commerce. The Federal Government does not control production, it does not direct the efforts of industry, and it does not involve itself in company affairs."[16]

During the first Nixon administration, 1969–73, the warning signs about possible pressures in energy supplies were becoming apparent, both domestically and abroad. This resulted in a considerable struggle within the administration to establish some coherence, authority, and urgency for the consideration of energy matters. The effort was bedeviled by many obstacles — inflation, ideological and personality clashes within the administration, environmental concerns, and conflicting demands by politically powerful groups, such as independent refiners. Coherence was not achieved before the 1973 embargo. However, three weeks afterward, Nixon announced the goal of energy independence by 1980: "Let us set as our national goal, in the spirit of Apollo, with the determination of the Manhattan Project, that by the end of this decade we will have developed the potential to meet our own energy needs without dependence on any foreign energy sources." But Watergate, not energy, was on the mind of the President and his crumbling administration, so the desire and ability to do much about energy were sharply limited.[17] Ironically, an administration strongly committed to the free market ended up regulating and rigidifying the petroleum market with price controls, with an allocation system that froze buyer-seller relations, and with an entitlements system that encouraged oil imports and inefficient refiners.

This was a heavy legacy for the Ford administration, which devoted much of its efforts to trying to unravel the controls and regulations. But, in response to Watergate, voters had sent a heavily Democratic Congress back to Washington in the 1974 elections. Some of the new members, highly critical of "big oil," wanted to roll back oil prices. With inflation still a primary concern, price decontrol was not exactly a popular issue.

Yet there were two highly significant achievements during the Nixon-Ford period. One involved a major environmental compromise. In the immediate aftermath of the embargo, culminating three years of debate, Congress gave a green light to the Alaskan pipeline. This made possible the single most important contribution to American energy supply in the 1970s, the buildup of Alaskan oil production to 1.5 million barrels a day. Without it, the second oil shock, in 1979–80, could have been a great deal worse. And second, in 1975, Congress set fuel efficiency standards for the automobile industry. By 1985, fleet averages would have to double to 27.5 miles per gallon. Since one out of every nine barrels of oil used in the world every day was burned as gasoline on American highways, such a change would have a major impact not only on America's but also on the world's oil balance.

Ford maintained the Project Independence theme with what might be called a "high production" strategy. In January 1975, he called for a ten-year program to build 200 nuclear power plants, 250 major coal mines, 150 major coal-fired power plants, 30 major oil refineries, and 20 major synthetic fuel plants.[18]

This strategy proved unrealistic for a variety of reasons. One of the most important was that the environmental movement had been gaining momentum since the late 1960s. The strongest impact had initially been on the strip mining and burning of coal, but in the mid-1970s it was even more so on nuclear power. By 1974, a national movement opposing atomic energy had taken clear shape. Environmentalism was not by any means solely responsible for the difficulties encountered by nuclear power; rather, it interacted with the economics — continually rising costs, inflation, and, later, high interest rates — to place major roadblocks in the way of nuclear's further development. Some of the increasing costs could be assigned to delays that resulted from the activities of the environmentalists, but some could also be attributed to poor management, to the government's changing specifications in midstream, and to a rapid scaling-up of plant size in the industry without sufficient operating experience. A renewal of concern about nuclear proliferation also contributed to the slowing of the development of nuclear power, which by the end of the Ford administration had, as far as ordering new plants was concerned, just about come to a halt in a de facto moratorium.[19]

There was another reason that the Project Independence plans would prove unrealistic: their inattention to the flexibility in energy use requirements. Such plans were based on the notion that energy demand would continue to grow at a relatively high rate. Expectations were basically extrapolations from the recent past with some lowering of curves due to price effects. These expectations were embodied in a number of forecasts in the early 1970s that projected substantial growth in U.S. energy consumption, tending to converge on a doubling by the year 2000. There were two reasons for these assumptions. As a U.S. Bureau of Mines study said, "This forecast of future energy consumption and supply is based on the assumption that existing patterns of resource utilization will continue."[20] The second was what seemed the altogether reasonable premise that continued high rates of energy growth were necessary to assure economic growth. This assumption was the commanding idea well into the middle 1970s.

But a change was at hand, and it was being expressed in a naive question — "Why?" Why did energy consumption have to grow so substantially? This question had been asked by a few dissenters years earlier; it had been posed by S. David Freeman within the administration at the very beginning of the 1970s and in a controversial Ford Foundation study in 1974.[21] But by the middle 1970s, the debate was in full swing.

Occasionally one encounters an acting out of a shift in commanding ideas. Such was the vivid case at this time in the tortured deliberations of the prestigious National Academy of Sciences' Committee on Nuclear and Alternative Energy Systems (CONAES), which had its beginnings in 1975.

"The historically dominant domestic view of the energy problem focused on its supply side," observed Jack Hollander, the staff director of CONAES, after the fact. "According to this view, the problem resides mainly in the threat of having 'too little' energy: energy demand growth is, a priori, desirable and necessary to nourish a healthy economy, therefore the principal solution to the growing domestic supply-demand imbalance should be found by increasing energy supply." But there was a growing energy conservation group, as well. CONAES provided an equal platform for advocates of both positions and, involving over two hundred people, became "a unique microcosm of the real-world energy debate":

The major issue dominating the CONAES study was unexpected at the start. The study had its beginning with a
traditional supply-technology orientation, originating from a
request by the Energy Research and Development Administration (ERDA) to the National Academy of Sciences to assess the
future role of nuclear energy, particularly the breeder reactor,
in the U.S. energy system. In view of this technological origin, it
came as a surprise to many that the issue which generated the
most debate, and which became central to all of the study's
conclusions, was not nuclear technology but rather an issue
with a large social component: growth in energy demand.[22]

While calling for an increased energy supply, the primary message of
CONAES was that "a vigorous and sustained drive to increase energy efficiency" was possible and should be the first item of energy policy.

This basic shift in ideas, gathering momentum in the mid-1970s, explains the considerable difference in emphasis between the Nixon-Ford
and Carter administration positions. In April 1977, less than a hundred
days into his term, Carter came forward with a National Energy Plan,
which put much greater emphasis on conventional coal use (to be doubled
by 1985) and on conservation, gave greater attention to renewables, and
sought to break out of the pricing impasse through a crude oil equalization
tax. In what was clearly a shift of attitude, the new President declared that
a major transition was at hand, no less significant than that from wood to
coal at the beginning of the Industrial Revolution and from coal to oil and
gas in the twentieth century. Moreover, reflecting the shift in the commanding ideas on energy, he raised conservation to first priority — "the
cornerstone of our policy . . . our first goal . . . the cheapest, most practical
way to meet our energy needs and to reduce our growing dependence on
foreign supplies of oil."[23] To be sure, the rhetorical commitment seemed
greater than the programmatic follow-through, and the imagery appeared
to emphasize sacrifice over efficiency.

The first Carter program was thoroughly worked over by a Congress
whose members reflected the many contradictory values and interests of a
diverse nation. The most significant achievement was the complex compromise on natural gas. There were also additional incentives for conservation and solar energy, and utility pricing reform. Because of a growing
concern about nuclear weapons proliferation, brakes were applied to both
the breeder reactor and nuclear fuel-reprocessing programs.

The Carter administration had sufficiently experienced the skepticism
and contradictory pressures engendered by the energy problem when
the second shock occurred in 1979. The President took a bold step in
April, when he finally broke the logjam on prices and announced that

domestic oil price controls would end by September 1981. He also called for a windfall tax on the additional revenues that would flow to oil producers.

Gas lines in the next two months, however, created national hysteria. While Carter was taking part in the Tokyo economic summit in June 1979, he received a memorandum from the head of the White House domestic staff, warning: "We have a worsening short-term energy crisis ... Nothing else has so frustrated, confused, angered the American people ... or so targeted their distress at you personally."

Thrown on the defensive by public and congressional clamor and panic, the administration fell back on a technological fix. In July 1979, the President offered a second energy program, built around an $88 billion synthetic fuels effort — primarily liquids from shale and liquids and gas from coal. The scale was reminiscent of the Nixon-Ford proposals. Essentially, it was a strategy to mobilize investment and reduce risk by channeling revenues from a windfall profits tax. The development of a synthetic fuels know-how and capacity was, at the very least, a reasonable insurance policy. But this program was very large. There were important questions about how rapidly such technologies could be perfected and scaled up, about costs, about environmental consequences. When would those technologies be available? Would the simple fact of the existence of such a program help to prevent further price eruptions or severe cuts in supply? Many concluded that, even if very successful, only a modest effect could be felt even by the early 1990s; and Congress proceeded to scale back the program to more manageable dimensions.[24]

But the election of Ronald Reagan clearly indicated that there was not yet a consensus on the energy problem in the United States. Indeed, the Reagan administration set its agenda in conscious opposition to that of the Carter administration. Three strands shaped its outlook. First, the government was seen as the real source of the energy problem. In 1978, at the time of congressional action on Carter's National Energy Plan, Michael Halbouty, subsequently head of the Reagan administration's energy task force in 1980, declared:

> There is no question that the public is confused about the energy situation. I would like to clarify a flagrant misconception by making it perfectly clear that there is no energy crisis in the United States. This country has a tremendous amount of energy potential. But there is a very very serious energy problem — in fact, the problem is a crisis — namely, Washington has politically manipulated, interfered, and imposed dictatorial controls and regulations which severely stymied discretionary productive efforts by the energy industries.[25]

Second, energy efficiency was seen not only as a sure recipe for a no-growth economy but also as a rejection of the American experience. This point of view was argued as the "policy theme" of the Reagan administration's transition team report:

> Much has been done. But what has been done is to impede production and curtail consumption. The government has acted on the principle that the way to deal with energy is to do away with it. Instead of unleashing the resources of a wealthy nation, we have, in the name of saving energy for some unspecified time in the future, tucked energy away like a rare bottle of wine ... The price of energy failure is not just economic stagnation, but social upheaval.

This view was pungently summed up by Reagan himself a month after his election. Conservation, he said, means "we'll be too hot in the summer and too cold in the winter."

These two strands both amounted to an assertion that the United States had not outrun its geological base and that it was possible to return to the energy economy of the 1950s and 1960s. In other words, here was an explicit reassertion of the belief in essentially unlimited abundance. Obviously, as well, such views emphasized the positive role of energy producers — quite a change from the dominant views of the preceding years.

Yet there was more to the stance of the Reagan administration. Energy policy was also shaped by a broader shift in national attitudes concerning the role of government in American society. In the latter part of the 1960s and throughout the 1970s, government, especially the federal government, had become more and more active throughout the economy and society. The 1980 election was seen as a mandate to roll back government influence, intervention, and regulation in almost every realm of American life. This certainly applied both to direct government intervention in the energy area and to government's much expanded role in the related area of environmental pollution. Those holding this view in the Reagan administration were not inherently hostile to energy efficiency; indeed, they favored "economy" and regarded conservation as a relatively cheap and attractive option. They emphasized the need to "free" the market to make whatever adjustments made the most sense on a decentralized basis. "This country could in the year 2000 be using a hundred quads of energy," observed one senior administration official, "or it could just as well be using eighty quads or a hundred and twenty quads, and it would still be a productive, efficient society, so long as we get there through a market process."[26] There was little confidence in the ability of government to, as variously described, "pick winners" or choose among the competing energy "products" whose advocates sought government assistance. Market prices would, they believed, lead to the optimal adjustment.

These views led to a rejection and repudiation of much of what had happened in the preceding ten years. The Reagan administration virtually eliminated conservation and solar budgets, greatly reduced support for synthetic fuels development, and eschewed planning for emergencies and disruptions. It speeded up the leasing of public lands and offshore areas for energy development and sought to roll back a decade of environmental protection programs. In the one major exception to its oft-repeated commitment to the market system, it actually increased government support for nuclear power. Symbolizing its drive to downgrade the importance of the energy question, it sought to eliminate the Department of Energy and disperse its functions among other agencies, though there was no evidence that this would save money or improve efficiency in government (and good reason to think just the opposite). It did resume the filling of the Strategic Petroleum Reserve.

The administration certainly had a very persuasive point about the importance of prices in the adjustment process and the incongruity of trying to encourage efficiency while repressing price signals, as well as about the excesses of regulation. But significant questions could be raised about its assertion that "market principles" were the be-all and end-all of the matter. So much of one's evaluation depends on one's mental timetable. The Reagan approach made sense if there were no clock ticking, if there were no reasons to expedite the adjustment to a rapid change in the world energy balance. But the bulk of the evidence in the early 1980s suggests that it is a risky gamble to assume that it is possible to return to the energy economy of the 1950s and 1960s and to restore the era of energy abundance, at least as it had been in the past.

Even with a reduced role for government, there were considerable imperfections in the market. Some of them were the result of entrenched subsidies provided in the past; others, the result of inadequate access to capital and information. Also, world market prices for oil still do not capture the total social cost of oil. In other words, there is a gap between the price the individual or firm sees and that borne by society as a whole. Moreover, one of the consequences of the Reagan administration's approach was a breakdown of the integration between the foreign and domestic elements of the energy problem. The defense secretary, Caspar Weinberger, declared that one of the most important "geopolitical realities" for the United States is "our dependence on foreign oil sources." He added, "The umbilical cord of the industrialized free world runs through the Strait of Hormuz into the Arabian Gulf and the nations which surround it. That area . . . is and will be the fulcrum of contention in the future." Even in the very unlikely case that U.S. oil imports declined to zero,

Table 4.3   Asymmetry in the Reagan Administration

Oil Import Bill	Rapid Deployment Force	Nuclear Power	Conservation	Solar	Strategic Petroleum Reserve
$77.3 billion	$4 billion (5.2%)	$1.6 billion (2.1%)	$22 million (.03%)	$83 million (0.1%)	$2.3 billion (3%)

Note: 1981 oil import bill; fiscal 1983 administration budget requests, submitted to Congress in January 1982; % indicates percentage of oil import bill.

the security problem would still be most serious because of the oil dependence of the rest of the world. Still, high U.S. oil imports greatly aggravate the security problem and make its management more difficult. Thus, the reduction of these imports, based on sound economics, is a relatively inexpensive but vitally important contribution to America's overall security. Yet James Edwards, the first secretary of energy in the Reagan administration, declared that oil imports were likely to remain at the same high level — 5 to 7 million barrels a day throughout the 1980s — but that federal encouragement of conservation and renewables would no longer be necessary because "market principles" would suffice.[27]

In other words, U.S. energy issues were divorced from their global context. But other industrial nations remained acutely aware that the United States, as the world's largest importer of oil, had a major impact on the world market, and they were alarmed by what they saw as a new American "energy nationalism," indeed, "energy isolationism."

The lack of coherence, as well as the lack of prudence and priority on the energy question, is suggested by Table 4.3, which compares various Reagan administration energy-related budget proposals against the annual U.S. oil import bill.

In contrast to its predecessors, the Reagan administration tended to ignore the macroeconomic costs of the two oil shocks as well as the potential cost of further disruptions. Its theory of inflation simply left out the effects of the two jumps in oil prices. In consequence, it saw much less urgency in trying to protect the economy against further shocks and also, therefore, operated on some misapprehension in dealing with overall economic problems.

Finally, while the basic principles had been articulated earlier, the Reagan policy was actually implemented during a period of slack in the international oil market, as the world economy struggled with the effects of the second oil shock. This apparent calm reinforced a widespread sense of complacency. In fact, the administration's approach depended upon a whole series of optimistic assumptions about both energy demand and supply. If they are borne out by events, then the administration could claim a considerable success. If, however, there were a significant shortfall from expectations, as did occur with economic policy, then the costs imposed on the economy and society could prove much greater than need have been the case. The Reagan administration proceeded as though the adjustment process were more or less complete. What it did not pursue was a prudent, diversified energy policy aimed at minimizing a broad range of risks.

The approach of the Reagan administration contrasted no less with that of the Nixon and Ford administrations than with that of the Carter administration. After all, it was those earlier Republican administrations that had seen a "crisis," controlled prices, and promoted large government subsidies to capital-intensive projects. Although the Ford administration had certainly sought to do so, it was the Carter administration that had actually initiated decontrol.

These lurches in policy demonstrate certain key themes that characterize America's response to the energy problem: sharp shifts from sense of crisis to complacency; a strong desire to return to "business as usual"; a drive to find a technological fix, a miracle solution; considerable difficulty in balancing energy concerns with equity and environmental considerations; a basic debate over what the sources of the "energy problem" are and over the role of various actors; a denial of the significance of America's integration into the world market; and an inability to weld domestic energy issues to fundamental foreign policy and security considerations. In consequence, no firm consensus has yet emerged. Perhaps the problem just was not yet severe enough. As one analyst has written: "The embargo of 1973–74 caused temporary inconvenience to many, but it fell short of being a disaster. That is to say, it was just irritating enough to give rise to a search for scapegoats without being so painful as to create a nationwide front in the face of adversity."

Consensus is made more difficult by fragmentation, for it is inescapably true that the energy problem is really many different problems that have converged in a short period of time and that are experienced and viewed differently by different groups. No matter what the specific issue, many competing interests are affected. Thomas P. O'Neill, Speaker of the U.S. House of Representatives, aptly described the energy issue as "perhaps the most parochial issue to hit the floor of the House."[28]

The conflicts over energy policies, priorities, and consequences took many forms. But the most significant and overarching conflict was over price — which was really a battle over income distribution, political power, visions of America, and conceptions of America's relation to the rest of the world.

The Battle over Prices — and Their Meaning

Throughout the 1970s, energy pricing — particularly that involving oil and natural gas — was a subject of great disagreement in the American political process. Indeed, no other aspect of the energy problem, not even nuclear power, generated as much political controversy as the debate over the pricing of these two hydrocarbons, which together provided three quarters of the energy for the American economy.

The basic debate centered on a single question: "What is the appropriate, sensible, and fair way for energy prices to be determined?" The question could be posed in many forms. Should price be related to costs of production, to historic costs, to a fixed return on investment, or to the varying values ascribed by the marketplace? Should prices be internally generated within the United States or should they be responsive to the world market? What kind of prices are necessary to generate sufficient investment in new exploration and production and alternative energy sources? What effect do prices have on consumption? How should the costs and benefits of the pricing system be distributed, whatever the system?

Price controls on oil and natural gas were in effect even before the 1973 embargo. Natural gas price regulation went back to a 1954 Supreme Court decision, which led to a system of price controls in the interstate market but not in the intrastate market. Domestic oil prices had been stabilized since the 1930s by production limits set by the Texas Railroad Commission. In the late 1950s, the Eisenhower administration instituted a quota system to protect domestic oil against a flood of cheaper foreign oil. But oil prices themselves were not directly set until August 15, 1971, when Nixon froze them as part of an economy-wide wage and price control system. Most price controls were lifted by 1973, but not on the oil industry. After the 1973 embargo, Congress mandated a complex system of price controls, refiner entitlements, and allocations. In 1975, Congress extended oil price controls at least through September 1981. In other words, what was meant as an emergency program was apparently turning into a permanent feature of American life. The result was a composite price for crude oil that was about 15 to 20 percent below that of the world market in the years 1976–78 (see Table 4.4).

The reasons for the price regulation of natural gas are easier to understand; the fixed pipelines that deliver gas are seen as utilities, enjoying natural monopolies. But the persistence of price controls on petroleum in the 1970s actually, on first consideration, seems quite odd. After all, the United States is probably the most "market-oriented" of the major industrial nations, and adjustment through market prices is a basic tenet of such a society. Free movement of prices — obviously upward — would, a priori, coax out more supply and reshape demand. What happens when there is a

Table 4.4   Refiner Acquisition Cost of Crude Oil
(dollars)

	Domestic	Imported	Composite	Composite below Imported
1976	$ 8.84	$13.48	$10.89	19%
1977	9.55	14.53	11.96	18%
1978	10.61	14.57	12.46	14%

Source: *Monthly Energy Review.*

bad frost and the orange crop is damaged? The price of oranges goes up. People may grumble, but many pay the extra cost. Some buy apples. Farmers, meanwhile, are encouraged to increase their orange crops the next year, and, when supplies become more abundant again, the price goes down. That is a principle of the system. Why doesn't it apply to oil?

The question of whether to insulate domestic consumers against higher world prices has bedeviled all industrial nations.[29] But the issue has taken on a special importance and intensity in the United States. The reasons for the persistence of controls start with the schizophrenia that arises from the peculiar case of being a significant oil producer as well as the world's largest consumer. Out of that arises a series of endemic tensions, expectations, and suspicions. As Thomas Schelling has observed, this led to a bitter debate as to whether price is part of the problem or part of the solution.[30] Some have argued that free pricing alone can solve the problem. Others warn about the dangerous economic impact of higher oil prices; many OPEC meetings have been preceded by the pleas of leaders of the industrial nations against price increases. When Americans rate inflation as one of the most important problems facing both the nation and themselves personally, it is difficult to see high prices as a "good." And even strong free-market advocates in various administrations have, when faced with the possibility of a surge in inflation, wavered in their commitment to remove controls.[31]

The truth, as paradoxical as it has sounded to many on both sides of the debate, is that price is both a significant part of the problem and a significant part of the solution. An increase fourteen times over in the price of domestic oil — as occurred between 1965 and 1981, six times over in constant dollars — does draw away people's income and inflict economic pain on segments of the public. Insofar as those higher prices are paid to foreign producers in the context of large price shocks, they inflict considerable damage on the national economy. Yet those upward movements in price also reshape consumption patterns and encourage additional domestic production of oil and other energy sources. In summary, higher prices "today" reduce American oil imports, thus increasing the slack in the world petroleum market and so helping to prevent much higher and much more damaging price increases at some unspecified "tomorrow." Why did this argument prove so difficult for so many to accept?

To begin with, major questions of income distribution are involved. Wealth is transferred from the United States to foreign oil producers. That is a national loss. But the United States is also a significant producer as well — the third largest producer of oil and the largest producer of natural gas. Thus, considerable transfers of income also take place within the United States — from sector to sector, from region to region, from social group to social group. This shift of wealth means that there are winners and losers; some gain, some give up. The stakes are very large. The value of U.S. oil and gas reserves was about $200 billion in early 1973 at world oil market prices. Those same reserves today, even allowing for depletion, are worth $2 trillion at world oil prices.[32] So the battle over pricing has been a struggle over how to distribute the income generated by the production of these hydrocarbons.

The sectoral transfer — from such energy-consuming industries as chemicals to energy producers — can have a significant and negative impact on the ability of the former to mobilize investment, including investment in more energy-efficient capital stock. Obviously, higher energy prices mean higher costs for firms. Nevertheless, the business community has generally favored decontrol, for the commitment to market pricing and reduced government regulation is deep-seated and because decontrol seems more likely to assure reliable supplies over the longer term.[33]

The major axis of tension on energy prices has run along a regional-class split. Lower-income people expend proportionally more of their income on direct energy expenditures. In 1980, low-income households spent more than a quarter of their income directly on energy — two and a half times as large a share as middle-income households (see Table 4.5). Low-income families have the least capacity to adapt, the least discretion in what they buy. They are locked into a way of life without access to the capital or means to change. Moreover, many see energy as a basic necessity, like food. In a modern industrial welfare state, there would be a drive to assist those so seriously affected. Price controls came to be the chosen instrument, although there were also programs of direct assistance. But two further points need to be made. If one includes indirect consumption of energy, as embodied in goods and services, then the distribution is less regressive.[34] Second, even when the benefits of price controls are measured only in direct expenditures, the groups that in absolute terms receive by far

Table 4.5 Estimated Average Annual Expenditures by Income Class, Fiscal Year 1981

Estimated Household Income	Estimated Average Home Energy and Gaso- line Expenditures	Percent of Income	Percent of all households
Less than $7,400	$1140	23	15
$7,400–$14,791	1550	14	21
$14,800–$22,099	2010	11	19
$22,100–$36,899	2580	9	28
$36,900 +	3230	6	18

Source: Congressional Budget Office, *Low Income Assistance,* June 1981, p. 7.

the largest benefits are the middle classes. One might well suspect that it is in response to them that controls were maintained. As a staff member of the Energy and Power Subcommittee of the Commerce Committee of the House of Representatives explained the matter (in the context of a discussion about increasing the prices of gasoline):

> With regard to constituencies, I've heard a lot about the fact that this isn't an issue of the poor and thus there should be support because it won't be perceived as hurting the poor. That, of course, is the exact reason why these programs have problems. These programs that would increase the price of gasoline affect the middle class, those who vote . . . They are the ones who yell and scream. Congress relates to anecdotes. They relate to letters from people who can write descriptively.[35]

Moreover, there is a very substantial redistribution of income among regions as a result of changes in oil and gas prices, taxation, and investment flows. Ten states produce over 90 percent of the nation's domestic oil. According to one estimate made during the windfall tax debate, these states will collect in excess of an extra $128 billion in increased revenues as a result of decontrol, through tax receipts and royalty income. Those living in energy-deficient regions pay more for energy. For instance, it was estimated that in 1980–81, it cost residents of the Northeast about 30 percent more than the national average to supply adequate energy to their homes.[36] Thus, energy has greatly aggravated tensions between the Sunbelt, the South, and the West, on the one side, and, on the other, the Frostbelt, the established industrial states of New England and around the Great Lakes. Legislators from the latter group of states tended to exert the main pressures for continued price controls, while those from the Southwest were most vigorously opposed. But the geographical distinctions are not absolute; there are discrepancies within regions as well. Officials in Iowa who had to cut the state budget three times in one year can be just as angry at Montana's severance tax on coal as at Texas' tax on oil.

But the debate has gone beyond the redistribution question. It is also concerned with a fundamental disagreement about the validity of prices. To grasp the nature of the disagreement, we must pose a basic question: What do prices do in a market economy? In sum, they constitute an intricate network by which supply and demand are balanced through decentralized decision-making. They provide signals about relative values and relative scarcities. This provides incentives for production and services, which causes shifts of resources from sector to sector, region to region, group to group.

The most important part of the debate has been a bitter disagreement about the very existence of the greater scarcity value that higher prices are "telling" consumers about.

One reason for the disbelief is the long habit of cheap, easily available oil. During the period of energy abundance, consumers experienced declining real prices, which they saw in price "wars" among gas stations, which seemed to sit on three of the four corners of every major intersection. Indeed, costs were so low that the price of gas was hardly a consideration when buying or operating an automobile. All the relevant experience for many years seemed to reinforce the deep-seated belief in abundance.

In fact, it is curious to observe how powerfully some of the fiercest adversaries in the price debate have clung to the basic American belief in abundance. Both sides have tended to argue that only a malevolent force stands in the way of high oil production. Of course, they have disagreed on who is malevolent. On the one side, some producer interests have said that the only thing holding up the great surge of extra production is the coalition of consumers and "extreme environmentalists" along with government intervention and price controls. For instance, in 1977, Senator Russell Long of Louisiana declared: "We have reserves of oil, gas, coal, shale oil, geo-pressured methane and various energy sources sufficient to last us for hundreds of years. The sad fact, however, is that misguided government policy, bureaucratic indecision, and an increasingly dangerous 'no growth' attitude on the part of some segments of our society have caused great disincentives to expand energy production." As late as 1981, the director of research at one of America's most successful high technology firms could declare: "Since the deregulation of oil and gas in the U.S. enough new reserves have *already* been discovered to last for several centuries."

Table 4.6 The Oil Companies as Villains

Which of these two statements is closer to your opinion?

There is a real shortage of petroleum in the world and we're just going to have to learn to live with less gasoline.	28%
There is plenty of petroleum in the world; the only reason we're having shortages is because the oil companies want an excuse to raise prices.	61%
Don't know	11%

Source: Cambridge Reports 22.

Yet some of the keenest "price controllers" seem to have believed no less strongly that there are vast additional supplies that could be produced — were it not for the oil companies, who, they seemed to think, manufactured the apparent higher scarcity of oil. This view has been widely shared by the public. In October 1979, more than half the respondents to a poll asserted that the oil companies had fabricated the shortage; a quarter thought that the government should take over the companies and run them. Table 4.6 captures the lopsided belief in the villainy of the oil companies.

That the suspicion is pervasive and deeply held has been demonstrated in many surveys.[37] Why? Such suspicions have strong roots in American history, back to the secrecy with which John D. Rockefeller assembled and operated the Standard Oil Trust. It has been reinforced by the very scale of the major companies and the absolute size of their sales and profits. Many Americans tend to see the oil industry as a monolith. On the contrary, it has many competing interests, one of the sharpest being between the majors and the independents. And the independent refiners and Texas producers can be as outspokenly critical and suspicious of the majors as any consumer activist from the Northeast. Nevertheless, for a large share of the public, even if a conspiracy could never be pinpointed, a central perception was inescapable: One domestic group (the oil companies) was doing well (measured in higher profits) in a situation in which most people thought themselves worse off (because of higher prices). In 1974 and 1980–81, gas lines disappeared and the press talked of a "glut," strengthening the idea of manipulation.

A chronological association made matters worse. The Watergate investigation dominated the media during the first two years of the oil shock, and some of the most dramatic revelations involved the illegal payments made by some oil companies to the Nixon campaign. This exposure tended to reinforce the notion of manipulation. What was difficult for many people to grasp was how the bargaining position of the major oil companies had eroded in the 1970s vis-à-vis that of the producing countries: that the companies were much less capable of controlling the market than they had been in the 1950s and 1960s, the time of declining real prices. Instead, the companies were perceived as having intimate relations and virtually congruent interests with the producing countries.

Thus, decontrolled oil prices were viewed as both painful and illegitimate. "The energy crisis is perceived," public opinion analyst William Schneider has observed, "as an abuse of power. Events such as the reports of oil company profits serve more nearly to confirm than to contradict this belief." Many consumers simply did not believe that oil had become fourteen times more valuable or that the earlier prices were misleading.

So pervasive was this conviction that in 1975, the head of the Federal Energy Administration felt constrained to advise President Ford to sign into law a bill extending price controls — a bill that ran directly counter to the administration's own policies — in order to show that "the oil companies do not control Washington."[38]

General trends, however, have tended to speak otherwise to the persistent belief in unbridled abundance — that, while the United States has hardly entered an era of scarcity, the historic abundance has run into constraints. But the significance of the peaking of domestic U.S. oil production in 1970 was hardly understood, matched as it was against the deeply held belief in American abundance. Moreover, in contrast to Japan or Europe, America had had little experience in what integration into and dependence on a world market could mean. Compare, for a moment, the situation in Japan. Both American and Japanese society go about converting the same kind of energy sources with the same kind of technology. Yet Japanese society has proceeded with an assumption of scarcity and insecurity of supply conditioned by its history and position. Recent events have only confirmed this assumption, so there has been a far less divisive debate in Japan about whether there is a problem and, if so, what it is and what to do about it. In contrast, it has been much more difficult to arrive at a consensus in the United States, where society begins with the assumption of abundance. Many Americans simply rejected the idea that a group of foreign oil producers could effectively set domestic oil prices — even though America's reliance on imports had clearly grown to the point where it was the largest buyer of oil in the world market. Here was a view that would deny reality — that the OPEC countries were the source, not only of the marginal barrel, but of many millions of barrels a day.

Both the erratic movement in the real price of energy and the character of the public dialogue during the period between the two oil shocks could only reinforce the public's skepticism. For prices went up and the gas lines disappeared. The dominant message was that there was no real problem to be overcome; some claimed, for instance, that at between $6 and $8 a barrel, everything from shale oil and coal liquefaction to Arctic gas would be economic.[39] It is not surprising that a public subjected continually to that kind of message would have been, at the very least, skeptical and suspicious of the source of higher prices.

Yet by the end of the decade, the debate on price controls had pretty much moved toward resolution. This occurred first in the case of natural gas. In his April 1977 energy program, President Carter proposed natural gas pricing reform, followed by an exhausting year-and-a-half battle on Capitol Hill. "I understand now what Hell is," said James Schlesinger, secretary of energy, during the congressional debate. "Hell is endless and eternal sessions of the natural gas conference."[40] Finally, in the autumn of 1978, a compromise was reached that provided for a merger of the interstate and intrastate markets and quasi decontrol by the mid-1980s. As difficult as it was, certain pressures pushed toward the compromise. For one thing, the distinction between interstate and intrastate markets was creating severe distortions. Natural gas pricing was also essentially a domestic matter, without the additional complexity of reliance on foreign producers.

That was not the case with oil, which was clearly next on the agenda. The Ford administration had tried hard to work free of price controls on oil, to no avail. Yet, on April 5, 1979, six months after the natural gas compromise, President Carter announced that he would allow oil price controls to expire in September 1981.

This development leads to a most puzzling question. When the White House was considering oil price decontrol in late 1978, it was very much within the context of world oil prices, then around $13 a barrel. The possibility that oil prices would rise above $20 a barrel was regarded as beyond reasonable speculation and was not even seriously considered in the internal discussions. Yet by the time the President announced his decontrol plan, world prices were already on a trajectory that would take them over $30 a barrel. Why was decontrol accepted so easily, especially given the passions and years of political controversy, the renewed concentration on and intensified suspicion of the oil companies, and the fact that decontrol was being set in motion during a time of sharply rising oil prices? After all, it was in March 1979, on the eve of the decontrol decision, that a White House official had warned privately of the "darts and arrows coming at us from all directions — from people who want to get rid of the regulatory structure, from people who care about inflation, from people who want a sexy and affirmative [energy] program, from people who don't want the oil companies to profiteer, and generally from people who want to make life miserable for us." Yet, for a number of reasons, decontrol was accepted in this quite inhospitable environment.

Table 4.7   The Attitude of Californians toward Energy

Believe present energy situation is:	July 1977	August 1978	August 1979	April 1980	October 1980
Extremely serious	29%	24%	41%	41%	50%
Somewhat serious	46	40	40	40	35
Not too serious	15	20	11	10	9
Not at all serious	7	12	7	5	5
No opinion	3	4	1	4	1

Source: Field Institute, California Opinion Index.

To begin with, there was a strong element of simple exhaustion. After the natural gas battle, both Congress and the administration were reluctant to slide into another war on energy prices. The fall of the shah and the rise of the ayatollah's Islamic Republic also seem to have a considerable educating effect on the public — underlining the United States' integration into the world market and emphasizing that the centers of power in this market were the producing nations, not the oil companies. To "deny" OPEC prices was to deny the reality and consequences of America's position. The producers set the prices. Moreover, these events began to integrate energy and security considerations in the mind of the public and establish energy as a serious problem (see Table 4.7).

Still, the tendency to believe in more or less unlimited abundance has not disappeared, as indicated by the results of an opinion poll, shown in Table 4.8, that was taken in 1981.

There was a growing consensus that price controls were highly inefficient. For one thing, they retarded domestic oil production, although by how much was the subject of considerable dispute. More clearly, they subsidized consumption, in particular the consumption of imported oil, and thus encouraged larger imports than would have been the case with decontrol. The seriousness of this cost was evident in the alarming growth of U.S. oil imports during the period of price controls — from 3.9 mbd in 1971 to 8.2 mbd in 1979. The difficulties with the allocation systems in the first half of 1979 also affected attitudes. Moreover, the overall costs to the economy of continued controls were seen as far outweighing the equity costs to some consumers. To put the matter more graphically, most Americans would be worse off in a very sick economy with high inflation and high unemployment than with decontrol.[41] The cost of price controls in terms of misleading signals was already becoming vividly clear in the case of the automakers. Environmentalists and those interested in energy efficiency and renewables also came to support decontrol as necessary if en-

Table 4.8 U.S. Oil Import Needs

Do you think the United States has enough oil to meet its needs, or do you think we have to import oil from other countries to meet these needs?

U.S. has enough oil	49%
We have to import oil	37
Don't know	13

Source: Union Carbide/Cambridge Reports, 1981.

ergy conservation and renewables were to be able to compete fairly in the market.

There was also an increasing interest in the notion of "social" or "true" costs of imported oil. It was argued that even if imported oil were priced on the world market, price would still not represent all the potential costs to the society in terms of extra inflation, lost GNP, and various political and strategic losses. In other words, if anything, it would take a premium on top of the world price — in the form of either a tariff or extra consumer taxes — to capture the "true" cost of the imported oil. At the beginning of 1978, this idea was regarded as quite odd. By the middle of 1980, the point was becoming accepted, at least among many energy specialists.[42]

The windfall tax, proposed by Carter together with decontrol, seemed to provide a way to respond to the equity issues. At the same time, there was a growing sense that assistance should be targeted directly to the lower-income people most painfully affected rather than using price controls as a very blunt and inefficient welfare system that subsidized everybody — and everybody's consumption.[43]

One further reason for the acceptance of decontrol was that a strong mood of opposition was developing against the whole thrust of federal regulation that had grown up since the late 1960s. This opposition was one of the main themes, in 1980, of the successful Reagan campaign, with its pledge to "get the government off the people's backs." The complex, jerry-built system of oil price regulation was an obvious target — quite a change from 1973.

Thus, by the beginning of the 1980s, world prices were sufficiently regarded as representing reality — a much higher scarcity value — and to be accepted as legitimate and appropriate, even if not very pleasant. The result was that there was little protest when, as one of the very first acts of the new administration, complete decontrol was moved up from September 30, 1981, to February 1, less than two weeks after the new President's inauguration.

The Process of Adjustment

In the middle 1970s, the dominant drive was to contain prices and repress their effectiveness. By the early 1980s, this view had been succeeded by a new belief — prices do everything.

While true prices are a vital part of the adjustment process, they are only part of it. After all, prices are very abstract pieces of information. They themselves do not achieve greater energy efficiency; their effect is felt only as mediated through human attitudes, actions, and investment decisions. Their meaning can be amplified — for instance, by information that persuasively clarifies the domestic supply system or by mechanisms that make capital available. Their meaning can also be blunted or even undercut — for instance, by the inability of consumers to obtain information on the relative efficiency of appliances, by difficulty in finding competent personnel to retrofit homes, or by the nation's leaders' constant refrain that there is no problem, and never was.

The price message can be confused by temporary changes responsive to short-term forces. Prices do not readily internalize the potential political "accident" that may — or may not — be waiting just around the corner. In other words, prices alone do not make an adjustment. To assume otherwise, to assume that prices are a be-all and end-all, is to make a very mechanistic assumption about human society.

Overall, we can understand the process of adjustment and adaptation as emerging from an interaction of forces and changes in six realms:

1. In *energy markets,* where the price and availability of various fuels alter decision-making by both producers and consumers of energy, with consequent changes in both supply and demand for various fuels.

2. In *attitudes, perceptions, expectations, and values,* on the part of both producers and consumers, of their own individual interests, needs, and prospects and of the larger economic and political setting in which they operate. These are affected by complex information flows.

3. In *government policies and programs,* which can generate regulations (fuel efficiency standards for automobiles, fuel switching); subsidies and incentives (depletion allowances for oil producers, price controls for consumers, tax credits, grants); new infrastructures (research and development programs, state and local energy offices); penalties (taxes); distribution (allocations); and information flows. All can affect decision-making and actions.

4. In the *technology* for producing and consuming energy and providing the desired service, be it comfort, illumination, motive force, or mobility.

5. In the realm of *capital availability.* Some would see this as part of the first realm. However, it can be considered a separate category because of

the rapid change in price and availability, possible constraints ahead, and the problems with generating investment in the current economic climate.

6. In the vast *political and economic realm* outside energy, in which other considerations — the general state of the economy, inflation, environment, equity, national security — interact with energy considerations.

The relative power of the various players in the various realms obviously affects outcomes. But power may take many forms — mobilizing financial assets, capitalizing on organizational and technical skills, making one's voice heard in Congress, intervening in the regulatory process.

The effect of these interactions can be seen on the supply side. For instance, in oil and gas exploration, the number of crews engaged in seismic exploration and the total footage of wells completed both increased more than two and a half times between 1973 and 1981. Coal production increased 34 percent between 1973 and 1981, while coal consumption increased 21 percent.[44] Moreover, for the first time in years, the amount of oil reserves added in 1980 almost equaled the amount produced.

But the far more dramatic impact has been on the demand side, where a considerable process of adjustment has taken place. This process began in the early 1970s, when industrial consumers first noted a tightening in natural gas prices. But certainly it took on its real momentum in the aftermath of the 1973 embargo. Growth in demand for energy slowed, and in recent years has actually flattened.

In part, we are now seeing the consequences of decisions and infrastructures that began to be laid down half a dozen years ago. Yet it also appears that the second oil shock has had an even more profound impact than the first. Oil imports in 1981 were 5.7 mbd, down 30 percent from the peak (though still the largest import level of any country in the world).

The change becomes quite apparent by comparing energy use per unit of GNP in 1973 with that in 1981 (see Table 3.3). In 1981, the United States consumed 35.1 million barrels a day of oil equivalent — the same level as in 1973. In the meantime, the U.S. economy has grown. If every unit of GNP generated in the United States had required the same amount of energy in 1981 as in 1973, then the country would have consumed 7.3 million barrels a day of oil equivalent, more than it actually did in 1981. In other words, the United States was about 17 percent more energy efficient in 1981 than in 1973.

To many, this kind of adaptation comes as a surprise. "Conservation of energy was never regarded all that seriously as a primary method for cutting the nation's appetite for imported fuel," *Business Week* noted in 1981. The magazine quoted one analyst as saying, "Conservation has taken hold much faster and produced larger savings than anyone could anticipate."[45]

Certainly the dominant view after the 1973 embargo did focus on in-

creasing conventional supply and tended to dismiss efficiency. This ac-
corded with the fact that most of the more prominent figures in the field,
naturally enough, came from the energy supply industries. But it was also
in line with the powerful traditions of the American experience. Turner's
frontier was to be opened yet again. As it turned out, however, the real
frontier has proved to be on the demand side. Moreover, as we have observed, there was a well-articulated body of thought that had focused on
demand for a number of years, and it had consistently pointed out that the
greatest potential for speedy adjustment was on that side of the equation.[46]

Many also assume that the process is automatic. But it is one thing for
the potential to exist, another for it to be acted upon. How it happened —
the interaction of the six realms — is well illustrated by the considerable
change that has occurred in the American automobile fleet, which is a
major force in world oil consumption.

In the period 1974–78, the real price of gasoline in the marketplace had
declined by 11 percent, with the result that the real price of gasoline at the
pump in January 1979 was the same as in 1960 — and actually lower than
it had been in the 1950s. Price controls held prices lower than would otherwise have been the case. This downward movement in prices reduced the
consumer interest in efficiency that had been piqued by the embargo. Manufacturers, responding both to their own perceptions of future prices and to
consumer tastes, also paid less attention to fuel efficiency. However, there
has been a tendency in recent years to blame all that has happened to the
auto industry on price controls. Since decontrol in the middle 1970s might
have added about 10 cents to the price of a gallon, it is not at all clear that
this would have brought the dramatic shift in consumer preferences that
the events of 1979 and 1980 did. After all, the real price of imported crude
oil for the United States declined by 13 percent between 1974 and 1978. In
other words, price controls reinforced an overall trend in the price message.

In 1979, in the aftermath of the shah's fall, the price of gasoline began to
move up sharply. But the price message did not stand alone. There were
also problems of availability in the marketplace, as dramatically evidenced
by the gas lines.

In 1979 and 1980, the attitudes, perceptions, expectations, and values of
the American consumer seem to have undergone a profound change. Even
if he continued to be suspicious of oil companies, the consumer concluded
that there was a real energy problem. Moreover, there was a movement
away from regarding conservation in negative terms — as deprivation —
to regarding it in a positive light — as efficiency, as a way to adjust to a
changed situation in order to preserve certain goals (distribution of expenditures, comfort, mobility). Thus, in 1979, one poll found that for the first

Table 4.9 Changing Attitudes toward Conservation

If the United States continues to use energy as we do now, ten years from now we will be very low in energy resources and will face severe cutbacks in our life-styles.

Strongly agree	30%
Agree	35
Disagree	17
Strongly disagree	7
Don't know	11

Source: Alliance to Save Energy/Cambridge Reports, 1979.

time, a majority of people (65 percent) regarded conservation as a way to avoid "severe cutbacks" in life-style (see Table 4.9). Recent polls have confirmed the positive shift and support in favor of conservation.

There seems to have been a shift in values as well. For many years, the American car had been a status symbol, a powerful way to proclaim individuality and achievement through size, design, and speed. As a California state senator declared, "A man's auto is almost as sacred as his castle."[47] But the public seemed to shift in what it expected; a car was becoming more of a tool, a convenience, a device to get from here to there, and less an assertion of ego.

Price and availability certainly affected the consumer's outlook. But so did the external environment. The fall of the shah, the turmoil in Iran, the rise of the ayatollah, the taking of the American hostages, the Soviet invasion of Afghanistan — all of these images, powerfully communicated on television news, dramatized the nature and dangers of American dependence on foreign oil and so undercut the traditional belief in American abundance and self-sufficiency. Public attention to the hostage crisis reached a near record high for any news event in the years 1974–80, and in March 1980, 90 percent of the public agreed that "excessive dependence on foreign oil is a danger to the nation's security."[48]

These shifts in attitude facilitated another key change, an erosion of the "first-cost mentality" that had traditionally underlay investment decisions in the United States. What had mattered was the initial cost, not the life-cycle costs, not the operating costs — including those for energy — over the years. This was reasonable as long as these costs were relatively insignificant. This attitude certainly prevailed in the residential and commercial sectors, where structures were built with the goal of minimizing construction costs and therefore selling prices. Little attention was given to the goal

Table 4.10   The U.S. Market Share for Automobiles

Year	American Cars Millions	% of total	Japanese Cars Millions	% of total	Total
1970	7.16	85	.31	4	8.44
1972	9.25	85	.63	6	10.87
1974	7.36	84	.60	7	8.77
1975	6.95	81	.81	9	8.54
1976	8.32	83	.93	9	9.99
1977	8.97	81	1.36	12	11.06
1978	9.16	82	1.34	12	11.16
1979	8.23	78	1.76	17	10.56
1980	6.58	73	1.91	21	8.98
1981	6.90	73	1.86	22	8.53

Source: Data Resources, Inc., Lexington, Massachusetts; *Ward's Automotive Reports.*
Note: Other imports make up the balance.

of minimizing long-term energy consumption and costs. Much the same was true in the purchase of automobiles, where selling price, performance, size, and design all mattered much more than operating costs.

In 1979 and 1980, however, the consumer began to be much more conscious of the life-cycle cost, which involved not only perceptions of current prices, but also expectations about future prices — and the world in which those prices would be generated. The result was a violent shift toward fuel-efficient imported cars, particularly Japanese cars. (This shift was amplified by a belief that Japanese workmanship was of higher quality than American.) The American automobile industry was caught mostly unprepared for this violent shift, and it suffered a significant loss of the market (see Table 4.10).

Once the proudest symbol of American industrial leadership — with half the total world automobile production as late as 1960 — the American auto industry is now widely viewed as a sick industry. It would have been in even worse shape, however, were it not for *government policy* — the fuel efficiency standards of the Energy Policy and Conservation Act of 1975, which was one of the most important responses to the 1973 embargo. Fuel efficiency standards were generally opposed by the auto industry, and as late as January 1979, Henry Ford II was calling for their relaxation. Yet, six months earlier, Ford himself had said, "I think it's fair to say also that the law requiring greater fuel economy in motor vehicle use has moved us faster toward energy conservation goals than competitive, free-market forces would have done."[49] Government policies thus had helped prepare the auto industry and helped expedite the process of adjustment.

The need for higher fuel efficiency has in turn stimulated a vigorous effort to use the existing *technology* better and to develop new technologies and improved materials. But the availability of *capital* constitutes one of the greatest questions hanging over the adjustment process of the American automobile industry. Much higher rates of investment are required — General Motors estimates that its rate of investment in the first half of the 1980s will be twice that of the 1970s — but it is difficult for the industry to mobilize such capital when it is experiencing substantial losses (as high as $4.5 billion in 1980). Moreover, the *external environment* also has its impact, for auto sales have proved to be highly sensitive to interest rate fluctuations, which in turn means that the national effort to control inflation can severely retard the industry's ability to mobilize the capital for adaptation.[50]

Life at the Lower Bound

Thus we can observe that America is in the midst of a moderately paced, hard-won adjustment to a rapid change in the realities of energy supply. It should also be clear that this adjustment is not by any means an automatic process. Its speed and range — and, indeed, its success — depend on a complex interaction of forces. The nature of that interaction depends on an array of forces. Indeed, gains in one year can turn into losses the next if the signals (e.g., prices, perception of an oil glut) become murky or if the instruments of change (e.g., access to capital) are not available or if the external factors (e.g., social or regional tensions) are not ameliorated.

If there were decades in which to adjust, then the question need not concern us. But the world could well experience continued pressure on energy supplies, in turn creating acute time pressures. The costs — political and social as well as economic — of inadequate adjustment may be much too high.

The prospects can be highlighted by putting them in the context of the Lower Bound.

Scenario I

Let us assume that the supply scenario is what Robert Stobaugh has described as the "Lower Bound — zero growth in energy supply through the year 2000 for the OECD countries." As he explains: "This scenario is based on what might well occur if things do not improve over the pattern of 1973–80, in other words, if things do not go very well." Note that this is far short of the worst imaginable case, the "foreboding scenario" of a major shutdown in the Arabian/Persian Gulf area.

The consequences of the Lower Bound are not inherent in the actual supply situation. Rather, they emerge out of an interaction of the supply picture with the consumption pattern of the OECD countries, including the United States. We now come to Scenario I, in which we assume that energy efficiency will continue to proceed, but in this case not at a rate sufficient to provide a bridge between the requirements of economic growth and available energy supplies. This means that there will be continued energy stringency and, in particular, that the world oil market, the locus of energy supply, will be susceptible to continued pressure — whether evidenced in "crunches" or "squeezes," whether instigated by political events or simply by rising demand. Whatever the case, the consuming world will face chronic destabilizing hikes in oil prices.

The economic consequences of this type of price movement would be similar to those that followed the 1973–74 and 1979–80 shocks. This assumption arises, not from a mechanistic application of the "lessons of history," but rather from the observation of systemic continuities. At the same time, we should note a difference so obvious that its significance may be lost. This would not be the first price shock, not the second, perhaps not even the third. So its impact would be felt on a system already considerably weakened by incomplete recovery from the earlier shocks. Consider, for instance, the current condition of the American automobile or the savings-and-loan industries.

The initial effect would be a sharp acceleration of inflation. The impact would be felt first in the price of energy and energy-derived products. There would soon be a catch-up effect by other prices, and inflation would be diffused through the consumer price index. Workers would seek to catch up. A severe recession would follow for two reasons: the abrupt withdrawal of a large amount of purchasing power from the economy — the "OPEC tax" discussed in Chapter 3; and the deflationary steps taken by the government in order to try to reduce inflation. Uncertainty as well as recession would significantly slow the tempo of investment. Energy-intensive industries would find that the capital available for investment was greatly curtailed. The residential sector would not find easy access to capital either. Unemployment would increase significantly, especially in two areas: sectors that are particularly vulnerable, such as automobiles and steel, and the marginally employed, such as minority youth. The survival of major enterprises — and, indeed, of some industries — would be thrown into question. Significant sums of income would be transferred from some regions and groups to others.

What happens? There are two considerations: How the American system responds to the energy issue, and how the stresses engendered by energy affect the system.

One immediate response might be more direct state intervention in the economy — direct bailouts for companies and sectors hard hit. Beyond that, there might be a renewed immobility as the system tried to apportion the pain. The hard-won compromises over price could collapse, as various segments of society once again sought shelter from the rigors of the world market. Price controls might be reinstituted, once again distorting and misleading the consumer. On the other hand, a ratcheting of price might have devastating short-term impacts on the society. The pressures would accentuate the fragmentation of authority in a quasi-crisis period. Regional conflict would be intensified, as some states benefited — and were seen to benefit — from the rising energy prices. Social conflict would be increased as a result of the sharpening of the divisions between the dual labor market of white and black. There would be a renewed impulse to find "miracle" solutions and a strong drive to find a villain to blame.

Working their way through the economy, the consequences would take a heavy toll in lost economic growth. The type of chronic and potentially precipitous stringency suggested here would probably lead to significant loss in income, at least for many groups.[51]

Here we come to the nub of the matter. "In the past generation, electoral politics throughout the industrial world, and beyond it," political scientist Fred Hirsch wrote, "has been increasingly dominated by the economic problem — gross national product, personal disposable income, and the rate at which these indicators of material prosperity grow." And with good reason. For whatever the limitations encountered in calculating economic growth, this growth has played a critical *political* role. It has, in a sense, provided the solution to the grave crises that afflicted industrial societies in varying ways in the 1920s and 1930s. The conflicts and competitions among groups in each society — and among the nations themselves — could be resolved through the distribution of the growth dividend. Lose that dividend, and America becomes a "zero-sum society," in which the main task of the political system becomes, in Lester Thurow's words, "loss allocation."[52] And that is what happens in this case, for constraints on imported oil and other energy supplies undercut the process of economic growth. Continued high dependence in the strait of stringency becomes an engine of "ungrowth." As conflicts can no longer be adjudicated, various groups seek to defend their positions. What results is sharp budget-cutting, eroding various social objectives, heightening domestic tensions, and possibly creating conditions for upheaval. While such cuts may be seen as necessary to offset inflation and deficits, they also intensify conflict within the society. Protectionist impulses become much stronger, temporarily insulating some industries from the pressures of the world market, but also penalizing consumers, retarding adjustment, undermining critical Western alliance relations, and lowering overall incomes. This protectionism really constitutes a dangerous competition among Western political systems about how lost economic growth will be apportioned. Depending on the public perception of the sources of the energy problem, there might also be strong pressures to seek a solution through foreign policy or military measures.

The new shock would accentuate a host of negative attitudes. Reduced expectations would take a heavy toll in terms of loss of confidence in the future, which would undermine both the economic and political foundations of society. Tensions would increase among social and ethnic groups as they experienced or warded off the loss allocation. Certainly, bitterness and suspicion would become rife, as would a sense of betrayal, that somehow some groups in society had sabotaged abundance. Already, there is evidence of how energy, more than other issues, engenders a sense of helplessness. In a survey of people's sense of helplessness and power, on a scale running from 1 (most helpless) to 10, the median for financial control was 5.3; in health, 6.7; and for energy, 1.6.[53] In other words, people feel that they may have much less control over energy than over other parts of their experience, and that is a precondition for rage. As one observer has noted:

> Energy is, above all else, power; and the symbolism of many energy-consuming devices that make up the pattern of modern consumption is that of power or potency ... Moreover, when a shortage of energy or power threatens, the initial psychological impact is to create a feeling of powerlessness, of anxiety, of death — as though that which has been sustaining us is exhausted. The prospect of everything running out can excite the strongest and most primitive psychological fantasies of desertion and destruction ... Add to this the observation that the prospect of the increasing shortage of things we value replicates the classic sociological conditions that are conducive to a panic response. Those conditions include a threat, to be sure; but even more, they include a condition in which escape or access is neither completely open nor closed, but closing and threatening to close entirely.[54]

And where would that leave the political system? Perhaps able to muddle along, but perhaps sufficiently paralyzed that it could muster neither the cohesion nor the purpose needed to respond to the domestic and international challenges. Expectations would not be met, discontent would rise, and contradictory claims would be made on the political system. "Two conditions of stable democracy are that its internal conflicts take place within a framework of general consensus about the legitimacy of the system of government, and that the level of conflict not rise above that which the mediating political processes can sustain."[55] In this scenario, the basic legitimacy would be in question, and the conflict might not be easily mediated. It is only one step further to observe that such a system, already made unstable by the political changes of the 1970s, might become politically chaotic and vulnerable to demagogues seeking scapegoats and offering firm authoritarian solutions or that the system itself might change in a more authoritarian direction.

There is a natural enough tendency to assume that such things simply cannot come to pass in the United States. But it is always a mistake to assume that what happened to happen inevitably had to happen. Franklin Roosevelt was quite on the mark in his comment: "When I became President in 1932, it looked as if any extremist group could gain a tremendous following among the underprivileged and unemployed. Who knows? The United States might have had a dictator if the New Deal hadn't come along with a sensible program."[56]

The intent here is not to present an unreasonably negative case, but rather to emphasize the seriousness of the challenge posed by energy, and how that challenge reverberates throughout the social, political, and economic spheres of society. The point is that the challenge might be on the same order, have the same political magnitude, as the Great Depression. While American political culture may have great durability, problems arising from energy can subject it to stresses beyond its capacity to give. Because this has not happened so far does not mean that it cannot happen. As the economist Kenneth Boulding has reminded us, "One of the most dangerous of all illusions is to mistake good luck for good management."[57]

The foregoing may seem to paint an extreme picture; yet it is also an imaginable outcome of a period of energy stringency. It is a conceivable situation, but hardly the only one.

Scenario II

We can equally well imagine an accelerated program of adjustment that does prove sufficient. Perhaps it would be a 3 percent annual improvement in energy efficiency — allowing for 3 percent growth, with zero increase in energy demand. This would not be a heroic assumption, as the average annual rate of improvement between 1973 and 1980 was almost 2 percent. But adjustment so far has involved the easier things. Still, recent research into energy use, as well as the practical experience of the last several years, points to a very considerable and even expanding potential for energy efficiency. A growing body of research suggests that substantial economic growth could occur with flat or even declining energy consumption.[58]

What is most striking in considering how U.S. society might become more efficient is that it need not look very different from the present society. Economic growth would be maintained — indeed, stimulated — by the new investment. The changes that did occur would be incremental, rather than sharp, making implementation more steady. They would also emerge in interactions with other factors — smaller families, lower rates of population growth, saturation of various consumer goods, high interest rates, and the merging of telephone, television, and computers in the home.

For instance, changes in family size and the financing of home ownership, as much as energy, are already leading to a new housing stock that uses less energy, a shift from the free-standing suburban house to condominiums and cluster housing. But energy considerations themselves hardly need dictate the demise of the comfortable suburban home. Innovative work in North America has shown that a house can be made far more efficient without any loss of comfort. A dramatic example is found in the Canadian province of Saskatchewan, where houses are being built that pretty much resemble conventional homes but average almost 90 percent less energy for space heating than houses around the corner — with only 5 percent added to the construction costs.[59] And both the wind and the cold are inescapable for much of the year in Saskatchewan.

Further shifts to alternative modes of transport, such as carpooling, could occur. Yet, even if the single-occupant automobile remains the favored mode of transportation, average fuel efficiency could move well beyond 50 miles per gallon.[60] Improvements in existing operations and the development of new products and processes will promote further conservation in industry. Electronics can play an increasing role, either as an alternative to activities now based on conventional energy sources or as a way to make these uses more efficient. Some examples are telecommuting, home computers, and sophisticated controls for auto engines and building systems.

In this scenario, efficiency proceeds at such a rate that possible energy constraints do not impede economic progress significantly. The difference between these two scenarios is a matter of degree: the difference in the rate and steadiness of the process of adaptation. The outcome will also be affected by the contingent and unpredictable, whether it be the negative effects of a major oil interruption or the positive benefits of a significant technical breakthrough.

Adjustment on the demand side does not preclude or make unnecessary new energy supply options. Rather, it responds to the probability that there will be important constraints on existing energy supply systems, and it can actually be understood as a strategy to buy time. It permits a more secure voyage through the strait of stringency; that is, it allows more time for the development, sorting out, and spreading of options from a wide variety of technologies that can meet the requirements of the twenty-first century.

The Matter of Time

How, then, can we speed up the process of adaptation and adjustment? Let us consider the question in the context of our six realms.

With the arrival of the year 2000 prices in the early 1980s, the *energy markets* are giving pretty clear signals.

No one can confidently predict today which *technologies* will be victorious in two or three decades — perhaps those that are already highly visible, perhaps those known only to a few readers of an obscure scientific report on early research. We can allow ourselves a more general confidence, however, because innovation responds to need, and the need is clearly perceived. But the innovation and deployment of new technologies on a scale large enough to be meaningful in American society simply cannot happen overnight. It takes not only money, organization, and consistency, but also time. And the time is provided by greater energy efficiency.

The *external realm* can make its influence felt in many ways, varying from environmental problems that disrupt or impose unexpected heavy additional costs on energy supply, to international trade conflicts that slow down the adjustment process, to political upheaval and military conflict in the Middle East, leading to a major oil disruption. Obviously, the highest priority must be given to those steps that help prevent or minimize upheaval and conflict and to the emergency measures that can reduce the costs of a disruption, should it occur.

In contrast to, say, Japan, there is little consensus on the role of *government policies* in helping in a graceful way to correct market imperfections and to create an environment in which time horizons extend beyond quarterly results. The rebellion against regulation has obscured the important nonregulatory functions that government can perform — in stimulating research and development, diffusing information, assisting capital flows, and in helping strengthen market signals. Canada and the United States are two countries facing similar energy issues. Yet the federal government of Canada, with one tenth the population, is spending twenty times more for energy conservation than is the federal government of the United States.

The energy-conscious homes in Saskatchewan did not spring unassisted from the prairie soil. Considerable inertia in the building sector had to be overcome. Initially, government-supported research investigated the question of whether and how such homes could be built. Then, despite skepticism among contractors about building such homes as well as about their market potential, the government provided market-oriented incentives to help those builders who were willing to take the risk. Then the government assisted in diffusing the know-how in a highly fragmented industry across the country. Thus, a small but well-targeted and well-executed program has significantly speeded up the process of innovation and investment in building efficiency. In contrast, one would have to say that current government policies in the United States have become a drag on the process of energy efficiency. For instance, in 1981, the administration sought to suppress the Solar Energy Research Institute report that sketched such efficiency possibilities for the United States.

Capital availability is a considerable obstacle. One study of energy end-use has estimated that $700 to $800 billion of incremental investment may be required to fulfill the current potential for efficiency. The study does go on to observe that such sums, to be expended over two decades, are only somewhat more than double the nation's current bill for energy expenditures in the single year of 1980. Still, it is a great deal of money. There is much evidence for lack of access to capital even for currently very cost-effective energy efficiency investments. Despite rapid paybacks, high interest rates and inflation have forced many firms to set one-year payback requirements for efficiency investments. Moreover, investments in efficiency must compete with the very substantial demand for capital for energy supply that was outlined in the preceding chapter, and the channels for investment in supply are much better organized than those for efficiency.[61]

Finally, we come to the realm of *attitudes, perceptions, expectations, and values.* What Americans believe about the nature of the energy problem, its impact on their lives, and their ability to cope with it — these will have a profound impact on how the nation ultimately responds. For several years, there was what proved to be a misguided effort to control prices and so deny reality. This view has been succeeded in some circles by a passionate belief that prices will do everything. This view also denies reality — the considerable market imperfections, the gap between short-term private interests and long-term national interests, the time pressures, geological and technical constraints — and the possibility for disruptions and catastrophes that can simply overwhelm the market. Moreover, blind insistence on price as the be-all and end-all can result in a breakdown in the balance between energy and equity concerns, setting the stage for renewed domestic warfare in the years ahead. The effort to deny reality takes another form as well: the passionate desire to believe that temporary calms in the oil market, a temporary leveling of prices, mean long-term economic and political security.

Certainly the possibilities for further intense conflict are inherent in the difficulties that arise out of problems with energy. The American people may be moving toward a deeper appreciation of how much and how rapidly the energy supply system has changed. Yet this acceptance has a very tentative quality to it. There is still a great deal of skepticism, confusion, and anger as well as uncertainty — which reflects the uncertainty that hangs over the entire energy picture. Moreover, severe tension can emerge between an acceptance of the need for adjustment and the character of American culture — a culture, it has been observed, that "has long been dominated by a growing psychology of entitlement and a greater emphasis on the freedom of the individual. One result of the new psychology of entitlement has been to catch Americans in a strange cross current of pressures between cultural trends that tell the individual 'you have a right to a greater freedom of choice, even in material well-being,' and economic trends that signal the individual American that 'the good times may be coming to an end.' "[62] The inflation and unemployment that owe so much to the energy problem accentuate that clash, and in that clash resides potential for even greater political and social upheaval. Accelerated energy efficiency, speeding up the adjustment process, offers a reasonable way to mediate between these two trends.

For the real danger to American society lies not in adjustment, but in the failure to adapt successfully and in a timely and prudent fashion. Ironically, the American society of the future is more likely to resemble our present society if adjustment proceeds apace. Without such adjustment, the architecture could weaken and even fail from the political and social stress engendered by the loss of economic growth.

5

The Remarkable Adaptation of Japan's Economy

by Teruyasu Murakami

A historian writing in the twenty-third century might regard Japan as an experiment in an extreme form of industrialism: A large and very advanced industrial economy, closely woven into a network of international interdependency, attempted to survive the political turbulence of the late twentieth century without owning either substantial indigenous sources of energy or a colonial empire. Japan's dependence on imports is among the highest of the industrial countries: In 1981, Japan relied on imports for 85 percent of its energy needs. Moreover, three quarters of its primary energy demand was met by oil, most of which came from the Middle East OPEC countries.

This high dependence comes from two contradictory but basic characteristics of the Japanese economy. First, Japan's manufacturing sector is one of the largest in the world, and it is relatively more concentrated in oil- and energy-intensive heavy and chemical industries than that of most other industrial countries. Second, Japan is extremely poor in oil and energy resources. It does not have British North Sea oil, German coal, French diplomatic capability, or American political and military power. Its only valuable resource is its 120 million rather well educated people. Japan's energy policy has tried to bring these characteristics of its economy into balance. Thus, Japan needs to enhance its adaptability to outside changes, especially to abrupt changes in energy prices and supply.

Here we shall examine how Japan has adapted so far and consider whether it can maintain this adaptability or whether it will find a new kind of strategy for its survival in the 1980s — and beyond.

The First Oil Crisis and the Japanese Economy

In retrospect, we can see that the Arab oil embargo of 1973, with its grading based on a standard of "friendliness," was almost irrelevant to the physical workings of the Japanese economy. The oil embargo affected Japan for only a very short period: from October 17, 1973, when OAPEC cut supplies by 5 percent, to December 25, 1973, when Japan was declared a "friendly" country and could, therefore, import as much as it wanted.

Despite the small restriction on oil supplies and the very short duration of the embargo, the first oil crisis of 1973 brought about widespread social hysteria. Such petroleum products as heating oil became the target of panic buying by Japanese housewives. So did laundry detergent, toilet paper, and even soy sauce. The real economic crisis began when the oil import bill had to be paid. Overnight, Japan faced an unprecedented quadrupling of oil prices. It took five years for this shock to be absorbed by the Japanese economy. Let us look at the adjustment process of the Japanese economy to the first oil crisis in order to analyze the macroeconomic and structural implications of oil crises likely to arise in the 1980s.

In understanding this process, it is important to observe that three fundamental tendencies were already changing the nature of the Japanese economy.

First of all, following an average 10 percent annual economic growth in the 1960s, concern about a possible trend toward a lower growth rate developed at the very beginning of the 1970s. The oil crisis ensured that the possibility became a reality, and surveys of businessmen's sentiments have confirmed this expectation. It was given a seal of approval by the reduction of the target growth rate in the government's medium-term economic plan, which is often used by economic planners. The five-year plan of the Miki government in 1976 reduced the target rate to 6.25 percent from the Tanaka government's 1973 target rate of 9.4 percent. Then, in 1979, the Ohira government's seven-year plan set a target for the years to 1985 of 5.7 percent, which was reduced, at the end of 1981, to 5.1 percent. This transition brought a sharp reduction of capital investment, an unavoidable supply-demand gap, and a large current account surplus. Even without any other disturbance, this was burden enough to the Japanese economy. These structural changes not only altered the character of the economy, they will affect its response to further oil crises in the 1980s.

Second, as a result of the high growth rate during the 1960s, the Japanese economy had become, by the beginning of the 1970s, second only in size to that of the United States among non-Communist countries. Therefore, despite the low-profile foreign policy of Japan, even a slight fluctuation of the Japanese economy exerted a large influence on other countries.

Yet Japan did not properly understand what neighboring countries expected of it in the changing conditions of the international economic scene. Unfortunately, the Japanese were thrown into such a panic by the first oil crisis and were so exercised by its traumatic effects, they could hardly believe that their country still had an important part to play in world economic affairs.

In the third place, before the first oil crisis, Japan was sharing in the synchronized upturn of the economies of the major OECD countries; indeed, the Japanese economy was at the climax of an upturn of twenty-two successive months when the oil crisis came. The government continued to follow a lax monetary policy out of a concern for the deflationary effect of the "Nixon shock" in mid-1971. The rate of increase in M2 (money and quasi money) in 1971 was 24 percent, 25 percent in 1972, and from January to October 1973, it was running at over 20 percent per annum. The Japanese economy, therefore, was already set for the explosive inflation of 1974.

The real issue of the first oil crisis was, as already suggested, not a cut in the quantity of oil, but an unprecedentedly rapid price rise combined with a crisis mentality created by a latent fear that Japan's oil supply might end.

The oil crisis imposed an internal shock to the system, severely disordered it, and introduced four major disequilibria: extremely high inflation, a large balance of payments deficit, very low economic growth, and high unemployment. The economy was forced to recover from this disorder and restore the equilibrium on all four fronts. All industrial countries had to aim basically at meeting the same problems.

The characteristics of the Japanese adjustment process were as follows: Instead of seeking a simultaneous solution to these four problems, which came to the Japanese economy virtually simultaneously, a step-by-step approach was followed; the four policy targets were to be achieved one by one. The effort to restore equilibrium began with inflation, then the balance of payments, then growth, and finally ended with the labor market.

The Japanese government initially perceived the first oil crisis as a problem of inflation rather than of employment or of anything else. It therefore concentrated on policies to combat inflation to the virtual disregard of the other three difficulties. One of the underlying reasons for this choice was the fact that the first conspicuous symptom of the oil crisis was an extremely high rate of inflation, which came to be called *kyoran bukka,* the "inflationary craze." It was particularly traumatic because the sharp price rises were accompanied by panic buying. As a result, the social and political reasons for trying to calm down this explosive inflation were more important than the purely economic ones. In addition, the Japanese economy had already gathered a great deal of inflationary momentum before the oil crisis. Even without oil price increases, drastic treatment would have been required to reduce future inflation. Fortunately, fears of unemployment were mitigated by Japanese labor-management practices, based on life-long employment, the seniority wage system, and a company trade union system — often called the Three Sacred Treasures of the Japanese employment system. Thus, the fourth target of diminishing unemployment could be put aside until a later stage of the adjustment process.

To begin with, the Bank of Japan followed a very stringent monetary policy, and the official discount rate was raised 2 percent at a stroke, to 9 percent, and tight window guidance was used to restrain private bank lending. In addition, fiscal policy was tightened, and price controls were introduced in certain categories of goods, such as heating oil. The stringent monetary policy, coupled with the competitive market, eventually reduced the inflationary expectation. Bargaining between suppliers and users of raw materials became tough. An upstream sector could pass on to downstream sectors only a part of the higher production costs caused by the oil price hike. The proportion depended inversely on the level of excess supply in each sector.

Nevertheless, the effect of the tight monetary policy on wage rates was not immediately felt, and they continued to increase — by 26 percent in 1974. Consequently, the corporate sector was forced to bear the initial burden of adjustment in the form of a sharp reduction of profit between 1973 and 1975.

The second target, balance of payments equilibrium, was achieved as inflation calmed and economic growth fell sharply. The sharp decline in industrial activity was reflected in a sharp decline in imports, measured in real terms, in 1974 and 1975. On the other hand, exports kept increasing partly because of the weakening of the yen and a strong export drive. The balance on the current account turned to surplus in 1976, and sharply increased in 1977, stimulating protectionist sentiment on both sides of the Atlantic.

The third target was growth. In 1974, during the struggle against inflation, the Japanese GNP actually declined for the first time since the end of the war. In 1974 and 1975, exports were the only buoyant factor in the economy. Until 1978, private capital investment and private consumption did not play an important role. Between 1976 and 1977, public investment dominated the demand. The expenditure on public works was kept low in 1974 and 1975, but after inflation had receded, it increased massively, beginning in 1976. As a result, the Japanese economy grew rapidly during 1976, 1977, and 1978.

In contrast to the rather smooth achievement of the other three targets, the employment situation continued to deteriorate. The unemployment rate climbed from 1.3 percent in 1973 to 2.2 percent, 1.24 million unemployed, in 1978. Although employment-related indicators improved in 1979, they did not recover to 1973 levels. In fact, the number of regular workers employed in the manufacturing sector decreased by 11 percent from 1973 to 1979. This deterioration is partly explained by the refraction of the growth trend.

So the Japanese economy recovered from the shock of 1973 in three stages:

The first phase extended from the beginning of the oil crisis to 1975. During this phase, the fight against *kyoran bukka* (the inflationary craze) was the sole target of Japanese economic policy. This was the period of economic decline, with only exports remaining strong.

The second phase ran from 1976 to 1977, when external balance was restored and the economic growth rate rose into the target range of 5 to 6 percent as a result of a massive fiscal injection. In this phase, Japan learned a most important lesson: If a mismatch between the pace of its recovery and that of other OECD countries arose, protectionist sentiment would gain ground and increases in exports would be checked by political pressure.

In the third phase, in 1978 and 1979, the Japanese economy restored the equilibrium and the normal economic growth pattern was regained, led by private consumption and private investment.

Changes in Energy Demand and the Supply Structure

The first oil crisis led, not only to the repeatable and reversible changes just discussed, but also to structural changes of a nonrepeatable and irreversible nature. Since the Japanese economy is so dependent on imported oil, it was natural that a great deal of effort was made to increase the level of efficiency of energy use, and in this process many structural changes were introduced.

Imports of crude oil declined from 5 million barrels a day in 1973 to 4.2 mbd in 1980, despite the fact that GNP had grown about 30 percent. The total final energy demand increased at an average annual rate of 0.75 percent between 1973 and 1980, so that energy demand elasticity to GNP was only 0.20 in 1973–80 in contrast to 1.17 between 1965 and 1970. In 1973, Japan was the largest importer of oil in the world, but in 1980, Japanese oil imports were only 71 percent of those in the United States.

The following factors contributed to Japan's improved efficiency in the use of oil.

First, Japanese companies — especially in high energy-consuming industries such as aluminum, steel, cement, sheet glass, petrochemicals, and pulp and paper — have made considerable economies in energy use. For example, Nippon Steel reduced the rate of energy input per unit of production by 10.4 percent from 1974 to mid-1978 by introducing direct rolling, improving waste heat recycling, and by using furnacetop pressure-recovery turbines to generate electricity as well as by taking other conservation measures. The cumulative gross saving of energy was equivalent to 160 billion yen — almost equal to the net profit of the company in 1978. Nippon Steel now plans to achieve a further 10 percent energy conservation by 1983.

Second, the manufacturers of such goods as televisions, air conditioners, and cars are also improving the energy efficiency of these products. The electrical consumption of a color TV (with a 20-inch screen) and an air conditioner of one Japanese company were cut by 26 percent and 38 percent respectively between 1973 and 1979. The average fuel efficiency of a Japanese 1600cc car was improved by a third — from 25 to 34 miles per gallon — between 1975 and 1978 by reducing the weight, increasing efficiency in power transmission, and by using radial tires.[1] The increased use of automotive electronics systems has contributed to the recent improvement of fuel efficiency.

Third, using conventional technology, switching fuel from oil to coal or LNG (liquefied natural gas) played an important role in reducing oil consumption in a short time, especially in industries such as cement and steel manufacturing. In 1980 alone, eight of the nine major cement manufacturing companies completed their fuel-switching program almost simultaneously. As a result, the coal consumption of this sector increased dramatically, from about 1 million tons to more than 5 million tons at a stroke, while oil consumption was cut correspondingly.

Fourth, the Japanese economic system as a whole adapted to the high relative price of energy. Since 1973, Japan's industrial structure has changed in three main directions. First, the share of manufacturing industries in total production and employment decreased, while that of the tertiary industries increased significantly. Second, raw and intermediate materials industries, such as steel and aluminum, decreased in importance, and processing industries, such as machinery and pharmaceuticals, became relatively larger. Third, structurally depressed industries emerged, in need of adjustment and restructuring.[2] This whole process can be seen as movement from a high energy-consuming economy toward a more energy-conserving and more energy-efficient system, forcing a painstaking adjustment in the structurally depressed industries. This change in itself shows the highly adaptable and flexible structure of the Japanese economy.

In addition, the Japanese government is trying to improve the energy situation through the Sunshine Project, dealing with new energy development, and the Moonlight Project, aiming at long-term energy conserva-

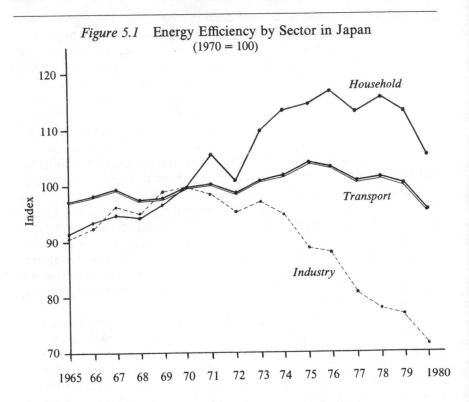

Figure 5.1   Energy Efficiency by Sector in Japan
(1970 = 100)

Note: Energy efficiency index = $\dfrac{\text{Final consumption of energy in each sector}}{\text{GNP in real terms}}$

Global Insecurity

tion.[3] In fiscal year 1980, the government budget allocated 741 billion yen to energy-related expenditures, a sharp increase of 31 percent despite only a 10 percent increase in the total budget.

However, there are some worrisome aspects to the Japanese energy picture. Although energy efficiency in the industrial sector has improved greatly, energy consumption in the transport sector has changed little in the last few years, and energy consumption per unit of income of the household sector did not show a decline until the end of the decade (see Figure 5.1).

Another concern is the change in how Japan gets its crude oil. The share of petroleum supplied by the major oil companies decreased dramatically in 1979. This reflects the changing structure of the world oil market that was triggered by the Iranian revolution. In the fourth quarter of 1978, 66 percent of crude oil was provided by the eight major oil companies; by the third quarter of 1980, this figure was reduced to 43 percent. This reduction means that the demand has to be met in other ways from OPEC countries (at least as long as other consumer countries do not share in the austerity by offering their oil to Japan). If not, the Japanese economy will suffer from a genuine oil shortage. This basic necessity for the Japanese economy — to adjust to the new reality of the world oil market — should have been remembered when the political stance of the Japanese government was criticized vis-à-vis sanctions against Iran and when possible government restrictions on purchases in the Rotterdam spot market were discussed. Japanese general trading companies, Sogo Shosha, tried to fill the gap. As a result, the share of the direct purchase of crude oil from national oil companies in oil-exporting countries, including direct deals and government-to-government deals, increased from 20 percent in the fourth quarter of 1978 to 45 percent of the total oil supplied to Japan in the third quarter of 1981. The pricing principle for these oil sales is not fully known yet, but it seems likely that in direct purchases, oil exporters would add premiums over the government price whenever the oil market became tight. Therefore, the greater the need for the direct purchase of crude oil, the higher the average cost to Japan. However, direct purchases could be the first casualties of a serious oil shortage. Thus certainty of supply under direct purchase in a tight market is in question.

A Range of Options for a Japanese Energy Policy

In the 1970s, Japan faced many economic problems but was unable to find effective solutions to those in three major areas: the distribution of income in both the domestic and international spheres, defense and national security, and energy.

During the 1960s, Japan was able, thanks to rapid economic growth, virtually to disregard social welfare and unemployment problems. However, the lower growth trend that emerged in the 1970s forced the government to face these questions. In addition, popular discussion of the aging of the population will lead to a serious conflict over intergenerational income distribution in the 1980s. At the same time, the concept of a "new international economic order," which demands a redistribution of income on an international scale, was widely discussed in the abstract. However, Japan will have to respond in a concrete form in the 1980s. Both the quantity and quality of Japanese aid to developing countries must be considered. The Japanese market must be opened to newly industrialized countries, and probably the industrial structure of Japan will move onto a high-technology- and knowledge-intensive structure.

During the 1980s, Japan will be asked to make a greater contribution to global security. Recent challenges to détente and the superpower conflict can only accelerate this trend.

Yet, while facing these difficult problems, Japan must still deal with the energy problem. During its rapid postwar growth, Japan took advantage of cheap energy by shifting into energy-intensive industries. Now, however, Japan is paying the unexpected costs of this policy and has to reverse direction for its industrial structure while trying to minimize the damages caused by short-term political contingencies in the Middle East.

Bearing these conditions in mind, we can identify three basic approaches toward solving the energy problem (see Table 5.1).

Crisis Management

In crisis management, there is no fundamental solution to the energy problem. An oil crisis is accepted as fate. Hence, the most appropriate policy is to minimize the damage to the Japanese economy should a crisis occur. The first priorities are thus to keep the oil stockpile at an appropriately high level and to take emergency measures, such as government restraint upon oil, natural gas, and electricity consumption as well as a mandatory gas rationing scheme.

The bargaining positions of consumer countries vis-à-vis OPEC have been strengthened since the first oil crisis by increasing stockpiles (targeted at 90 days' use) and by the emergency sharing system among the twenty-one countries of the International Energy Agency (IEA). At the time of the first oil crisis, Japan had an oil stockpile of merely 49 days, but now the private sector and government stockpile together is running at more than 100 days. To understand the resilience that oil stockpiles give consumer countries as a whole in the face of supply cuts by OPEC and OAPEC countries, see Figure 5.2. For each level of stockpile, the length of time that the need to reduce oil consumption can be resisted depends upon both the extent and length of the oil supply cut. The parabola CC represents the amount by which the oil supply is cut and the duration of the cut that a 90-day oil stockpile can offset without any reduction in oil consumption. This diagram is well worth some attention, for if it is well understood, panic in the years ahead can be avoided.

It is very difficult to measure the amount of oil that can be drawn from the stockpile in an emergency. A part of the stockpile could not be used although it is included in official figures. However, in an emergency, we could use oil in tankers already on the ocean when the supply cutoff starts. To simplify matters, therefore, let us assume (1) the net stockpile for emergency purpose equals the officially stated stockpile, at a 90-day level for all

Table 5.1 Profile of Three Approaches toward Energy Problem-Solving

	Crisis Management	Structural Adaptation	Political Actions
Type of approach	Preparing only for minimizing damage to the economy when a crisis occurs	Reducing the dependency upon OPEC oil by reducing the use of oil and developing alternative sources of energy	Establishing agreement between consumer countries and OPEC on stable supply of oil and reasonable revenue
Measures	Stockpiling Emergency measures	Privately led Energy conservation Shifting fuel source away from oil Government controlled Accelerated development of new sources of energy Energy consumption control	Collective dialogue Industrial development of collective bargaining of oil pricing between oil producers and consumers Military option
Nature of approach	Internal adjustment to external crises	Internal adjustment to external crises	External solution to external crises
Timing	Short lead time Immediate effect	Long lead time Long time to take effect	Discontinuous Little time to have effect if successful
Perception of the problem	One of the relatively important problems but not the most important	The most important and immediate problem for the whole society	A political problem of an international nature

Figure 5.2 The Effectiveness of the 90-Day (Net) Stockpile

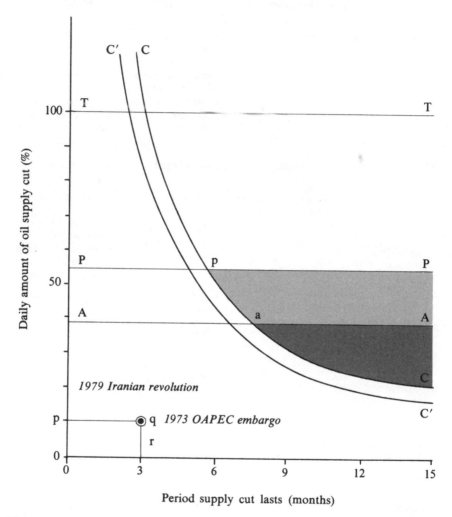

Period supply cut lasts (months)

Notes:
1. Daily amount of oil supply cut is expressed by the percentage of crude oil supply lost in total oil consumption of the IEA countries plus France.
2. Lines PP and AA move up or down every year. P and A indicate the share of crude oil exported from OPEC and from OAPEC countries respectively in the total crude oil consumption of the IEA countries plus France. This figure uses the number for 1979.
3. The stockpile in this figure is 90 days of effective oil stockpile.
4. See pages 146–50 and 161 of text for further explanation.

IEA countries and France; (2) an emergency sharing scheme works effectively; and (3) the mix of petroleum products in the demand is roughly the same as that of the oil stockpile. The area to the right of CC is the crisis area, which represents cuts that the stockpile cannot totally cover. However, in reality the crisis zone is not the whole area to the right of CC. So long as non-OPEC oil can be secured, we can disregard the area above the line PP, which indicates the share of OPEC oil in the total supply. If we believe it is only Arab member countries (OAPEC) that pose a real threat of a supply cut, the crisis zone is further reduced to the area below line AA, which represents the Arab oil exporters' share of the total oil supply.

In retrospect, the 1973 OAPEC embargo (plotted as ⊙ on the chart) was far from the crisis zone if we had had 90-day stocks. In the same manner, if we assume that the supply disruption in 1979 caused by the Iranian revolution was equivalent to the loss of the whole production of Iran (that is, 5.5 mbd for two and a half months, plotted as * in the chart), then it would have been at about the same level of crisis as that of 1973. If it is supposed that this supply cut continues, it would take two years to eat up the whole 90-day stockpile of consumer countries under the strict condition of no change of total demand whatsoever.

The Strait of Hormuz has a bad connotation in Japan, as it does in most other oil-consuming countries. The closing of the strait often represents the entrance to Armageddon. However, it is important to remember that a 90-day stockpile can delay our arrival in Armageddon by seven months, even in the absence of any increase in oil or substitute energy production and of any reduction in oil consumption. In theory, if the world oil consumption is cut by 25 percent and non–Gulf oil producers increase oil supply by 20 percent — which is not completely unrealistic — then the 90-day stockpile will last more than four and a half years. It is highly improbable that, faced with this long-lasting predicament, oil-consuming countries would be unable to mount effective political or physical countermeasures. Therefore, it is reasonable to assume that a 90-day stockpile provides consumer countries with a significant degree of resilience in the event of a single oil supply cut.

On the other hand, emergency measures must be imposed to avoid panic, which is, for social, political, and economic reasons, one of the most serious possible effects of an oil cut.

In retrospect, the type of crisis that the Japanese imagined when faced with the 1973 embargo, which was much inflated by the prevailing crisis sentiment among the general public, was far removed from the type of crisis OAPEC countries could in fact impose. In the 1980s and beyond, therefore, it is vital that the illusion of crisis be separated from the reality and that the magnitude and nature and limitations of disturbances that a real oil crisis can generate be clarified. Reliable government emergency measures, prepared beforehand, will help prevent social panic, which occurs if the market mechanism starts functioning in its crudest form during a crisis.

The coordination of oil stockpiling and emergency measures will be very effective in a crisis situation and thus can prevent its further exacerbation.

Structural Adaptation

A policy of structural adaptation tries to increase the countervailing power of consumer countries by reducing the dependency upon OPEC oil. This approach requires a massive shift of resources into energy, especially conservation, and into the development of alternative energy sources. This implies replacing expensive energy by relatively less expensive means, using the resources of capital, technology, and skilled labor. This approach can take two forms:

Privately Led Structural Adaptation

In the private sector, price mechanism plays a vital part, and the government limits its role to ensuring that the price mechanism works without distortion. In practice, government encouragement, through education in saving energy or through tax incentives that encourage investment in energy development, will play an important role in triggering structural adaptation. As the cost-benefit ratio of oil-substitution measures moves against oil, companies as well as individuals will respond by switching their fuel away from oil or by conserving energy. The higher the relative price of oil, the quicker structural adaptation advances. Such a process of change toward a highly energy-efficient economy is relatively smooth — provided markets work efficiently to allocate resources according to relative prices.

Government-Controlled Structural Adaptation

The shift away from an oil-based economy may be perceived by Japan as an absolute requirement to ensure its long-term economic security. In that case, whatever the change of the relative price of oil, the government will exercise strong initiatives in order to mobilize all available resources to accelerate change and to save oil. This requires that the government control the allocation of resources as well as prices. At the same time, energy consumption will be limited by the government, often by rationing. In its extreme form, such a regime is similar to a wartime economy, if not that of a Communist country of the Soviet type. This approach makes the development of new sources of energy feasible even if a commercial calculation does not justify the investment.

Political Action

Political action is based on the belief that the source of oil crises is actually the hostile relationship between OPEC and consumer countries. Improving this relationship and establishing an acceptable agreement on the stable supply of, and reasonable return from, oil should, therefore, be the ultimate objective of energy policy. This approach aims at removing OPEC-generated crises through collective political actions of consumer countries and takes two forms.

Collective Dialogue

Collective dialogue attempts to establish an institutional framework for collective bargaining on oil prices between oil producers and consumers. The producers will be represented by OPEC, consumers by the IEA. However, in order to make this framework work, the membership of OPEC and the IEA will have to be altered, for at present OPEC may not represent the interests of such important oil-producing countries as Mexico, China, and even Britain. The IEA may also have to expand so that it reflects the interests of developing countries.

Since this approach will reduce the ability of individual OPEC members to exert political influence through oil pricing and production control, it will be difficult to persuade OPEC countries that are politically minded, or are simply major producers, to support this idea. Indeed, the divergence of interests among consumers as well as among producers is so great as to make any effort at collective dialogue extremely unwieldy.

Military Option

If the economic consequences of an oil price increase or a production limit became intolerable to the consumer countries, there is a danger that some would be tempted to exert political influence underpinned by military force. This could vary from a crude military threat to suppress OPEC's demand for an excessively high oil price to intervention in a chaotic situation. Although the cost of this option is not necessarily high, the probability of its success may be relatively low. This is especially true given circumstances in which the target country has an option to rely on the Eastern bloc.

Although in theory Japan can choose one of or even all of these three approaches, in fact, the range of options that are immediately effective and that are open to Japan is limited.

At the current time, Japan must rely upon crisis management and gradual privately led structural adaptation. How profoundly structural adaptation, either private or governmental, takes place will depend upon the state of the energy market Japan faces in the future, and it will require several years before structural adaptation affects the energy market at all significantly. There is no real chance that Japan will be able to take the initiative in political action, at least in the first half of the 1980s. Collective dialogue requires political and diplomatic influence as well as extraordinary efforts at coordination; and the military option needs a military capability, which Japan lacks.

The Future of the Japanese Economy under Energy Constraints

In the 1970s, it looked as if energy forecasting had established itself as a viable industry. Enormous quantities of financial as well as intellectual resources had been allocated to it, most of which were employed on the basis of a common methodology, often called the "notional gap approach." This approach first assumes economic growth rates and energy coefficients. Having thus computed energy demand, it forecasts probable non-OPEC energy supply and possible OPEC oil supply, and so obtains the notional supply shortage as a result.

As long as it is used merely as a device to remind us of the seriousness of the energy problem, it is useful. However, once one tries to use the forecasts obtained by the notional gap method for any practical purpose, one cannot extract much from them. Energy demand is forecast in isolation from energy supply and hence does not predict actual demand, but represents a theoretical limit to the range of such predictions. At the same time, it is very difficult to make a reliable projection using a variable oil supply.

This study starts from the realistic assumption of an oil supply and attempts to analyze the macroeconomic and structural implications for the future of the Japanese economy. The Atlantic Institute approach, therefore, starts where a notional gap approach leaves off.

Two Oil Shortage Scenarios

The basic assumption of future energy supplies is given in Chapter 2, by Robert Stobaugh. His analysis is summarized in Table 2.8, which presents the world supply of each type of energy for 1978, 1985, 1990, and 2000. Stobaugh sets up two realistic possibilities for energy supply: an Upper Bound and a Lower Bound scenario.

The same table includes the aggregate energy demand figures for OECD, OPEC, and other developing countries for the two cases. Here, total energy demand always equals supply and, therefore, indicates the amount of energy each group of countries is theoretically allowed to consume in each year. The potential of economic growth is basically constrained by this allocation of potential energy consumption.

Two contrasting scenarios of the oil supply-demand situation, based on Stobaugh's energy scenarios, are set up in this chapter, focusing on oil instead of on total energy. In this analysis it is the oil supply-and-demand relationship rather than the total energy supply that makes the great difference in the development of the Japanese economy.

Scenario A: The Upper Bound Case

The Upper Bound case that Stobaugh set up forecasts a 2.6 percent per year increase in total net energy supply to the free world and a 2.0 percent per year increase of energy consumption in OECD countries. The same forecast does not detail energy consumption by country.

Starting with these figures, one option open to some countries, probably including Japan, is to continue economic growth by increasing their share of total oil imports on the basis of their competitive strength and their relatively high purchasing power, at the expense of weaker countries. In other words, this survival of the fittest option aims at changing the oil distribution pattern in favor of economically strong countries. However, it is not too naive or idealistic to argue that this option is virtually impossible to implement in such a highly interdependent world as the present one. Therefore, it is more realistic to assume that the distribution patterns of oil and energy imports will not change drastically in the years ahead.

In this case, if the maintenance of a 0.6 energy coefficient, which was discussed at the Venice Economic Summit, is at all feasible, it makes 3.3 percent per year economic growth in the OECD possible. Consistency with an unchanged distribution pattern of oil imports implies that the pattern of distribution of economic growth potential is also unaltered. In that case, an average growth rate for OECD of 3.3 percent per year is perfectly consistent with 5 to 6 percent per year economic growth in Japan.

However, the assumption of a constant 0.6 energy coefficient is very stringent in practical terms, because a constant energy coefficient entails an accelerated improvement in energy efficiency (energy consumption per unit of production). For example, it implies that companies in energy-intensive industries in Japan must continue to improve energy efficiency at the same accelerating rate, having already made the most easily available economies between 1973 and 1978. Failure to do so means a decline in economic growth. Entrepreneurs who seek to maintain the momentum of corporate growth in these circumstances would certainly perceive a shortage of energy, particularly of oil.

Second, this scenario presumes a drastic change in the pattern of energy consumption through a reduction of the share of oil in total energy consumption. In the Upper Bound case, the share of oil will be reduced by 5 percent from 1978 to 1985 and by another 5 percent from 1985 to 1990. This reduction would not be as simple in reality as we might imagine. In practice, a shortage of oil may be felt when the speed of reduction of the share of the oil supply is faster than the speed of adjustment of the economy to the new energy supply situation.

On the whole, the main feature of the Upper Bound scenario is an oil shortage, although only a minor one, at least in the earlier part of our time horizon. For example, one can visualize the shortage as being less than 3 mbd in 1985, or 6 percent of the total oil supply. However, economies are cyclical, so at a peak of the business cycle the oil shortage will be felt more acutely than the average indicates.

Scenario B: The Lower Bound Case

The Lower Bound case postulates zero growth in possible energy consumption in OECD countries. In this case, as Stobaugh rightly states, economic growth would be dependent solely on the ability to use available energy more efficiently. And, as was noted above, continuous improvement of the energy coefficient and efficiency becomes increasingly difficult as time passes.

The pattern of energy consumption changes in roughly the same way in the Lower Bound case as in the Upper Bound one. It means countries are exposed to the problem of a mismatch between the speed of reduction of the share of oil in the total energy supply and economic adjustment.

In the Lower Bound case, energy becomes a genuine constraint on economic growth, and oil production in all OPEC countries will keep declining after 1978 for the rest of the century. This scenario envisions a significant, absolute shortage of energy, especially of oil. One may quantify such a shortage at more than 10 mbd in 1985, or 26 percent of total oil supply.

The Implications for the Japanese Economy

Scenario A: The Upper Bound Case

The recovery from the second oil crisis will be marked by the emergence of the crisis management approach in Japan, with the stockpile running at well over the 90-day level. What is required is a sound consensus in both government and the private sector that the oil stockpile is not merely a buffer stock for smoothing a short-term fluctuation in price but a contingency reserve to secure the long-term security of the economy. The Japanese government is reported to have been formulating emergency measures, including a gasoline rationing scheme.

The future of Japan will be very different depending upon whether Japan regards crisis management as the ultimate solution to the energy problem or whether such an approach is accompanied by other policies. Since the crisis is perceived to be relatively minor in terms of quantity, structural adaptation will be delayed. However, such perceptions are often deceptive. The second oil crisis teaches us that a shortfall of, say, 3 mbd is more than sufficient to raise oil prices dramatically.

Renewed economic growth will come with recovery from the slump that followed the second oil crisis. World oil demand, which had sunk to a low ebb by 1982, will thus be stimulated again, setting the stage for new increases in the real price of oil. The consuming economies will again then enter a synchronized recession that will be accompanied by tight monetary and fiscal anti-inflation policies. Once inflation is controlled, a new upturn occurs partly as a result of government-induced reflation. Once more OPEC raises the oil price, which, thus, is repeated in a series of discontinuous steps. Moreover, if OPEC members are able to control their oil production strictly, then the arrival of crises could occur independently of the business cycle in consuming countries.

Given sufficient oil stockpiles and the imposition of emergency measures, panic will be avoided. Paradoxically, perhaps, this lack of panic may postpone structural adaptation. Adjustment is likely to be centered upon the manufacturing sector's effort to use less oil, and more coal and gas, and upon energy conservation by the household and transport sectors.

Inflation

In the absense of panic and because the Japanese have learned much from past experience, the bottleneck-induced inflation caused by panic buying in 1973–74 is unlikely to recur. The concept of "homemade inflation," which has recently been taken up by Japanese businessmen, politicians, and even trade unionists, reveals how Japan will fight the consequences of oil price rises.[4] When the price of oil increases, the crucial action is to break the link between imported and homemade inflation, by restraining unjustifiable increases in wages and unit profit. At the same time, average productivity is improved through the cooperation of labor and management on cost-reduction measures. In line with the worldwide burgeoning of monetarism, a tight control of monetary aggregates will receive greater support. A crucial factor in this approach is the flexible and cooperative relationship between labor and capital in Japan, in which workers volunteer improvements in productivity.

The rise in the relative price of oil and oil-related products in the domestic market will not be restrained by government action. The more the real domestic price of oil is allowed to rise in consumer countries, the more the repercussions of the international oil price on their domestic inflation are limited. This occurs because the link between the oil price and prices of other factors of production is uncoupled. At the same time, the more the real price of oil increases, the faster the price of oil approaches the cost of alternative energy, which, in theory, forms the upper limit of the oil price increase. In this sense, from an energy economist's point of view, a good anti-inflation policy, including a workable monetarist policy, is a good energy policy. Detaching the oil price increase from the rise of general prices and thus raising the relative price of oil is very important in setting limits to increases in the international price of oil itself. However, until the price of oil reaches its elusive upper limit, the increase in price, coupled with a possible temporary weakening of the yen exchange rate, could add a further push to inflation. However, since the Japanese people want to avoid the impact of these influences on homemade inflation, it is likely that it will be contained within acceptable limits.

In this it will be assisted by the employment situation. Reactions to past crises reveal that the employment cycle lags behind the business cycle. The result is that when a hike in the price of oil hits, the labor market is still relatively weak and below the peak of its own cycle. Yet it is at this very time that deflationary policies are imposed, which reduces the next peak of the employment cycle.

Balance of Payments

In the Upper Bound case, the oil price will be raised in steps. Each step will more than recover the intervening fall in the real oil price. Each sharp increase, of course, implies a rapid rise in the oil bills of consumer countries. Japan faces the most severe problems of any developed country, for it imports about twice as much oil as do both France and West Germany. In terms of price, Japan is also poorly placed. First, it has substantially increased the proportion of direct purchase of crude from national oil companies of OPEC countries; the price of oil in these sales tends to depend upon relative bargaining positions. Weaker buyers pay higher premiums in the form of "key money" or "signature bonuses." Japan's energy supply structure and political bargaining position mean that it cannot negotiate lower premiums than other countries. Second, the scramble for spot oil, seen in the second half of 1979, is likely to happen periodically in the future. Although this effect should not be overestimated, there is no doubt that it will have a negative effect on Japan's terms of trade.

The negative impact of a recession, induced by a sharp rise in oil prices, upon other commodity prices will only partially offset the sharp increase of Japan's oil bill. As the world economy slips into recession, newly industrialized countries will increasingly pressure Japan to open its market to their exports. However, such exports tend to be less energy intensive and so will not improve Japan's balance of payments as far as energy is concerned.

In order to pay for these increased imports, exports must also be increased. But such action is severely constrained. Steep increases in oil prices have imposed strong inflationary pressures on all countries. After the second oil crisis, all major OECD countries, but in particular the United States, tightened their monetary policies and kept interest rates at a high level, both to control domestic inflation and to finance balance of payments deficits. This is causing a recession throughout the world, whatever the future effect on inflation, with the result that export markets for all countries are being eroded. At the same time, there is an increasing uncertainty about OPEC markets due to political instability, the emergence of a sentiment against modernization in oil-producing countries, and the recent decline in oil revenues.

Since the impact of the oil bill on Japan is larger than that on other countries, Japan is under the most pressure to increase its exports. If it seeks to increase its exports to existing markets in the OECD, this can only be achieved at the expense of market shares of domestic firms in the targeted market. In these countries this will add to the strains that already exist between trade unions and management and will, therefore, stimulate protectionism against Japan; this may take the form of demands for the replacement of exports by direct investment by Japanese concerns. In the short term, this will have a negative impact on the balance of payments.

All of this will result in a weaker yen, and because of already shrunken world markets, the adjustment of Japan's current account imbalance will be difficult. One strategy would be to concentrate upon the export of those products that are in an early stage and that have no competitors yet. Alternatively, Japan may attempt to develop markets quickly in which it is already present but has yet to fully exploit, such as the Communist bloc. It is, however, unrealistic to suppose that either course will yield rapid results.

Therefore, Japan's only choice is capital account adjustment, but before this is possible, the Tokyo capital and money markets must be internationalized.[5] This will involve yielding to international pressure for further liberalization of the Japanese monetary system. However, Japan will not be alone in seeking inflows of capital. We may, therefore, see a competitive scramble for oil money similar to that already experienced for spot oil in the autumn of 1979.

The internationalization of capital and money markets merely means that the institutional framework for capital account adjustment is set up. For such adjustment, the Japanese economy must be attractive to overseas investors in order to induce capital inflows. For the short term, high interest rates form the incentive. However, in the long run it is vital to preserve confidence in Japan's unique adaptability to crisis situations. This means achieving a continuous and high level of investment centering on energy conservation, new energy development, and on technology- and knowledge-intensive industries as well as on channeling a part of the inflow of capital to developing countries in the Pacific region through Japanese financial institutions. The hardship in the initial stage of containment of inflation will only be justified if this long-term confidence in Japan as a whole is maintained and reinforced.

Economic Growth

In an attempt to reduce inflationary expectations at the outset of a crisis, the government will deflate the economy, which will reduce activity. However, once inflation is under control, the outbreak of an oil crisis itself will not reduce the long-term growth targets of the Japanese government. This contrasts with the experience of the 1973 oil crisis, in which the long-term growth rate was halved as a result of coincidental structural change of the Japanese economy. Thus, after the initial recession, recovery will wait for resurgence of capital expenditure.

In recovering from the first oil crisis, exports and public investment played a crucial role. However, in the future the importance of these factors as driving forces of the Japanese economy will be reduced. We have already described the limitations of exports in this role. Public investment will not be an important prime mover of economic growth because for some time the public deficit will be dominated by provisions for social welfare, partly due to the rapid aging of the population, thus reducing room for discretionary use of government investment. This leaves growth wholly dependent upon private investment and consumption.

Private consumption will continue to exhibit its stable growth because of the greater expenditure on services, resulting in a rapid increase in the importance of tertiary industry.

A more significant and important role, however, will be played by capital investment. If business and government policy follow the precept of "preventing homemade inflation," it will certainly force a higher relative price for oil and oil-related products. The present prices do not yet justify most of the large-scale development of unconventional sources of energy, such as solar heat and tidal power generation and nuclear fusion, but much oil and energy conservation investment has already become economic, and when the relative price of oil rises further, projects leading to a shift of fuel sources away from oil will become increasingly viable. In fact, this process is already gaining momentum in the corporate sector. The switch in fuels from heavy oil to coal in the Japanese cement industry in 1980 is a good example of such a dramatic change. Thus, privately led structural adaptation is under way in Japan.

More capital will be invested in industries with lower energy input per unit of output, such as microchip-based manufacturing, social systems industries (new transportation and new medical information systems), and those based on new technologies (optoelectronics, life sciences, and industrial ceramics and carbon fibers).[6]

On the other hand, since 1973, private investment in factories and machinery in Japan has been increasingly dominated by replacement rather than expansion. By 1978, replacement investment formed 50 percent of the total, and between 1973 and 1978, the average age of capacity rose from 6.8 to 7.6 years. As a result, in the future the economy will see a more cyclical investment pattern. It is still too early to judge how important these developments will be in the 1980s. However, in light of them, and in view of this analysis of Japan's adjustment to future oil crises, there is little to indicate that the Japanese target growth will be reduced below the present level of 4 to 6 percent.

Employment

Employment indicators were the slowest of all macroeconomic variables to respond to the 1973 crisis, for Japan's lifetime employment system postpones the time it takes for an economic shock to be reflected in unemployment.

The pattern of Japanese economic growth is in the process of changing from a continuous upward trend to a more cyclical path. As a result, the amplitude of the oscillations in growth are increasing. Lifetime employment and the seniority wage system were well adapted to a situation of continuous growth, but the capacity of firms to adjust to a drastic decline of business activity in the short term is limited. The principle of lifetime employment is secure, but the seniority wage system will probably be adapted to new circumstances.

The Second Oil Crisis and After

The Iranian revolution has brought about several changes in the world oil market, such as (1) the decreasing role of the major oil companies in the distribution of oil; (2) an increased importance of spot market, direct deal, and government-to-government deal oil; and (3) the collapse of the normal pricing system. The addition to the oil bill of consumer countries was larger this second time than in the first oil crisis. However, in contrast to 1973, the price increased in a series of sharp steps through 1979.

One should recognize that the 1979 oil crisis came at a moment when the Japanese economy was in far better shape than it had been in 1973. The business cycle was in the early stages of an upturn rather than at its peak, the rate of inflation was decelerating, the money supply was well controlled, and the labor market was weak despite a short-lived upturn in 1979.

There was no panic partly because of the step-by-step nature of the price increase and partly because of the experience of the first oil crisis. The recent rise of inflation in Japan has not, therefore, been the result of panic buying. Although inflation seemed to be accelerating at the beginning of 1980, one should have carefully looked into the details of price increases. There seems to be a clear contrast between price structure in 1974 and 1980. Raw material prices increased by the same magnitude in 1974 and 1980. However, final prices increased only a third as fast in 1980 as in 1974. As of April 1980, the wholesale price index was rising at 24 percent annually. Compare this with 31 percent in 1974.

However, this time, raw materials increased 68 percent, intermediary goods 30 percent, and final goods 7 percent, and the rise in consumer prices was 8 percent. In contrast, in September 1974, raw materials rose 72 percent, intermediary goods 29 percent, final goods 22 percent, and the corresponding consumer price index rose 24 percent. The tight monetary policy pursued since the beginning of the second oil crisis ought to be credited for this well-restrained price rise in final prices. The wholesale price index hit a peak in April 1980 and in May started to decline.

This time inflationary expectations fueled by the second oil crisis were neatly contained, and Japan did not see a price explosion such as the 1974 *kyoran bukka*. The 1980 spring offensive limited wage increases to about 6 percent during a period when productivity in manufacturing was rising at more than 10 percent per year.

We can see, therefore, that Japan's adjustment to the second oil crisis has so far basically conformed to the Upper Bound scenario. This will change only if there is a breakdown of international relations or inflation fails to be contained. The Japanese economy has taken the first steps of adjustment to the cycle of oil crises that awaits it in the 1980s.

Yet, as noted earlier, the shortage of oil will be increasingly felt as the end of the century approaches, and the pressure upon the economy to seek alternative sources of energy will be further intensified, although not as urgently as under the Lower Bound scenario.

Scenario B: The Lower Bound Case

The Lower Bound case assumes zero growth in energy supply to OECD countries, which will certainly place a tight corset around the growth of the economy. The primary factor that brings about this stringent scenario are the constraints that arise in the supply of oil, presented in Table 1.1. The distribution pattern among countries will probably be kept virtually unchanged in the case of oil, but probably not in the case of other energy supplies. The constraints determined by uncontrollable political developments upon the supply of non-oil energy is significantly less than those upon oil, with the exception of internationally traded natural gas and perhaps nuclear energy. Everything depends upon how far a country can mobilize national resources to this end. If the Japanese economy is to maintain a rate of economic growth consistent with social and economic stability, two basic conditions must be satisfied. First, the consumption of oil must be controlled so that demand matches restrained supply with the minimum hazard to the economy and the whole society. Second, the use of energy must be improved, and the supply of non-oil energy must be increased as rapidly as possible.

The important difference between the two scenarios is that under Scenario B, a stockpile policy does not help substantially in dealing with the oil shortage problem. The examination of the past two oil crises in Figure 5.2 dealt only with single crises, not with a succession of supply crises. The development of the Iranian revolution since early 1979 and the outbreak of the Iran-Iraq war underlines the importance of the latter case.

In a single crisis, consumer countries reduce oil stock by the area of the rectangle Opqr in Figure 5.2. The line CC shifts downward in proportion to Opqr to a new line, C'C'. This shift widens the crisis zone. In order to keep the area of the crisis zone constant, consumer countries must shift the line C'C' back toward CC; i.e., build up stocks once more.

However, if crises come in quick succession, stocks cannot be replenished and C'C' shifts down farther, and the crisis zone is widened even more. This is the most serious situation and will continuously weaken the resilience of consumer countries to future crises. Scenario B does not represent a once-for-all cut of the oil supply, but rather poses a continuous reduction of the supply, and the magnitude of the supply cut is greater than that in Scenario A.

It is difficult to envisage actual circumstances in which this scenario will become reality. For many OPEC countries, oil is the only source of revenue; without it, social or economic development cannot take place. It is not in such countries' interest, therefore, to drive up the price of oil so as to threaten the consumer countries' ability to pay or to restrict supply to the extent that the world economy ceases to function, although, of course, consumers and producers might well disagree on the degree of tolerance. Left to OPEC's decision alone, therefore, any shortage caused by deliberate OPEC decisions is unlikely to exceed more than 3 mbd, which is the situation covered by Scenario A. We are driven to the conclusion, therefore, that Scenario B will come about only in an extreme crisis, independent of the policies of OPEC.

In the first oil crisis, Japan's oil imports fell by only 10 percent, which was followed by a fall in the economic growth rate from 6.5 percent to zero.

This experience suggests that the onset of Scenario B requires strict government control over oil use. The degree of severity of this control will vary according to the actual shortage. As a result, the energy situation becomes a real restraint on economic growth and thus represents an extension, over the long term, of circumstances that most consuming countries faced for only a matter of months in 1973 and 1974.

Such a situation cannot be maintained indefinitely. A consensus will quickly emerge to place a higher priority on energy conservation and the development of substitute energy sources. Public resistance to nuclear power plants and the reluctance to allocate funds to new energy development could soon be relaxed. Once a consensus emerges in Japan, society changes rapidly, so the lead time for alternative energy — usually stated as five years for access and four years for commercialization — could be shortened dramatically. Scenario B, therefore, implies that after a very difficult transition period, structural adaptation will occur. However, it will be controlled by the government rather than privately led.

Initially, the price of oil rises rapidly, but such is the damage done to consuming economies that OPEC cannot summon the courage to raise prices much in the future. Indeed, attempts to escalate the oil price rapidly over a long period may elicit very strong political action by consumer countries in the form of the military option.

Consensus Formation in the Initial Stage

When the absolute shortage of oil is perceived, some panic buying will be unavoidable, which, if taken to extremes, will have serious inflationary repercussions and will result in the government's taking a tight monetary and fiscal stance.

Scenario B implies that until alternative energy development becomes sufficient to eliminate the oil shortage, strict control of oil use will be necessary. This will sharply reduce industrial production of, and capital investment in, oil- and energy-intensive industries centering on the production of intermediate goods in Japan. The expectation that this will continue for a considerable time will reduce targeted economic growth. As a result, a negative accelerator effect will bite the recovery of capital investment.

Attempts to maintain strict control of energy use in the long run may spill over into other spheres, thereby producing an almost wartime economy which, while not acceptable over time, will quickly call forth a national consensus that leads to structural adaptation in the form of energy conservation, a shift from oil as a source of fuel, and the development of alternative energy sources.

It is generally agreed in Japan that the energy problem is, together with domestic and international income distribution and defense and security, one of the three most important issues to be faced in the 1980s, but it is by no means certain that the energy problem will be given first priority. Programs aimed at alleviating energy shortages will have to compete with other policy targets. Under Scenario B, both distribution of income and the question of security will be complicated because of the likelihood of a serious political casualty among the Middle Eastern countries. There will, therefore, be greater pressure for defense spending. The resulting economic deterioration will hit the developing nations in general and the poorer sections of both developed and developing societies.

However, it is clear that the rational sequence of decision-making for Japan is to give first priority to energy and so open a path toward solutions to other problems. Except for a slow but steady increase in defense capability, Japan can, at present, contribute to the maintenance of global security only through economic measures, such as offsetting payments for U.S. troops. Japan learned during the 1960s and 1970s that distribution problems are best solved through economic growth. However, under Scenario B, which implies a long period of crisis, it may not be possible to preserve this clearly rational priority. Attempts may be made to achieve multiple policy targets simultaneously, which could lead to serious instability. Will Japan be able to avoid this situation and to pursue a solution to the energy problem before all else throughout a long period of crisis? The answer depends largely on the continued stability of the political process and on the ability of the government to provide effective political leadership.

Structural adaptation requires a vast amount of money as well as consensus. According to the estimate of an influential business association, Energy Sogo Suishin Iinkai, the capital requirement for energy-related investment between 1979 and 1995 amounts to 155 trillion yen at 1979 prices (approximately $700 billion), a sum that requires heavy investment from both the private and the public sectors.[7]

The Transition Period

Once consensus is formed, structural adaptation begins. But the most difficult period that Japan will have to face under Scenario B is the transition period, before structural adaptation begins yielding its benefits. During this period, investment in energy development will have three major effects, which will emerge successively in waves, with one beginning as the other ends: energy-consumption effect, income effect, and energy-supply effect.

The effect of energy investment when it first emerges is paradoxically an energy-consumption effect. In order to make the initial investment essential for government-controlled structural adaptation, an increase in energy consumption is necessary. This is because existing technology, which is highly energy intensive, has to be used to develop alternative energy such as nuclear or solar power or gasification and liquefaction of coal.

Once construction of new energy capital begins, there will follow a stimulus to effective demand via the usual income-multiplier mechanism. This may again contradict the stringent energy constraint that the economy faces.

Only after a long gestation period will the energy-supply effect come through. The fact that energy investment has different effects at different times will add to the difficulty in meeting the challenge of structural adaptation under severe energy constraint. In addition, since Japan is more dependent on oil than other countries, it must undertake more far-reaching structural adaptation than the rest of the developed world.

Under Scenario B, as oil prices will not increase sharply after the initial jump, there is no continuous, rapid deterioration of the balance of payments as a result of increases in the oil bill. However, the accumulation of current account deficit will increasingly become a heavy burden, although imports will decline because of the deflationary effect of oil price increases. The energy-consumption effect will increase Japan's foreign exchange requirements just when exports, especially intermediate goods exports, are severely hit by energy constraint and potential protectionism. Thus, the current account will weaken. This effect will be intensified, if anything, by the energy-income effect. The resulting fall of the yen will mean that Japan will be able to finance its balance of payments deficit via the inflows of short-term capital only at the expense of higher short-term interest rates.

The only real solution to these balance of payments difficulties lies through inflows of long-term capital into Japan. Such capital is not highly sensitive to interest rates but moves in relation to the relative growth potential of countries and to the effectiveness of structural adaptation. However, the weakness of Japan's balance of payments and the reduction in expected growth trends will make it difficult for it to attract this kind of capital. It is almost certain, therefore, that Japan will continue to run into a balance of payments ceiling throughout the transition period. The Japanese who lived through the 1950s will have a feeling of déjà vu.

Government commitment to structural adaptation is very important because of possible reactions by OPEC. If OPEC tries to disrupt the planning of alternative energy by threatening a real reduction in oil prices and so increasing the risk to private investors, or indeed if market forces accomplish the same thing, then the government may have to take a direct stake in the project or provide subsidies or incentives. That question moved from the hypothetical to the real in the first months of 1982.

It has often been said in recent years that the Japanese economic system possesses a unique flexibility in industrial relations, which enabled it to accommodate the serious effects of the first oil crisis. This flexibility has been based on two factors: the short-term shifts of the distribution of income between capital and labor, as shown in the recovery process from the shock of the first oil crisis; and the long-term prospect of each party holding to its income. As a result, both labor and capital in Japan have been concerned about the medium- to long-range performance of the total system — that is, the business performance of the company. To that end, they have both been prepared to make short-term sacrifices. This has usually taken the form of creative and voluntary improvements in productivity. Such flexible industrial relations rest on intense competition among companies, the fragmentation of the union movement, and the identification of workers with their employers that is implicit in lifetime employment and that is strengthened by the seniority wage systems.

Under Scenario B, the supporting factors of this flexible situation will be challenged. Until the energy-supply effect is felt, deflation will increase unemployment. In the discussion of Scenario A, the seniority wage system was envisaged as modified to maintain lifetime employment. However, in Scenario B, even lifetime employment will be under attack. In addition, differences in economic activity among various sectors (e.g., steel and electronics) will be large; hence, vertical movements of workers between parent companies and subsidiaries will not solve the problem and horizontal movement between sectors will be necessary. Indeed, this implies a more profound change than the horizontal movements within existing industrial groups that were seen after the 1973 crisis. Because in Japan the costs of unemployment to individual workers is very high, a decline in lifetime employment may well lead to a growing militancy among trade unions. These pressures on the base of the system of flexible industrial relations are particularly traumatic because they continue throughout the transition period.

Political Action

In the event of certain crises, there will be a strong temptation on the part of consuming countries to consider the military option. It is unlikely that Japan would entirely change its traditional low-profile foreign policy and quickly restore the asymmetry between economic and military power, although symptoms of change can be detected. Therefore, in the foreseeable future, this option is not open to Japan.

However, this does not necessarily mean that other powers, politically more influential and militarily more powerful, will not be tempted to take strong political action. An oil crisis is caused by increased politicization of oil pricing by OPEC. Consumer countries have tried to cope with the problem so far only by mobilizing economic instruments. However, in time, excessive politicization of the oil economy could lead to the politicization of the means to solve the oil problem.

If this military option approach is attempted by other countries, Japan will be affected, not only in its oil supplies, but also on international political and economic relations in general. The stronger the action, the higher the risk of a supply cut and a price increase. The vulnerability of its energy supply will make Japan try to stay out of trouble, if at all possible.

What is crucial is the extent of structural adaptation in Japan at the time of political action. The worse the bargaining position of consumer countries taking political action, as measured by their dependence upon OPEC oil, the stronger that action will be, and the more difficult and costly it will be to maintain any equilibrium that is attained with it. Since Japan will not be prepared to participate in this tough business of maintaining a fragile equilibrium through military capacity in the 1980s, it might have to bear an enormous economic cost to pay for the anti-OPEC crusade.

Japan will never be in a position to promote or participate directly in political action in the early stages of its structural adaptation. However, in later stages, in which the resilience of the Japanese economy to an oil crisis will be greater, a high degree of commitment by Japan to political action through collective dialogue is likely.

The economic benefit of political action is the stability of supply and price of oil, not cheap energy. As a sweetener for OPEC cooperation or a means to maintain the equilibrium peacefully, it is likely that the real price of oil will have to rise, so that even in this case, the balance of payments burden will be heavy. This approach will accelerate the development of oil resources that otherwise would be exploited much later in the century. In this case, the solution to the oil shortage is regarded as the further development of oil, not of oil substitutes. However, it does not solve the long-term problem of the absolute limit of oil resources. Paradoxically, this approach could advance the date at which this absolute limit becomes a pressing problem for mankind.

*   *   *

Even the analysis of the minor oil shortage case of Scenario A shows that a series of sharp oil price rises is likely. As long as the crisis management measures are in readiness, the social panic seen in 1973 need not recur. The inflationary impulses will not be translated into homemade inflation, and the Japanese economy will exhibit a healthy resilience to oil crises of this type. In this scenario, the cycle of oil crises may become a cycle of protectionism. A synchronized deflation to prevent an outburst of inflation and possible protectionism will lead to a large and persistent current account deficit. In this scenario, the Japanese economy will maintain the present level of economic growth that is strengthened by active investment. Energy conservation measures and new sources of energy will be developed through privately led structural adaptation, and growth in less energy-intensive industries, notably those that are technology and knowledge intensive, will take place. In addition, the flexibility in Japanese industrial relations will be preserved, and it will support the general confidence of foreign investors in Japan. If the Tokyo capital and money markets are sufficiently internationalized, the restoration of the balance of payments equilibrium will be achieved in the form of capital account adjustment.

Scenario B presumes a major oil shortage. Moreover, it is not a single crisis but a series of successive crises. Therefore, the oil stockpile cannot cope, and structural adaptation is needed. In this case, government-controlled structural adaptation is required. Owing to a heightened perception of crisis, a general consensus for structural adaptation will be formed quickly. The transition to a more energy-efficient economy will be constrained by a double ceiling formed by the supply of energy and the balance of payments. It will be a very difficult and lengthy process. This scenario foresees strong deflationary pressure staying for a long time, the perception and expectation of which will lead to the reduction of targeted growth rates. The supporting factors of flexibility in industrial relations will be challenged.

The prolonged austerity implied by Scenario B may lead some countries into political action, probably in the form of the military option. Japan, unable to take any significant initiative at all, will merely pay the economic cost and accept whatever result the venture may bring.

Under both scenarios, Japan's industrial structure will shift toward low-energy-, high-technology-, and knowledge-intensive industries and toward the non-oil energy production sector. At the same time, the financial system will become freer and more international. However, the speed of these structural changes varies by scenario. As a result of the reduction in the flexibility of industrial relations, the resilience of the economy to an oil crisis will be reduced. It is clear, however, that the character of the Japanese economy in the 1990s will be colored by whether the erosion of flexible industrial relations occurs more quickly than structural adaptation, or vice versa.

Whichever scenario afflicts the world, Japan must attempt structural adaptation. Scenario A will allow Japan to take the least troublesome course to this end. The sooner structural adaptation begins in Japan, the greater will be the resilience of its economy to oil crises and the less will its relative position in the world economy decline. No matter which approach is taken, international cooperation and solidarity among consumer countries is vital to cope with the common problem of future crises. After all, the energy problem is global by nature, and naturally requires international collective action for its ultimate solution.

6

Japanese Society and the Limits of Growth

by Joji Watanuki

The energy issue is not merely an economic problem, but has broad implications for all of society. Our daily life consists of energy-consuming behavior. Working, commuting, eating, and resting; most activities are, in a sense, concerned with energy, and so the energy issue is in fact about ways of life. The Japanese have consumed and still continue to consume less energy than other industrialized societies. When considering recurrent energy problems in the future, it is important to ask certain questions: How has the pattern of life changed for the Japanese as their economy has grown, and what might be the future trends? Can any lessons be drawn from the Japanese example to help us become more energy efficient in daily life? On the other hand, the ratio of industrial use of energy to total energy consumption has been considerably higher in Japan than in other developed countries. Why and how has this happened, and what lies ahead? The transformation of the Japanese industrial structure to one that will be less energy intensive is a vital part of this problem.

Energy has become one of the central political issues, both internationally and domestically. How has the Japanese government been performing in this respect? Moreover, the political implications of energy issues reach beyond energy policy to broader questions about the stability and capability of the Japanese democracy as a whole.[1]

Japan: 1955–1973

In order to understand the social and political implications of the energy factor for Japan, we must first review the period of rapid economic growth

168

between 1955 and 1973. For Japan's strengths and weaknesses in coping with the energy crisis have been formed by this economic surge and its social and political consequences.

During this time, the OECD countries achieved high levels of economic growth and prosperity with the aid of abundant and cheap supplies of oil. The oil was provided largely by the major oil corporations, which managed the oil supply under the umbrella of the United States. Among these countries, Japan achieved the highest economic growth rate and the highest increase in energy consumption, especially of oil (see Table 6.1). This sustained growth was a considerable achievement not only domestically, by raising the standard of living of the Japanese people, but also internationally, by increasing world trade and supplying various industrial products worldwide.

On the other hand, this growth also brought with it a number of liabilities. Domestically, the concentration of industrial plants in the narrow flat country of Japan created the highest density of energy consumption in the world and seriously polluted the environment. The rapid rise in the standard of living and the large migration of population from rural to urban and industrial areas resulted in an erosion of traditional values and social structure. At the same time, Japan, a nation with few natural resources, found its dependence on imported energy, food, and other raw materials rising to perilously high levels.

Table 6.1 The Growth of Energy Consumption, Import Dependency, and the GNP of Japan, 1955–1973

	1955	1960	1970	1973
Primary energy (1955 = 100)	100	167	554	683
Oil consumption (1955 = 100)	100	313	1945	2626
Dependency on imported energy	24%	44%	84%	90%
Dependency on imported oil (share of total energy)	20%	37%	71%	77%
Amount of oil imported (mbd)	0.2	0.6	3.8	5.3
GNP (1955 = 100)	100	152	418	524

Source: On energy figures, calculated from Shigen-enerugi-cho (Energy and Resources Agency, Japanese government), *Sogo Enerugi Tokei* (Comprehensive Energy Statistics), 1981 ed. (Tokyo: Tsusho-sangyo Kenkyusha, 1981). Abbreviation for that source in the following is *SET*, 1981 ed. Also, Keizai Kikaku Chō (Agency of Economic Planning), *Gendai Nihon Keizai no Tenkai* (*Development of Contemporary Japanese Economy*) (Tokyo: Okurasho Insatsukyoku, 1976).

Let us examine more carefully several of these points, which are highly relevant to present and future energy issues.

The Role of Heavy and Chemical Industries

Four sectors — steel, electricity, coal mining, and shipping — were regarded as the basic industries in the 1950s. They were favored with low-interest government credit from the Japan Development Bank as well as with other preferential measures. Then, in the 1960s, the government's emphasis shifted to three newer industries — petrochemicals, machinery, and electronics. While steel and electricity continued to be regarded as essential, coal mining was shifted into the category of "declining industry" and coal itself was replaced by oil, especially after the liberalization of oil imports in 1962. Petrochemicals, machinery (including automobiles), and electronics developed rapidly in the 1960s as a result of government "guidance," company efforts, and fierce competition among private enterprises.

Since those heavy and chemical industries are highly energy intensive, their development resulted in a rapid rise in energy consumption for industrial use (see Table 6.2). From 1970 through 1973, the steel industry consumed 18 percent of total energy and 16 percent of total electricity in Japan. However, even at that time, the steel industry was improving its energy efficiency; steel production increased more than energy consumption.

The Growth of Energy Use in Transportation and in the Residential and Commercial Sectors

Although more than half of Japan's total energy was used by industry in 1973, that sector's share was declining (see Table 6.2), and energy use in

Table 6.2   The Growth of Industrial Energy Use in Japan, 1955–1973

	1955	1960	1970	1973
Energy for industrial use (1955 = 100)	100	172	529	647
Share of industrial use to total energy consumption	61%	64%	58%	57%
Energy used by steel industries (1955 = 100)	100	203	799	1017
Share of energy used by steel industry to total energy consumption	12%	16%	18%	18%
Production of crude steel (1955 = 100)	100	235	992	1268

Source: On energy figures, *SET*, 1981 ed.; on steel, Keizai Kikaku Chō, *Gendai Nihon Keizai no Tenkai.*

Table 6.3   The Growth of Energy Consumption in Nonindustrial
Sectors in Japan, 1955–1973

	1955	1960	1970	1973
Gasoline in Transportation (1955 = 100)	100	221	791	1028
Electricity in Commercial/Residential Sectors (1955 = 100)	100	174	384	775
Kerosene in Commercial/Residential Sectors (1955 = 100)	100	318	3593	5275
LPG in Commercial/Residential Sectors (10^{10} kcal)	—	361	3953	5714

Source: *SET*, 1981 ed.

other sectors was beginning to grow faster (see Table 6.3). In transportation, the number of registered automobiles increased explosively after 1965, nearly tripling between 1965 and 1970 — from 4 million to 11 million. The number of privately owned passenger cars in that period quadrupled — from 1.3 million to 5.3 million. This growth actually accelerated thereafter, with an annual increase of 1.5 million vehicles between 1970 and 1973. This meant, inevitably, a very rapid rise in gasoline consumption, although this increase lagged behind the increase in the number of vehicles. (Already there was a tendency to use one's car less, primarily in response to crowded roads, particularly in the big cities.)

In the residential and commercial sectors during this period, Japan experienced a very substantial rise in energy consumption because of a remarkable spread of electric appliances as well as major improvements in domestic heating and cooking. Electric washing machines, refrigerators, and black and white television sets had, by 1970, reached nearly 100 percent saturation. Color televisions, air conditioners, and electric cooking ranges were the appliances of the 1970s. Traditionally, the use of heat in homes and businesses had been extremely limited. Even coal or wood stoves were rarely used in private homes, except on the northern island of Hokkaido. The most common heating appliances were the brazier and the charcoal footwarmer. Therefore, the introduction of the kerosene stove in the late 1950s, and its spread in the 1960s, was really a revolutionary change, for it illustrated a new idea — heating a room (entire houses were not yet heated) instead of merely one's hands or feet. Another major change was the introduction of LPG (propane gas), which was put into small cylinders for cooking and other household use. In the cities, pipelines had provided coal gas or natural gas for many years, and that supply had been expanding steadily at an average annual rate of 3 percent. However, for those in small towns and villages, the availability of LPG in cylinders brought a tremendous improvement in convenience and comfort, and its use grew rapidly in the 1960s. Yet one must also note that in spite of the sharp increase in energy consumption in the transportation and residential-commercial sectors in the 1950s and 1960s, Japan's energy consumption per capita in those sectors was still far below that of any other OECD country in 1973. In spite of slight increases in absolute energy consumption in those sectors in recent years, Japan's per capita level has stayed relatively low and will remain so in the future, as will be discussed.

Labor Unions, New Political Parties, and Citizens' Movements

During the years of great economic growth, Japanese labor unions had developed a strategy called the "spring struggle." This meant a nationwide coordinated demand for wage hikes each April, which usually succeeded in winning wage increases of about 5 percent more than the inflation rate. The spring struggle attracted more and more unions. Neither workers nor management was unhappy with the wage increase, for the increases were below the annual economic growth rate and were less than the increase in productivity. Thus, in spite of the increasing participation in the spring struggle, both the duration of the strikes and the total number of days lost by strikes actually became shorter. The labor unions did not oppose the introduction of new technology or the transfer of workers to new jobs and to new plants. Both management and workers were prepared to retrain workers and to develop the necessary skills when a new technology was introduced and a new plant constructed. Thus, management and labor unions and workers had, in this period, built a sense of common interest and mutual credibility.

Economic growth meant an expansion of the labor market in industry, and as a vast number of people migrated from rural to urban industrial areas, companies and labor unions helped these new workers settle down and adjust. Still, many newcomers — particularly those working in small shops and factories without labor unions or those who had operated their own small shops after several years as employees — felt alone and alienated in their new environment. The nascent political forces of the 1960s that attracted these people were the Komei party, the political arm of a Buddhist sect, and the newly revived Japan Communist party, both of which made conspicuous gains in the big cities, each to a 15 to 20 percent share of the vote. The labor force engaged in primary industries had decreased continuously throughout this period — from 41 percent in 1955 to 19 percent in 1970. In addition, remote towns and villages lost population. Still, agricultural and fishery interests, and all other local interests, were heavily protected by the politicians of the governing Liberal Democratic party, who depended to a considerable extent on those constituencies.

Huge subsidies to domestic rice production, import barriers on fruit and beef, preferential distribution of public works — such local interests were protected by the governing party. In this regard, the Liberal Democratic party was, and still is, a party of local interest, a party of the periphery, and yet at the same time a party of the center, closely aligned with big business and bureaucracy. This has, as we shall discuss, significant implications for the future.

Another phenomenon observed in Japanese society since the late 1960s has been the emergence of new values similar to the postindustrial values in other advanced societies. Environmental movements began to appear in the mid-1960s, and they attracted wider and wider attention in the late 1960s and early 1970s. Mingled with them were growing movements that demanded more "participation" in the decisions of local bodies. Various protest actions, including those by students on campuses, became rampant in 1968 and 1969. In the 1970s, the student protest movement diminished. Environmental movements also became less conspicuous because the government and business took various antipollution measures. Still, the new values and movements — emphasizing participation, environmental protection, and other humane values — did not by any means completely disappear. Actually, the government and local bodies have adopted some of those values, and many other groups pursue them vigorously.

Government and Bureaucracy under Incremental Budget-Making

During this period of economic growth, government budget-making was relatively easy. It relied on a natural increase of revenues due to a sustained economic growth of over 15 percent in nominal terms. "Incrementalism" had become central to budget-making, which still left room for the yearly addition of new expenditures in the budget.[2] Thus, vested interest could be protected while new demands were accommodated. Moreover, because of the lack of large military expenses, the post–World War II Japanese government was relatively small, and the tax burden remained comparatively low in regard to either GNP or per capita income. According to one estimate, as of 1965, the ratio of government expenditure to GNP was about 20 percent in Japan; it had reached 35 percent in Western European countries and nearly 30 percent in the United States and Canada.[3]

This incremental budget-making under the umbrella of rapidly growing revenues fits not only the character of the governing party — representing various interests, both central and peripheral — but also the sectionalism of Japanese bureaucracy, whereby each ministry demands more of the budget and insists on a fair share or "balance" among ministries.[4] In its functional aspect, Japanese bureaucratic sectionalism was accompanied by a sense of mission based on each ministry's belief that its interests were linked tightly to the interests of the nation as a whole.

The Ministry of International Trade and Industry (MITI), in particular, had built up a record and pride of being the "policy-oriented" ministry that guided Japanese industry. In the late 1960s, MITI had started to work out a long-range "vision" of Japanese industry. That vision was expressed in a document, released by MITI in 1971, which made the following argument about the future of Japanese industry:[5] In the 1950s and 1960s, development of heavy and chemical industries had been based on two criteria — productivity and income elasticity (i.e., expected growth of demand surpassing that of income). However, in the 1970s and beyond, two other criteria — environmental concerns, including conservation of energy and resources, and concerns about work safety, job satisfaction, and fulfillment — assumed a greater importance. MITI concluded that the desirable industrial structure for the future of Japan would be one "centered in industries with high knowledge intensity." The idea of "knowledge-intensive industry" was — and still is — not completely clear.[6] But an important point is that MITI anticipated, before the 1973 oil crisis, the need for a ceiling on massive energy- and resource-consuming heavy industries. In addition, in 1973 MITI created the Agency for Resources and Energy. So even before the first oil crisis, the Japanese had created a mechanism to deal with possible supply interruptions.

1973–1979: The Impact of the First Oil Shock

But there were neither institutional nor psychological preparations for such a crisis as that of 1973, and the initial response was confused, belated, and ineffective.

The Economy, Private Enterprise, and Consumers

Just before the 1973 oil crisis began, the Japanese economy had been overheated both by the boom of "rebuilding the Japanese archipelago," proposed by Prime Minister Kakuei Tanaka, and by the sharp currency fluctuations that resulted from the introduction of floating exchange rates. (The Japanese yen had jumped by 13.5 percent in relation to the U.S. dollar.) Land speculation was rampant, and commodities were oversold and in short supply. OAPEC's announcement of its price hike and cut in supply of oil in October immediately triggered further price hikes as well as an unwillingness by producers and merchants to sell oil and oil-related products and many other commodities. The mass media carelessly reported the spread of panic buying and predicted imminent price hikes for commodities, thus exacerbating the mood of panic. Agitated consumers rushed to hoard such commodities as toilet paper, laundry detergents, and kerosene. The result was the *kyoran bukka* or "inflationary craze," the very high rate of inflation described in the preceding chapter.[7] Among advanced countries, Japan experienced the largest and quickest price rise from the 1973 oil embargo.

The Response of the Government and Bureaucracy

The government was not totally unprepared. The new Agency for Resources and Energy had been launched in July. A White Paper on energy had been released in September, in which some reference was made to the overall insecurity of the oil supply and the necessity of measures to cope with an emergency. Immediately after OAPEC's move of October 18, cutting oil supplies by 5 percent, MITI and its Agency of Resources and Energy started to discuss the oil situation. But it took a month for the government to develop "An Outline of Measures to Cope with an Oil Emergency" and to form an interministerial task force to deal with the crisis. And, although after November 16 various administrative measures were taken to restrain oil consumption and to smooth out oil distribution, effective laws were not enacted until December.

Two pieces of legislation — the Law for Proper Supply and Consumption of Oil and the Law for Emergency Measures to Stabilize People's Daily Life — gave the Japanese government the power to set the price of commodities, to order the selling of inventories, and to introduce rationing of oil and oil products if necessary. This was the first time since the end of World War II that the government obtained rationing powers as well as price-control powers for specific commodities. The Price Control Law of 1948 was still in effect, but it was too comprehensive in this context.

Although the enactment of these laws may appear slow, by Japanese standards it was exceptionally fast. Indeed, five weeks — three weeks of drafting the bills in the bureaucracy and two weeks of discussion in the Diet — set the record in recent years for speedy lawmaking. Moreover, these laws received unanimous approval by all the political parties, including the Japanese Communist party, though only after some minor revisions proposed jointly by four opposition parties (the Socialist, Kōmei, Democratic Socialist, and Japan Communist parties) were accepted by the governing party.

On December 22, 1973, both laws were approved by the Diet and enacted immediately; under the Law for Proper Supply and Consumption of Oil, an emergency declaration was made, which remained in effect until

July 31, 1974. On January 14, 1974, standard prices for kerosene and LPG for residential use were set, based on the Law for Emergency Measures.

Since July 31, 1974, no additional measures based on these laws have been taken. However, these laws do provide the necessary legal framework should the government have to cope with future emergency interruptions of oil supplies. Emergency plans for varying degrees of oil supply interruptions are updated annually and deposited in the safe of MITI. However, MITI's strategy for a slight undersupply of oil is to rely on "administrative guidance" rather than on the laws, for MITI is effective in providing such guidance to industries and it wants to avoid the panic among consumers that might result from prematurely applying the laws.

The Aftermath: Economic Adjustment in 1974–1975

Inflation hit Japan quite fast. The price rise continued in 1974, with the year ending with a 25 percent inflation rate. However, economic adjustment to inflation took a very peculiar form in Japan. The average wage increase in the spring struggle of 1974 was an unprecedented 33 percent, substantially higher than inflation. So workers were happy in spite of both the inflation and the anxiety engendered by the oil crisis.

Most companies had accumulated large profits before the oil crisis and did not want to jeopardize their harmonious relationship with the labor unions. The inflation rate plus a substantial increment had become the norm for wage hikes during the spring struggle. Since they could afford to maintain this norm in 1974, they did not want to fall short of union expectations.

Thus, a fundamental problem was deferred to the following year. If inflation continued, demand for wage increases in the next spring struggle would be intensified, and it was doubtful that companies could afford them. The government, in order to terminate the rampant inflation before the spring struggle of 1975, persuaded private corporations to cut their profits. Nobuyoshi Namiki argues that Japanese corporations learned a lesson: Their behavior just after the outbreak of the oil crisis — raising prices — had triggered inflation and wage increases and eventually led to the loss of profit.[8] In other words, they learned that inflation would hurt them, and therefore it was wiser to refrain from behavior that would stimulate inflation. This lesson would be applied effectively in 1979.

As a consequence of cutting corporate profit in 1974, government revenue from corporate taxes dropped sharply from 1975 to 1977. The government chose to cope with this decrease of revenue by issuing government bonds, without cutting government expenditures. The share of bonds in total government revenue had jumped from 11 percent in 1974 to 25 per-

Table 6.4   The Use and Efficiency of Energy in Manufacturing and Mining, 1973–1980

	A Total energy (1973 = 100)	B Oil (1973 = 100)	C Production index (1973 = 100)	A/C	B/C
1973	100.0	100.0	100.0	100.0	100.0
1974	96.9	94.8	90.3	107.4	105.0
1975	92.5	94.6	86.3	107.2	109.6
1976	96.4	90.2	95.6	100.8	94.4
1977	94.1	86.8	98.6	95.4	88.0
1978	95.1	88.8	105.6	90.1	84.1
1979	98.8	86.0	115.4	85.6	74.5
1980	95.3	70.8	120.7	80.8	58.7

Source: *SET*, 1981 ed.

cent in 1975 and 29 percent in 1976. Moreover, in 1977 and 1978, the government adopted a policy of stimulating business through increased public works expenditures. As a result, despite the recovery of revenue from corporate taxes, total government expenditure has continued to grow faster than revenue, and more bonds have been issued. The share of reliance on bonds in the total revenue has exceeded 30 percent, thus causing what is known as a "financial crisis of government" and creating a serious challenge for government in the 1980s. This continued growth in the budget was congruent with bureaucratic incrementalism and the governing party's accommodation of various interests — those practices and habits established in the era of great economic growth. For these reasons, the growth in expenditures was welcome.

The pattern of energy consumption in Japanese industry after 1973 is outlined in Table 6.4. Both the total energy and the oil consumption by industry (manufacturing and mining) decreased. However, until 1976, the main reason for the decline was lower production due to a recession. Energy efficiency measured by A/C or B/C was lower in 1974 and 1975 than in 1973. A MITI report says that this resulted from the reduced operation of production facilities because of the recession and continued measures to protect the environment.[9] However, energy saving by industry became more effective after 1976, lowering the ratio of energy consumption to production continuously, especially in the case of oil.

A Changing Life-Style?

The 1973 oil crisis delivered a strong blow to the attitudes formed during the period of economic growth embodied in such slogans as "Consumption is a virtue" and "Material affluence has the top priority." These values had been promoted during the 1960s by commercials for consumer goods. After the oil crisis, the traditional Japanese values of frugality and savings regained their position. According to a public opinion survey conducted nationwide in July 1978, 85 percent of the respondents said that they were trying to be frugal in daily life.[10] The breakdown of data showed a clear correlation with age: the older the people, the more conscious they were of thriftiness. This was especially true of people over fifty, who had been brought up under prewar indoctrination, which emphasized frugality, and who had experienced the austerity of the war and early postwar period. The survey also showed that frugality correlated inversely with level of education, indicating that this value comes out of the Japanese tradition.

Another opinion poll, which asks the same question every year, found an increase in respondents after 1976 who said that "spiritual richness is more important than material affluence."[11] In this case, breakdown by age was random, but that of education showed a positive correlation: the higher the level of education, the greater the emphasis on spiritual richness. In contrast with the value of frugality, this observation indicated that the emphasis on spiritual richness is related to the so-called postindustrial value rather than to traditional values. Various evidence suggests that, in the case of Japan, the oil crisis seemed to generate an alliance between the traditional values of frugality, the simple life, and a respect for nature with the post-industrial values that decry materialism and industrialism and work toward more humane life. This alliance nourishes attitudes favorable to energy conservation.

However, attitudes are only attitudes, and they proved different from people's actual energy-consuming behavior. In the transportation and residential sectors, consumption stagnated in 1974 because of the increase in the price of energy and the psychological impact of the oil crisis. However, from 1975 on, as the people became accustomed to the higher prices and as the price of oil stabilized, energy consumption started to increase (see Tables 6.5 and 6.6).

In the transportation sector, the growth in the number of automobiles has accelerated since 1973, with registrations increasing at a rate of 1.5 to 2 million annually between 1975 and 1979 (more precisely, to March 1980; these statistics are determined by the Japanese fiscal year, which ends in March). The increase in privately owned passenger cars was particularly explosive, with the number growing by nearly 1.5 million annually. One reason for this increase is the growing need of residents in small towns. For

Table 6.5 Energy Used in the Transportation Sector, 1973–1980

	Energy (1973 = 100)	Registered automobiles (millions) (index: 1973 = 100)	Gasoline consumption (1973 = 100)	Private passenger cars (millions) (index: 1973 = 100)
1973	100	20.7 (100)	100	11.3 (100)
1974	100	21.0 (102)	100	12.9 (124)
1976	110	24.8 (120)	113	16.0 (140)
1978	120	28.5 (138)	126	18.9 (167)
1980	124	31.3 (151)	128	21.3 (187)

Source: For energy figures, *SET,* 1981; Unyushō Daijinkanbo Jōhōkanribu (Ministry of Transportation), *Unyu Keizai Tōkei Yōran, Shōwa 56nenban* (Handbook of Transportation and Economy, 1981) (Tokyo: Unyu Keizai Kenkyu Senta, 1981).

these people, having a car means much more convenience in commuting, shopping, and social activities. But it is also a response to the deterioration of public transportation in areas caught up in a vicious circle — fewer passengers, less service, still fewer passengers. The regional statistics for private ownership and use of cars in Japan shows a reverse correlation to urban-rural factors: the more rural a community, the higher the percentage of ownership and the greater use of cars for various purposes. At the same time, the increase in gasoline consumption has lagged far behind the increase in the number of automobiles and, indeed, in 1980 there was no growth at all in gasoline consumption. This is due partly to greater fuel efficiency but also to a decrease in mileage per year per car. This is especially true of privately owned cars in big cities.[12] In considering the future of the automobile in Japan, especially of privately owned automobiles, we may assume that ownership will reach a saturation point far below that of the United States as well as below those of Western European countries. A number of factors, including the price of gasoline, the tax on car ownership, the degree of development of public transportation, and urban density, are likely to bring this about.

In the commercial and residential sectors (see Table 6.6), kerosene for heating and LPG for cooking seem to have almost reached a saturation point by 1976. This does not mean that the level of heating in the Japanese

Table 6.6   Energy Used in the Residential and Commercial Sectors, 1973–1980

	Total Energy (1973 = 100)	Kerosene (1973 = 100)	Electricity (1973 = 100)	LPG (1973 = 100)
1973	100	100	100	100
1976	113	122	126	111
1980	126	107	155	109

Source: *SET,* 1981.

home has reached that of Western Europe or of North America. Central heating in private homes in Japan lost its appeal before it could spread very far. The amenity that did start to spread in the 1970s was the kerosene- or gas-powered fan heater. This is a stove with an electric fan and thermostat that warms a whole room, but not a whole house. The kerosene heater has not driven up home use of kerosene, probably because room temperatures are still kept low. In a public opinion poll conducted in February 1980, the majority of the respondents said that they are keeping the temperature in their home below 64 degrees Fahrenheit; that is, what the government recommends.[13]

The increasing consumption of electricity is a grave problem, as shown in Table 6.6. Between 1973 and 1978, Japan experienced a 50 percent increase in electricity consumption in the commercial-residential sector. Since common electrical appliances such as refrigerators, televisions, and electric washing machines reached a saturation point of over 90 percent before 1973, this increase can be attributed mainly to the spread of air conditioners. In fact, since the 1970s, the seasonal peak of electricity use has shifted from winter to summer, indicating a rapid increase of consumption of electricity for air conditioning. Another factor pushing up electricity consumption in homes is the perpetual introduction of new appliances, such as electronic ranges, and the increased use of familiar appliances, such as the addition of a second television set or longer viewing hours.[14] The use of air conditioners has risen from 13 percent in 1973 to 30 percent in 1978 and, during the summer of 1979, reached a level of 36 percent.[15] However, in the summer of 1980, because of the 50 percent increase in the price of electricity, unusually cool weather, and, to some degree, a government plea for conservation of energy, air conditioners were used less than in the previous year. Remarkable improvements are taking place in the energy efficiency of electric home appliances. For instance, some new refrigerator models consume only half as much electricity as similar models of a few years ago.

Still, the use and supply of electricity will remain a key problem in Japan's energy future.

A Damage to Political Leadership?

Between 1973 and 1979, four cabinets came and went in Japan. In November 1974, after two years and four months, Kakuei Tanaka was forced to resign because of rumors of his involvement in a financial scandal. His successor, Takeo Miki, became a target of intraparty factional strife and resigned in December 1976. The next prime minister, Takeo Fukuda, gave up his position in December 1978, when he was defeated in the leadership election of the governing Liberal Democratic party (LDP). Meanwhile, when Tanaka's suspected involvement in "dirty money" became connected to the Lockheed scandal, he was arrested and put on trial. Not only had factional strife within the LDP been rampant throughout the latter half of Miki's tenure, it also created a political vacuum for the forty days following the general election of October 1979, when the LDP could not decide on the candidate for the premiership, and two candidates — Ohira and Fukuda — ran for it.

Still, policies were formed and implemented by the bureaucracy, legitimated through advisory commissions (*shingikai*), and finally approved by each successive cabinet.[16] In 1975, MITI's huge advisory commission, the Overall Investigative Commission on Energy, with eight subcommittees, presented a report, "Stabilizing Energy Supply." It followed up in 1978 with another, broader report, "Energy Strategy for the Twenty-first Century," which established the basic lines of Japan's energy policy for the future. In the field of industrial policy, again MITI and its Advisory Commission on Industrial Structure have played a leading role, publishing in March 1980 a report called "A Vision of Trade and Industry Policies in the 1980s." As for an overall land-use plan, "The Third Comprehensive National Development Plan" was prepared in 1978 by the National Land Agency and its Advisory Commission on Land Use, which emphasized developing smaller cities and towns, striving to create "a living environment where nature, living, and production environments are in harmony." The Economic Planning Agency and Advisory Commission on Economy made economic plans for the seven years from 1979 to 1985, assuming a 5.7 percent annual growth rate.[17]

On the other hand, in 1978, Prime Minister Fukuda wanted to create a Ministry of Housing and a Ministry of Energy. However, his attempt was opposed by bureaucrats, especially by the MITI bureaucrats, and failed. In addition, the LDP's energy policy is virtually identical to the government's; that is, to the policy of MITI and its overall investigative commission on energy.

Thus, in spite of this succession of short-lived cabinets, Japan did not lose political leadership or the ability to form policies — at least in the fields of energy policy, industrial policy, economic planning, and regional development planning — thanks to the bureaucracy and its decision-making system of using deliberation councils as legitimating organs. However, it is not so certain that the same mechanism can work as well in the future.

In another political trend in this period, cities and towns acquired an enhanced significance. Environmental and citizen movements that demanded more participation were vocal and effective in the late 1960s. Since then, however, they have declined, or at least have become less vocal. However, local bodies — not only those in big cities but also those in smaller cities and towns — have become more oriented toward residents, accommodating more participation from citizens, and are more conscious of harmonizing the environment with industrial development in their communities. This trend was quickly encouraged by the bureaucrats of the central government, especially those in the Ministry of Autonomy and in the National Land Agency, which are supposed to take care of communities. The governing LDP's politicians, most of whom came from local constituencies, also supported this because they were essentially oriented to their communities, so long as the local emphasis was not against central government or against the LDP. As a result, "the coming of age of the locality," or "localism," has become a favorite chorus both of the government and governing party and of environmentalists and some intellectuals, despite their disagreement over the content. The government and governing party are focusing mainly on ways to distribute government money and improve the infrastructures for living and industry, while the environmentalists and intellectuals are concentrating on such matters as more local participation, more environmental protection, and a more humane life. This difference tends to manifest itself on the issue of constructing electric power plants, whether nuclear, oil, or coal.

The Second Oil Shock of 1979 and Perspectives on the 1980s

The second oil crisis, triggered by the Iranian revolution, hit Japan hard, creating conditions that will constrain Japan's energy future. Also, other events occurred in 1979, such as the Three Mile Island accident and the

Tokyo Declaration at the economic summit, which will have a lasting impact on Japan. Let us consider how Japan responded and what the implications for its future may be.

Japan's Response

As we already observed, the oil crisis of 1973 resulted in immediate and sharp price increases for commodities by corporations and merchants and panic-buying behavior by consumers. The response in 1979 was quite different, despite the cessation of oil exports by Iran for three months, the cutoff in deliveries by major oil companies to Japan, and what ended up being a more than doubling of oil prices. Why did the 1979 shock not set off the same reactions as seen in 1973? First of all, unlike the conditions in 1973, the Japanese economy was healthy and stable before the crisis, with the added stability of prices. In fact, in 1978, the wholesale price rise was −2.3 percent due to the rise in the value of the yen. Second, consumers had learned that panic buying was harmful to their own interests. Third, corporations refrained from speculation. They too had learned that seeking profit through speculation not only resulted in public criticism, but also heightened the workers' demand for wage increases in the following spring struggle. Fourth, the mass media, taking care not to inflame the inflation consciousness of the public, refrained from sensational reporting and pessimistic forecasts of price rises.

However, it was unavoidable that a rise in oil prices and the floating downward of the yen would increase the price of imported goods and so push up domestic prices. The rise in wholesale prices was conspicuous in the latter part of 1979, reaching a 20 percent annual increase rate in the first half of 1980. However, as noted in the preceding chapter, the rise in consumer prices has been contained at one digit annually. This compares exceptionally well with all the major advanced countries except West Germany. Moreover, through the spring struggle of 1980, wages were raised by 6 percent, enabling workers to blunt somewhat the deterioration of their living standard against the rise of prices.

As for the supply and price of oil, Toyoaki Ikuta, director of the Institute of Energy Economics, has pointed out that Japan had some difficulty in acquiring enough oil in the first half of 1979.[18] Oil imports in the latter half of the year compensated for earlier shortfall. But, compared to other major oil importers, Japan had to pay more money for its oil because of several increases in the official price of oil in midyear, as well as because of premiums demanded by oil-producing countries for newly contracted direct-deal or government-to-government oil and some buying on the spot market at higher prices.

One consequence of the 1979 oil crisis for Japan's supply was the sharp decrease in the share supplied through major oil corporations and its replacement by the big Japanese trading companies.[19] Some say that Japan's go-it-alone purchasing policy will destabilize the supply of oil to Japan and weaken the bargaining power of oil-consuming countries vis-à-vis oil-producing countries. Others argue that the involvement of the Japanese trading companies in the oil business may have a positive effect, that these companies can not only stimulate exports to oil-producing countries but can also promote big development projects in the producing countries. It is thus argued that the trading companies' involvement may lead to closer economic cooperation between Japan and the oil-producing companies, resulting in a more stable supply of oil.

In 1979, energy conservation was promoted with vigor. The Japanese government, in accordance with the IEA's decision to consume 5 percent less oil in 1979, set guidelines on March 15, 1979, including such items as keeping room temperatures below 64 degrees Fahrenheit in the winter and around 82 degrees during the summer, stopping 20 percent of the elevators, reducing the use of automobiles for leisure, converting power plants from oil to liquefied natural gas or coal, and, in industry, making more efficient use of energy and converting from oil to other fuels.

What was the actual record? On June 1, 1980, newspapers carried a large government announcement, which trumpeted: WE ACHIEVED 5 PERCENT CONSERVATION OF OIL THANKS TO YOUR COOPERATION! According to official explanations, in the 1979 fiscal year ending in March 1980, oil consumption was 5 mbd which was 0.4 percent less than in the previous fiscal year and 5 percent less than the expected figure — and this despite a 5.5 percent growth rate in real terms. Certainly that was a fine achievement. Government buildings strictly obeyed the guidelines in heating, cooling, and elevator use, other offices more or less voluntarily followed the guidelines, and the public was very cooperative. For example, a letter to the editor of a newspaper suggested that the level of heating the room at 64 degrees was too much; 59 degrees, even 53 degrees, would be all right when a traditional small footwarmer (*kotatsu*) was used. Thus, said the letter, not 5 percent, but 10 percent conservation could be achieved.[20] Ironically, however, the person who wrote that letter must have already been following such a practice and thus would not have contributed to the decrease in energy use.

The 1980 fiscal year, ending in March 1981, recorded a further decline of almost 7 percent in oil consumption, to 4.7 mbd while 4.2 percent economic growth was maintained. However, some factors peculiar to that year contributed to this; i.e., an unusually cool summer in 1980 and unusually abundant rainfall for hydroelectric generation. Also, in spite of economic growth, such industries as petrochemicals and aluminum suffered downturns, resulting in less consumption of oil. A MITI analysis suggests that about half of the drop in oil consumption was due to those factors, but another half can be attributed to more enduring factors, such as switching to non-oil energy and more efficient use of oil.[21]

Japanese Society in the 1980s

There is a government plan for the years 1979–85, "The New Economic and Social Seven-Year Plan," which is intended to provide guidance for economic management.[22] According to that plan, the economy is expected to grow at a rate of 5.7 percent annually until 1985, and would thus be able to offer new jobs to a labor force increasing at an annual rate of 0.9 percent, reducing the need to reorganize the employment system. In other words, if this plan is successfully implemented, full employment would be achieved, living standards improved further, social stability maintained — thus leaving no serious problems within Japanese society. In December 1981, the target was slightly lowered, to 5.1 percent.

But what about the necessary energy supply? In order to achieve this goal, according to estimates made for the plan prior to the second oil crisis, Japan would consume 6.5 to 7.0 mbd of oil in 1985, along with a big increase in the use of alternative sources of energy. But this projection was thrown into doubt by the second oil crisis and was actually called into question at the Tokyo summit meeting in May 1979. After some negotiation, the Declaration of the Tokyo summit meeting set a somewhat lower target for Japanese oil imports in 1985: 6.3 to 6.9 mbd. In order to make the Seven-Year Plan compatible with the agreement of the summit meeting, revised estimates for energy supply and consumption, entitled "Provisional Estimate of Long-Term Energy Supply and Consumption," were released by the Japanese government in August 1979.[23] This estimate has become very famous in Japan and has been cited frequently in government documents, popular books, and experts' articles on energy.

Table 6.7 is somewhat simplified, calculated from these government projections. In spite of high conservation targets, the growth of total energy consumption continues to be strong; it is thought that by 1995, total energy consumption will double, based on an assumption of sustained annual growth of around 5 percent throughout this period. The estimate for imported oil was set, according to the Tokyo summit agreement, at 6.3 mbd in 1985. After that, it is expected to decrease very slightly.

However, as Toyoaki Ikuta, a participant in this effort, has warned, these estimates do not constitute a plan at all, but rather a target or a program, perhaps only a mere hope. As for a realistic prediction, Ikuta argues that first of all, it would be very difficult for Japan to get 6.3 mbd of oil from a

Table 6.7　Government Estimates 1979

A. *Energy Consumption and Oil Imports*

	1977	1985	1990	1995
1. Total consumption of energy	100	161	196	236
2. Conservation ratio		12%	15%	17%
3. Consumption after conservation	100	141	170	196
4. Imported oil	100	120	114	113
	(5.3 mbd)	(6.3 mbd)	(6.0 mbd)	(6.0 mbd)

B. *The Composition of the Energy Supply (%)*

	1977	1985	1990	1995
Imported oil	75	63	50	43
Alternative energy	25	37	50	57
Coal	15	16	18	18
Nuclear	2	7	11	14
LNG	3	7	9	9
Hydro	5	5	5	5
Geothermal	0	0	1	2
New energy	1	2	8	9
TOTAL	100	100	100	100

tightening world oil market in 1985. He estimates that Japan's oil imports in 1985 and in 1990 might be at most 10 percent higher than the present level or might even remain at the present level.[24] Based on the actual import levels since 1979, the official projection will be brought down to a level close to Ikuta's estimate, around 5.7 mbd. Second, he offers some doubts about the feasibility of developing alternative sources of energy as presented by the government, in spite of the strenuous efforts that will be made by government and private corporations. His realistic estimate, as shown in Table 6.8, is 10 to 40 percent lower than the government's estimate of major items of alternative sources of energy. Thus, he concludes, "The total energy supply of 1990 will be more than 10 percent less, probably 12–13 percent less, than the government's estimate."

However, as for the consequence of this energy constraint on Japan's economic growth, Ikuta writes: "This means that the economic growth rate of Japan in the 1980s would drop to around 4 percent, which is much lower than the government program. This would be Japan's first experience of

Table 6.8   Estimates of the Alternative Energy Supply in Japan, 1990

	Government Estimate	Ikuta's Realistic Estimate
Hydroelectric	26 gw	24 gw
Domestic coal	20 mt	18 mt
Geothermal	3.5 gw	0.5 gw
Domestic oil and natural gas	9.5 mt	5 mt
Steam coal import	53.5 mt	38 mt
LNG	45 mt	38 mt
LPG	26 mt	15 mt
Nuclear	53 gw	35 gw

Source: Toyoaki Ikuta, "Energy Problems of Japan" (mimeo), May 25, 1980, p. 16; revised December 1981.
Note: gw = gigawatts; mt = million tons.

low economic growth and there will be serious consequences in many fields such as the economy, politics, and society."[25] Even so, Ikuta may be optimistic. Japan's economy depends a great deal on exports, and in view of the intensifying trade conflicts between Japan and the United States and Japan and the Western European countries, its ability to increase its exports in the future is somewhat doubtful. Yet, despite 12 or 13 percent less total energy than estimated, Ikuta still expects the Japanese economy to grow at a rate of 4 percent. If so, this would reflect the tremendous vitality of Japan's economy and people and necessitate not only more strenuous efforts to conserve energy in every sector, but also a reorganization of the industrial structure toward less energy intensity and greater knowledge intensity.

Second, why do we regard 4 percent as "low economic growth"? Certainly, compared to Japan's average annual rate of more than 10 percent between 1955 and 1973, it is low. If it is low only in this comparative sense, then it is mainly a psychological phenomenon. Four percent would not be regarded as low in Western Europe or the United States.

There are several structural reasons why an annual economic growth rate of 4 percent is too low and will cause serious troubles. Japan's population will grow slowly from its 1980 level of 117 million at an annual rate of 0.9 percent in the 1980s, at 0.5 percent in the 1990s, and probably reach an equilibrium of about 140 million sometime around 2020. Thus, the Japanese economy of the 1980s needs to create about a million new jobs a year in order to maintain the present employment level.

However, Japan's population and employment situation in the 1980s will not be so simple because of the changing character of the population and of the labor force. First, over 90 percent of those in the 15- to 18-year-old age bracket are in high school. Nearly 40 percent of those between 18 and 22 are currently in colleges and universities. This decreases the labor force in the 15- to 24-year-old category but increases the availability of a labor pool with more education. Also, the expected gradual increase of the population in the 1980s and 1990s is based on increased longevity, not on the birth rate, which has already declined to the level of zero population growth. An increase of the population and the labor force, therefore, will arise primarily from a growing population of workers over 55 years old and of the elderly.

The Japanese employment system — with lifelong employment, a seniority wage system, and retirement around the age of 55 — is about to have to make room for the better-educated younger workers. In order to make a smooth transition, the unemployment rate must be as low as possible, and both public and private funds will need to be allocated to expand the employment possibilities for those over 55. A growth rate just corresponding to, or slightly greater than, the increase of the population would result not only in increased unemployment but also would create grave social and moral confusion. The aged would be hit most seriously and would feel that society has betrayed them. In 1978, the 7 percent unemployment rate among the population over 55 in Japan was already higher than that in other advanced countries: 4 percent in West Germany, and 3 percent in the United States in the 55–64 age bracket.[26]

A rapidly aging society will also bring increasing demands for such social welfare expenditures as pensions and medical care. In 1980, social welfare was the biggest item in the government budget, and it is expected to grow throughout the 1980s and even more rapidly in the 1990s just to maintain the present level of benefits. However, in order to avoid financial bankruptcy, the government is proposing to emphasize self-help and the solidarity of families, neighborhoods, and communities, thus creating a less expensive but more humane "Japanese-type welfare society."[27]

The government's Seven-Year Plan points out: "There is an ongoing change in the national consciousness from concern with quantitative expansion to improvement of the quality of life."[28] Consciousness about quality of life implies more concern for such nonmaterial things as purpose, meaningfulness, and self-actualization. Yet it means that the government must do a lot of very material work — building more convenient roads, more spacious housing, better sewage and waste disposal systems, more medical facilities, more parks, and more cultural and educational facilities. In a word, "quality of life" demands more public expenditure for social capital. The Seven-Year Plan estimates that a public investment totaling 240 trillion yen ($1.2 trillion) would be necessary to improve living-related social capital to the level that would satisfy the needs of the people. This level, which could be achieved by 1985, would not be superior to that of other advanced countries today. For example, the goal of housing is 80 square meters (864 square feet) per house and the percentage of the population to be covered by public sewage systems is 55 percent. The decreasing emphasis on material goods may contribute to a reduction of personal consumption of goods and energy. Moreover, the quality of life argument is related to the idea of the population dispersing from the big cities to smaller cities and towns, which ideally will result in a harmonious combination of nature, humane living, and production. The idea is beautiful, but more spending on social capital in smaller cities and towns will be necessary. That is exactly what the Seven-Year Plan and "The Third Comprehensive National Development Plan," a plan for land use and regional development, are saying.

Even now, the government's financial crisis is that of a small government with a large debt, and new demands for increased allocation of funds are emerging for the 1980s. Energy-related expenditure is one. As the international responsibility of Japan as an economic giant has been recognized, increased foreign aid, especially official development aid, is a must. At the Bonn summit meeting in 1977, the Japanese government made a commitment to double its overseas development aid within three years. It fulfilled this goal by increasing the budget for this item by 15 percent annually — again, larger than the average increase of the budget. This was another exceptional decision by the government in a budget-making practice that has been dominated by incrementalism and balance. In the 1981 and 1982 budgets, defense expenditure required exceptional allocation due to the collapse of détente and the urging of the United States that Japan strengthen its defense capability in Northeast Asia.

Theoretically, solving the financial crisis is not so difficult. Government revenue could be increased through the introduction of a value-added tax (VAT). In 1979, Prime Minister Masayoshi Ohira wanted to do just this. However, when he suggested it just before the October general election, it was rejected by his own party and also proved unpopular with the public. Although he withdrew his idea before election day, the poor showing by his party has been partly attributed to it. In the June 1980 general election that resulted from the dissolution of the Diet, Ohira and the LDP did not discuss a nex tax at all and won an overwhelming victory, though partly due to "sympathy votes" for Ohira, who had died during the campaign.

Even with 286 seats out of 511 in the House of Representatives and 136 seats out of 251 in the House of Councilors, the new prime minister, Zenko

Suzuki, and the LDP have no intention of bringing up the unpopular tax anymore. How can the government meet its financial crisis? The answer seems to be "low incrementalism": restraining the budget allocation to ordinary items at zero percent growth and to policy items by each ministry under 7.5 percent. Extra increases at the rate of the expected nominal increase of GNP of around 10 percent would be allowed for defense (especially), foreign aid, and energy-related items. Since the economy is growing by a rate of over 10 percent in current value, the government can expect a "natural increase" of tax revenue in current value in proportion to that growth rate and inflation rate, even without introducing new taxes or increasing the rate on existing taxes.

Therefore, by keeping the incrementalism of budget items at a rate lower than the expected growth of revenue in current values in general, and by allocating more of the budget to defense, energy, and foreign aid, the government can somehow decrease the deficit and the reliance on bonds by several percentage points. Actually, in the 1980 budget, Ohira's cabinet did lower the reliance on bonds from 40 to 33.5 percent. Suzuki's government followed the same course and may further cut the bond reliance in 1981 to 26 percent and to 20 percent in 1982. But this way of coping with the financial crisis testifies to a weak political leadership and becomes possible only with a combination of sustained economic growth and a moderate, perhaps one-digit, inflation rate.

Electricity and Nuclear Generation

Does nuclear power provide a likely way out of the quandaries that energy problems are likely to create for Japanese society? Since the nuclear power issue is related to the place of electricity in the total energy supply, we must briefly review Japan's electricity situation. Among the major industrial countries, Japan's use of electricity as part of the total energy consumption is the highest, 39 percent in 1980. Electricity consumption has grown rapidly, at an annual rate of 10 percent between 1960 and 1975. It is expected to continue to grow, although at a reduced rate — about 3 percent annually. The growth of industrial use of electricity is slowing down, but it is still high compared with those of other industrial countries. In 1980, two thirds of electricity was used by industry. To generate electricity, Japan is heavily dependent on oil, and it is imperative to replace it by other sources as much as possible. A substantial drop is occurring — from 66 percent of total generation in 1975 to 46 percent in 1980. This is the result of fuel switching and good operating results for nuclear and hydroelectric plants. Electricity consumption also has wide seasonal fluctuations, peaking in the summer, due to the spread of air conditioning, which requires electric companies to construct a much larger generation capacity than that which would otherwise be needed.

Whatever the nature of the power source, electric companies have been troubled by the problem of how to acquire the necessary sites for generating plants. In the face of opposition, winning cooperation from the residents of a potential damsite was very difficult. It has been even more complex with coal and oil generation. Persuasion must extend beyond the immediate landowners. Nearby residents, fishermen affected by the release of cooling water, localities — all have been a headache for companies. Often the land is owned by hundreds of people, some of whom are still farming. To buy farmland, it is still necessary to get the approval of the farmland committee of local governments — a legacy of the effort to protect postwar land reform. Even when the land for a planned site is publicly owned, some terrain is covered by the right of the villagers to use it as a result of historical practice that goes back to before the Meiji restoration. This makes negotiations necessary and requires compensation for the villagers. In the case of the rivers and ocean, the native fishermen also have historical rights that were institutionalized by a law on fishing cooperatives. In order to release the cooling water from a generation plant to a river or to the sea, electric power companies have to negotiate with the fishing cooperatives involved and, in addition to paying compensation, persuade them to declare the "disappearance" of their fishing rights. This requires the agreement of two thirds of their members.

All these factors apply, not only to electric plants, but also to the construction of any kind of factory or public works project. However, site acquisition for power plants faces several additional disadvantages. First, unlike factories, which bring new jobs, a power plant does not create much employment. Second, there is the possibility of air and water pollution, and, in the case of nuclear generation, there are further doubts about safety.

In spite of these obstacles, many factories have been built every year, and large numbers of public works have been completed, often taking a lot of time (eight years to open the Tokyo International Airport in Narita) and requiring patient negotiation, payment of lucrative compensation, and, in rare cases, legal expropriation as the last resort. The same has been true of power plants. Moreover, special laws were enacted in 1974 that give financial incentives to communities in which a new power plant is to be constructed in order to make acceptance easier. According to these laws, communities can get a huge grant (about $2 per kilowatt of the capacity of plant for five years), to be used for improving public facilities, from a special account funded by a new tax on the consumption of electricity. This measure provides an effective incentive for communities to approve or even to invite the construction of a power plant. However, other factors —

such as opposition from fishermen, the refusal of landowners to sell, and influence from environmentalists and/or antinuclear movements from outside — have been causing delays in plant construction, whether using steam or nuclear power.

In fact, public opinion on the development of nuclear generation has been rather favorable in Japan. According to polls conducted by the prime minister's cabinet,[29] favorable opinion for further development of nuclear generation was 50 percent in 1976 and 51 percent in 1978, while the share favoring a stop to the development dropped from 15 percent in 1976 to 6 percent in 1978. Another poll was conducted both in the United States and Japan by the newspaper *Yomiuri Shimbun* in 1979, after the Three Mile Island accident. At that time, only 35 percent of the Japanese respondents favored the further increase of nuclear generation — a marked decrease from the government poll of the year before. However, comparing the Japanese and U.S. responses, those definitely opposed to nuclear generation are far fewer in Japan (see Table 6.9). Moreover, the Japanese data show that the younger the respondent and the higher the level of education, the more favorable he is to developing nuclear power. Here Japan differs sharply from the Western European countries, where ecological and antinuclear movements are attracting younger, better-educated, and rather well-to-do people and, on this base, showing some electoral strength. In Japan, only 5 to 8 percent of the people are definitely against nuclear power. One directory of citizens' movements in Japan listed, as of 1979, forty-three antinuclear movements in addition to twenty-one movements against construction of power plants by oil.[30] However, these groups have little influence on elections, and although the Japan Socialist party, the largest opposition party, has opposed nuclear power, this stand does not seem to have helped it at the polls.

The governing conservative party, the Liberal Democratic party, has been in favor of nuclear power. This is hardly surprising since Japanese business and industry favor nuclear power and they have close ties with the

Table 6.9 Opinion on Nuclear Generation in Japan and the United States

	Develop further	Keep as is	Abolish it	Don't Know/ No Answer
Japan	35%	46%	8%	12%
U.S.	26	42	22	9

Source: *Yomiuri Shimbun*, November 28, 1979.

LDP. At the same time, the LDP, as a party of the "periphery," represents farmers and fishermen in towns and villages, including those whose land and fishing rights are affected by nuclear or oil plants. They express their antipathy to the "center" — big cities and industry that consume the electricity generated in their community — by asking, "Why should we be sacrificed?" The LDP, especially those politicians from the smaller constituencies, are quick to respond to these grievances by getting grants from the government and extensive compensation from electric power companies. But this kind of appeasement drives up the cost of construction and eventually the electric bills.

In addition, the assurance of safety is a vital problem of nuclear power — between electric power companies and residents, between government and antinuclear movements, among scholars who are divided on this issue, and among political parties, the LDP and Democratic Socialists favoring nuclear power and the Japan Socialist party and others opposing it. Safety is built into the design and construction of nuclear plants by the Atomic Safety Commission. After construction, operations are supervised by MITI and carried out by electric companies.

However, some residents are not satisfied with this arrangement and wish to participate in the operation of the plant. For individual citizens, it would obviously be almost impossible, but the role of local governing boards, especially that of the prefecture (Japan is divided into forty-seven prefectures) has been increasing.[31] Bigger than towns and cities, having larger staffs, and concerned also with the residents as the constituency of their governor and prefectural assemblyman, some prefectures have been making safety agreements with the electric companies that give the prefectures several important rights — access to information on safety in operating nuclear power plants, on-the-spot inspections, and the power to shut down the plants if necessary. The prefectures have also been monitoring the level of radioactivity in the areas near nuclear plants. Under the present laws and regulations on nuclear power, the prefectures have no actual jurisdiction over nuclear plants in their areas, but practice has preceded the law and, therefore, it has been argued that the laws should be revised in the direction of legitimating the reality and giving more say to prefectures in supervising the safety of nuclear plants.[32] A nuclear specialist with an international reputation, Rhukichi Imai, observes a recent tendency to acknowledge the right of local bodies to evaluate the impact of the plant on the environment, but he argues that safety issues concerning radioactivity should remain the responsibility of the central government.[33]

Another area in which local bodies are taking greater interest is the development of "local energies," such as mini-hydro (or low-head hydro) geothermal, wind generation, solar heating, cogeneration, and recovery of waste heat. Much of the expansion of the 1980 energy budget went to pre-

fectures to survey the potential of those local forms of energy. Many prefectures, cities, and towns are already doing experimental projects in the
field.[34] The place of local energies remains modest in the government's
total energy program, though their role will increase. But local sources of
energy cannot solve Japan's energy problem now or in the near future.
Regardless, they do have their place in the government plan, and the role
of local bodies in this respect will be increased.

The nuclear power issue, both domestically and internationally, has
been associated with the problem of nuclear weapons. The Japanese government's position on this point has been clear: Japan will not have nuclear
weapons; therefore, nuclear power and nuclear research are strictly for
peaceful use. However, in spite of the generally favorable opinion on nuclear power expressed in the polls, the Japanese public is sensitive to the
possibility that nuclear power might be used to develop nuclear weapons.
Asked about the image in their minds when they hear the words "atomic
power," 30 percent of those polled responded with the atomic bomb.[35]
Japan's powerful Council for Abolishing Atomic and Hydrogen Bombs
(Gensuikyo) — a peace movement that grew out of the memories and experiences of Hiroshima and Nagasaki — took a tougher stand on nuclear
power in its 1980 policy, calling for a moratorium on new construction of
nuclear plants unless a safety check is made on all currently operating nuclear plants in Japan. A radical group held a meeting in Tokyo on July 3,
1980, calling for "anti-nuclear power, anti-nuclear bomb, and anti-repro-
cessing (of uranium used in nuclear generation)."

The last point — antireprocessing — coincides with the U.S. policy on
exporting uranium to Japan from the United States and with the same
logic inherent in the American concern that reprocessing can produce plutonium, which can be used to make a nuclear bomb. From the start,
Japan's nuclear power industry had depended heavily on the United States
for both uranium and technology. The United States had been encouraging the export of both, though not without restriction. For instance, used
uranium originally exported from the United States could be moved only
with American permission. At the time, when Japan was reaching the technological level of enriching uranium itself and reprocessing used uranium
by itself, as well as planning to proceed with an experimental fast breeder
reactor program, it ran into President Jimmy Carter's antiplutonium policy
of 1977. The consequence was disagreement between the United States
and other uranium-poor advanced countries, such as Japan and West Germany, and the subsequent international fuel cycle evaluation. Japan's reprocessing test plant is currently operating on a provisional basis with tentative approval from the United States. Thus, the future of Japan's nuclear
power depends to a considerable degree on U.S. policy concerning the
nonproliferation of plutonium. Moreover, Japan's plan to dump low-

level radioactive waste from nuclear plants in the Pacific has caused protests from newly independent islands in the South Pacific, which can retaliate by prohibiting Japanese fishing in their territorial and offshore waters. Japan's nuclear power industry has already accumulated a huge volume of low-level radioactive waste (250,000 drums, according to newspaper reports) and will accumulate waste at a more rapid rate as nuclear energy continues to expand.[36] This is another obstacle to the future development of Japan's ability to generate power.

Nuclear power in Japan cannot be a panacea for Japan's energy problem. Even those people who promote nuclear power admit that the hard path relying on nuclear power cannot solve the energy problem. They are calling for the "flexible path," a combination of soft and hard paths.[37] On the other hand, nuclear power in Japan is not in a stalemate, unlike that in the United States, as described and analyzed in the Harvard Business School report.[38] The attitude of the public is more favorable; widespread alternative movements are lacking; center-periphery cleavage is mediated in a peculiar way within the LDP; and the increased role of local bodies can improve safety both objectively and psychologically. However, although it is not in a stalemate, it certainly is a rough road on which to travel.

A Worst Case Crisis

To prepare for shortfalls and/or partial supply cuts in oil, Japan has increased its stockpile of oil. If a serious supply cut should occur, the government would declare an emergency based on the Law for Proper Supply and Consumption of Oil, which allows it to limit the use of oil, to order the sale of oil, and to ration, if necessary, as well as to set the level at which supplies would be withdrawn from the stockpile. There is another law to control the prices of commodities essential for people's living, which would apply not only to oil and oil products but also to other goods.

Depending on the degree and expected duration of the oil supply cut, the government would decide how and when those measures would be implemented. If the economy could cope with the cut without those measures, all the better. This was the case in the first half of 1979, when the supply of Iranian oil stopped for three months. By not exercising the emergency measures, the government succeeded in restraining the spread of inflation psychology among both private companies and consumers.

However, in a situation that necessitated the government's exercising those measures, there would no longer be a psychological problem. Production would be hit, transportation would be hit, the economy would stagnate, and, however the government tried to control prices, inflation would creep in.

The shorter the crisis, the lighter the damage to the economy. But it is also possible that the lessons learned will be easily forgotten, and the development of alternative or new energy will not be accelerated. On the other hand, if the damage caused by the crisis were serious, and stagflation continued, both the public and the government would be caught up in the immediate problem. In the best possible case, the damage would be light, a consensus formed to act on the lessons learned, and the development of new energy and more conservation accelerated. But Japan's energy vulnerability would nevertheless remain.

If a serious oil supply cut should continue longer, and most or all of the stockpiled oil were used, then real austerity would be imperative. For the people and for the society, it would become a matter of survival. Since Japan is heavily dependent on imported food and feed, the interruption of these imports would result, sooner or later, in a serious food shortage. Because of fuel shortages, even coastal fishing would become difficult. Stock and poultry raising and dairy products would be curtailed rather quickly by the cutoff of animal feed. Even rice production, which has been subsidized by the government for many years, would decline if deliveries of chemical fertilizer, various agricultural chemicals, and machine fuel should drop.

There is a best-selling novel, *Yudan,* which in Japanese has a double meaning: "oil cut" and "unpreparedness."[39] Published just after the first oil crisis of 1973, its plot concerns the consequences of a total cutoff of oil from the Middle East that lasts for seven months. It includes the prediction that three million Japanese would die from starvation. Taichi Sakaiya, the author, is a former MITI official who actually participated in a simulation exercise of such a cut, held just before the first oil crisis of 1973, when Japan had only a 60-day stockpile of oil and was far less prepared for such a crisis than it is today.

The novel mentions that if Japan should be hit so seriously as to experience mass starvation, many other nations would also suffer similarly or more seriously. But some of those countries are powerful militarily and would be tempted to intervene in the area of trouble, which might result in either a better and swifter solution or, conversely, trigger another world war, with the nuclear annihilation of mankind. It is significant that the story, despite food riots and intense control measures taken by the government, does not mention a coup d'état or civil war in Japan. Indeed, the novel ends with Japan's political structure remaining unchanged. Notwithstanding the loss of human lives and the serious damage to the Japanese economy (and the world economy as a whole), Japan starts to recover somehow, just as it did after the devastation of World War II. The novel is reassuring in its implication that, in a vivid crisis, the Japanese people would have no reason at all to think that some other, more authoritarian form of government could ease the plight better than its democratic government, even though it would have to resort, democratically, to emergency measures. But here we have gone beyond the Lower Bound.

Future Energy Stringency

Where is Japan left, given the kind of energy premises that provide the starting point for this study? Stobaugh offers for discussion two possibilities for world energy supply to the year 2000, on the basis of which Teruyasu Murakami presents two scenarios for the Japanese economy in the future.

The Japanese government's own estimate of future energy consumption and oil imports is shown in Table 6.7. It assumes that Japan's total energy consumption will nearly double by 1995, even after serious conservation efforts. This does not accord even with the Upper Bound scenario, which suggests only 2 percent annual growth of energy supply for OECD countries.

Is Japan staking its future on an energy supply surpassing that of the Upper Bound, a condition not very probable? If so, it is a dangerous course, not only for Japan, but also for the OECD countries as a whole. For it would impel Japan both to become more aggressive economically in acquiring energy, which could stimulate price hikes by oil-producing countries, and to accelerate its export drive in order to pay for oil, thus intensifying trade conflicts with the United States and Western European countries.

However, the recent record of energy consumption, especially for oil, indicates that Japan will not need as much oil as previously estimated, and the government will probably lower its estimate for oil consumption considerably. However, the target figures for alternative sources of energy in Table 6.8 will probably remain the same. But achieving those targets for alternative energy sources will not be easy, in spite of strenuous efforts by both government and private corporations. The ironic result: Both total energy supply and total consumption will end up lower than the current government estimate and so fall within the range of Stobaugh's Upper Bound.

Japanese society seems to be able to achieve Murakami's Scenario A and can endure the burden caused under it. Being free of any serious ethnic or religious cleavages, and having become more egalitarian as a result of the postwar reforms and through the period of high economic growth, Japanese society is — and would remain — more stable than those of the other OECD countries.[40] Although the Japanese political system is not suited to producing strong political leadership, it is capable of cautious guidance and steady implementation through its bureaucracy. There is a virtually complete consensus on the need for energy conservation in the residential, commercial, and transportation sectors, more efficient use of energy in the industrial sector, and more use of alternative energy, along with the development of new energy sources. The issues in which divisions are clear include: (1) the promotion of or opposition to nuclear power, (2) government or private initiative, (3) centralization or decentralization, (4) a balanced allocation of government funds or ordering by priority, and (5) large or small government. The eventual choices will lie more or less between those paired poles, and, consequently, Japan will have fewer problems than other societies.

However, Japan is still in an inherently weak position in international affairs. Its exports have been causing trade conflicts with other industrial countries. And with no historical ties to the Middle East and no military power to offer, how can Japan contribute to a dialogue with oil-producing countries, a dialogue that will be essential for securing a stable supply of oil?

As we go down from the Upper Bound case, lower and lower, eventually reaching the Lower Bound case in world energy supply, we find anxiety and uncertainty and even despair setting in. The idea of zero energy growth, which would result in zero or even minus economic growth for more than a short period, is totally alien both to Japan's leaders and to its populace (except to a few environmentalists). Psychologically, the Japanese are totally unprepared.

Living would be, at a minimum, uncomfortable for all and difficult for many. People would become tense and angry. Conflicts among various interests would become manifest and acute, despite Japan's homogeneous and relatively egalitarian society. Labor-management relations would be hurt seriously, more so than in other countries. Japanese workers and labor unions have been accustomed to lifelong employment, and their loyalty to and cooperation with the company have been based on the expectation that their jobs are secure against the ups and downs of business. Thus, a large-scale dismissal of workers due to the continued stagnation of the economy would provoke fierce and prolonged struggles and would undermine workers' loyalty to the firm. The collapse of loyalty to the company would lead to lower morale and lower productivity, triggering a vicious circle: bad economic situation, lower morale, worsened economic situation. Initially, older workers would be hit most seriously, because they would be the first to be fired and the last to find another job, while social welfare for the aged would be curtailed. But eventually all the workers would be affected.

In such circumstances, Japanese society and politics would face the type of strains not felt since their defeat in the Second World War. There would be a need for stronger and more effective political leadership than that which has been customary, and for a stronger consensus on policies. Yet it is exactly under such conditions that greater divergence of opinions on policies and principles would likely emerge. New extreme political forces beyond the present range of the political spectrum — which currently extends from the Liberal Democratic party to the Japan Communist party — might gain wide followings and eventually win power. Some might want to transform Japan, internationally, into a major military power in order to secure its *Lebensraum,* as in the 1930s, and, domestically, into an authoritarian state, with the power to overrule opposition to its policies. On the other extreme, anti-industrialism, which calls for a radical change of way of life and return to nature, might become widespread. The result would be confusion, confrontation, less consensus, and weaker political leadership, thus initiating yet another vicious circle: from economic difficulties to weaker political leadership, resulting in further deterioration of the economy and a further weakening of political leadership. Such a pattern in one industrial country would affect international cooperation adversely among all advanced countries, causing further deterioration on such common matters as the world energy situation and world trade, and so accelerating the downward domestic spiral. We can see sinister symptoms of such an international vicious circle in current trade conflicts and in the trend toward protectionism among advanced countries.

If the world energy supply ends up somewhere between the Upper and the Lower Bound, Japanese society and politics could somehow adjust to the higher half of the range between the two bounds. But the Japanese people are not prepared for the lower half of the range.

But it is better, for Japanese leaders and for the public, to be preparing for the worse energy situations. The better prepared we are, the more likely we can prevent their actual occurrence.

7

Europe's Farewell
to Full Employment?

by G. F. Ray

This chapter, coming as it does after the sections concerning the United States and Japan, requires qualification. The United States of America consists of many states, and the individual states and regions have their own particular energy perspectives. Nevertheless, the United States can be regarded as one entity for this discussion of energy. The case of Japan is much the same. While Western Europe may be a convenient unit, it cannot easily be treated in the same way for our particular topic. The economic and industrial structures of the score of independent countries are different; their historical development and present social and institutional situations also vary, as do their endowments of energy resources. Consequently, their energy situations and policies also differ.

The existence of the European Community has so far not changed this situation much. There has been no lack of honest attempts on the part of the Brussels Commission — with more or less enthusiastic support from the governments of the member countries — to achieve some significant degree of harmony in energy policies, but it has not led to more than the recognition and better definition of the national differences.

The International Energy Agency (IEA) has been more successful in achieving the implementation among its members of a number of measures, such as the establishment of minimum oil stocks, but they do not add up to a European energy policy (which, in any event, is not one of the IEA's objectives).

As a matter of fact, given the seriousness of the problem, the individual European governments have come to grips conceptually with the energy situation at a haphazard and slow pace. There has been no real cooperative approach to the development of a coherent and cohesive European energy policy. A failure either to act in this matter or to recognize the impact of one country's actions on other countries could ultimately impose very heavy costs on all the national economies of Europe, with very serious political and social implications.

Import Dependence

In recent years, there have been only two countries in Western Europe whose production of primary energy has exceeded domestic consumption, Norway and the Netherlands. Next is the United Kingdom, which has recently become just about self-supplying. Yugoslav home production covered almost two thirds of its energy needs, but the rest of the European countries had to import more than half their requirements. Even Germany,* with its historically important and still significant coal industry, imported 55 percent of its domestic consumption in 1978; this proportion has been only slightly reduced since. Energy imports in the past year or two accounted for an even higher share of the needs of other European countries — from about two thirds in the case of Austria, to three quarters in France, Spain, and Switzerland, to the rest, who have hardly any energy resources.

Although European energy production covered about nine tenths of its energy requirements immediately after World War II, Western Europe was importing over 60 percent of its energy needs by 1973 (see Table 7.1). The rapid growth of energy requirements coincided with the abundance of relatively cheap oil, mainly but not exclusively from the Middle East. Light distillates met the steeply rising needs of the mushrooming motor vehicle fleet, while the medium and heavier products of the refinery process, which competed directly with coal, won an increasing share of the industrial markets because of price advantage and convenience. As a result, coal's share fell in gradual steps to 25 percent by 1973; in the same period, the share of oil increased fivefold, to about 60 percent. Natural gas, a relative newcomer in Europe, accounted for 12 percent by 1973, and the earlier small contribution of primary electricity doubled, to 4 percent. Coal suffered also as a result of major technological changes, such as the phasing out of the earlier exclusively coal-based carbonization process for making gas, the dieselization and electrification of the railways, and the new steelmaking processes requiring less energy.

The penetration of oil and the retreat of coal occurred even in the two major coal-producing countries, Germany and the United Kingdom (see Table 7.1). Since the first spectacular increase in OPEC's oil price, in 1973, the contribution of domestically produced energy has grown as a conse-

* "Germany" means the Federal Republic (West Germany) throughout.

Table 7.1 Energy Indicators of Western Europe, Germany, and the United Kingdom, 1950–1980

	Western Europe					Germany					United Kingdom				
	1950	1960	1973	1978	1980	1950	1960	1973	1978	1980	1950	1960	1973	1978	1980
	mbdoe														
Primary energy consumption,	7.5	10.9	20.1	20.6	20.7	1.6	2.7	4.8	5.1	4.9	2.9	3.4	4.0	3.8	3.7
Domestic production	6.7	7.4	7.7	9.6	10.9	2.0	2.4	2.2	2.2	2.2	3.0	2.7	2.2	3.3	3.8
Share in consumption:	*Percent*														
Domestic production	88	68	38	46	53	123	90	47	45	46	99	77	53	84	101
Import requirement	12	32	62	54	47	−23	10	53	55	57	1	23	47	16	−1
Coal (incl. other solid fuel)	86	65	25	24	15	96	77	34	32	34	91	77	39	37	39
Oil	12	29	59	55	53	3	21	53	49	45	9	22	47	40	36
Natural gas	—	2	12	16	17	—	1	12	17	18	—	—	13	21	23
Primary electricity	2	4	4	5	5	1	1	1	2	3	2	1	1	2	2

Source: *World Energy Supplies* 1973–78, United Nations, New York; author's estimates for 1980 are based on the *BP Statistical Review*.

quence mainly of North Sea oil and gas production and to a smaller extent of increasing nuclear power generation. The share of oil in total energy supplies has also turned down, while the contribution of natural gas, partly imported, has risen substantially. Coal's share at first continued to decline after 1973 (even in the two countries producing the bulk of European coal), though at a reduced rate, and more recently it has tended to stabilize.

Plans and Expectations to 1990

The most recent and publicly available plans and forecasts of seventeen European countries reflect the 1980 situation, when the average OPEC price was around $31 a barrel.[1] Oil prices then rose another 10 percent and have more recently fallen back. The outlook to 1990 is for slower economic growth than had been planned — or hoped for — in the late 1970s.

According to these plans, the energy:output coefficient* would decline in most countries and for Western Europe altogether, indicating that, for any additional output, relatively less energy would be required than in the past.

Western Europe also assumed strenuous efforts to maximize domestic energy production. The aims for coal, gas, and oil are shown in Table 7.2, and the expected rapid buildup in nuclear-generated electricity is shown in Table 7.3.

Coping with Needs

The forecasts have been eroded by these developments and by uncertainty, inertia, and a variety of obstacles. How do the different countries now plan to cope with their energy needs to the 1990s? In Belgium the further fall in domestic coal seems unavoidable, but the use of coal is expected to rise from imports. Natural gas and nuclear electricity provide the cornerstones of the future; these plans are fairly well based on current investment and contracts for gas from the Netherlands, Norway, and possibly other sources.

Until the recent modest finds of offshore oil and gas, Denmark was practically without domestic energy resources. These, and more imported coal, are planned to reduce its earlier almost total reliance in imported oil.[2]

In view of declining domestic coal, oil, and gas resources, France has built up a powerful nuclear industry and is working on the implementation

---

* The ratio of the percent change of primary energy use to the percent change in output. This is not a very reliable indicator, as explained later. It is discussed here because in quite a few announcements and publications its reduction has been treated as a "target."

Table 7.2 Plans and Forecasting: Coal, Gas and Oil Production in Western Europe, 1980, 1985, 1990

Coal (million tons)*

	1980	1985	1990
Belgium	6	7	7
France	22	18	13
Germany	135	137	140
Greece	7	9	12
Spain	19	24	25
U.K.	116	102	98
Yugoslavia	21	87	31
Others	6	5	10
TOTAL	332	329	336
CONSUMPTION	388	—	—

Natural Gas (billion cubic feet)

	1980	1985	1990
France	294	210	126
Germany	672	672	630
Italy	462	294	294
Netherlands	3150	2898	2814
U.K.	1344	1596	1764
Norway	924	1008	1008
Others	126	336	420
TOTAL	6972	7014	7056
CONSUMPTION	7770	—	—

Oil (Thousand b/d)

	1980	1985	1990
Denmark	—	40	80
France	40	40	60
Germany	100	100	100
Italy	40	40	60
Norway	522	642	642
Netherlands	40	40	60
Spain	240	40	20
U.K.	1606	2309	1907
Yugoslavia	80	100	100
Others	20	80	121
TOTAL	2490	3394	3052
CONSUMPTION	13,313	—	—

* Including the coal equivalent of lignite and peat.
Source: G. Ray and C. Robinson, *European Energy Prospects to 1990* (London: Staniland and Hall, 1982).

Table 7.3   Nuclear Power: Western European Capacity Installed and
            Planned, 1980–1990
            (thousand megawatts)

	1980	1985	1990
TOTAL	44.4	105.1	150.9

Source: International Atomic Energy Agency, *Power Reactors in Member States,*
Vienna, 1981.

of a nuclear plan, which, even after its considerable reduction from an overambitious original objective, will still build up a very large nuclear capacity if successfully finished. Dutch and Algerian gas are supplementing domestic supplies and will help — with additional imports from elsewhere (the Soviet Union) — to reduce the importance of oil.

Germany remains the largest coal producer in Western Europe. Hard coal production is to be maintained at 85 to 90 million tons, while lignite (brown coal) output exceeds 100 million tons; the latter is used mainly for electricity generation, for which it provides relatively cheap fuel (being strip-mined). Dutch, Norwegian, and Soviet gas are the backbone of the greatly increasing gas supplies, and the expansion of nuclear generation continues.

Offshore natural gas reduces Ireland's reliance on oil somewhat. Otherwise, it has only peat.

Italy's import dependence will hardly change; nuclear stations are being added to the small domestic oil and gas resources.

The Dutch stopped mining coal some time ago. Although the Groningen reserves of gas will last for some time, actual production will be declining; it may be partly offset by new offshore gas and perhaps also by oil. Still, probably well before 1985, the Netherlands will become a net energy importer again.

In the United Kingdom, it seems unlikely that the National Coal Board's target of 135 million tons of coal will be reached by 1985. Future oil production is subject to the government depletion policy; production may peak around 1985 and begin to decline around 1990. Gas from the northern fields (in the North Sea) and imports from Norway will more than balance the decline of gas output in the older, southern fields.

Outside the EEC, in Austria, a referendum decided against the operation of the single nuclear power station already built. Hence, with meager domestic resources, it cannot avoid becoming even more dependent on imports, although oil's share may fall as a consequence of the further exploitation of its hydroelectric power potential and of higher gas imports.

Despite the planned expansion of nuclear power and peat production, Finland will also remain highly dependent on imports.

The case of Greece is similar, even after allowing for the rising lignite production and modest oil and gas finds in the Aegean Sea.

Norway is in an entirely different and, for Europe, entirely unique position. It already exports considerable quantities of both oil and gas. Its cautious depletion policy will decide the size of future exports. There is plenty of scope for further exploration, with very good hopes for expanding the offshore production base. A sizable addition to hydroelectric power is also a possibility.

Portugal will invariably have to rely on imports.

Spain, on the other hand, initiated an ambitious energy program some years ago, aimed at raising coal and lignite production significantly and building up considerable nuclear capacity. Although imported oil requirements will mount, they will be at least partly offset by higher home production from offshore fields.

In Sweden, uncertainty surrounds the politically vexed question of the further expansion of nuclear power; otherwise, there is little scope for the further development of hydroelectricity, and only natural gas imports can reduce the reliance on imported oil.

Although the Swiss authorities are still considering a plan to build a large number of solar stations in the Alps for producing electricity, it will not affect the country's import needs before the 1990s. Meanwhile, only the planned moderate expansion of nuclear power and natural gas imports can restrain rising oil requirements.

Yugoslavia is in a better position than many European countries. Lignite production can be expanded, and there is still untapped hydroelectric potential. Intensive exploration has been going on for additional oil and gas deposits, with promising, though so far moderate, success. Oil shale deposits have been identified, though their exploitation depends, as everywhere, on technological advances and on oil prices. Nevertheless, oil requirements will grow, to be supplied via the recently built Adriatic pipeline, which is planned to be — and has partly already been — extended to Austria, Hungary, and Czechoslovakia.

The Structure of European Consumption

The energy use of the various countries is, naturally, influenced by a number of complex factors, among them climate, tradition, and economic structure. Most notable are the variations in industrial use, reflecting the structure of industry. Some countries — Denmark, Ireland, and Switzerland — have hardly any iron and steel industry. In others, that industry accounts for a large part, more than one tenth, of total energy use — in Belgium, Luxembourg, Germany, Austria, and Spain. Similarly, big differences can be found within the two main divisions of the chemical industry (petrochemicals and others), where the outstandingly high Dutch share is the most notable. There are smaller variations in the share of other industries.

We list the industries in order of their energy intensity, using a scale of 1 to 30 (see Table 7.4). The industries rated 10 and higher are the energy-intensive ones, led by cement and iron-steel.

Transportation use depends, to some extent, on the country's geography and population density as well as on the availability and quality of public transportation. The share of the residential (household) sector is markedly influenced by the climate, varying from 11 percent in Portugal to around 40 percent in Denmark and Switzerland. In Switzerland, the high share of household use is due, not only to the climate, but also to the relative absence of energy-intensive industries.

Approaching the Future

Energy is a world problem. The outcome of events in Iran was a shortage of oil supplies that proved to be temporary but pointed unmistakably to the unstable nature of the oil market.

Despite increasing oil production in European offshore waters and attempts to reduce oil use, petroleum will remain the most important source of energy for a long time to come. The bulk of this oil will invariably come from OPEC countries, and the prices set by OPEC will determine the prices for domestic oil and imports from other sources, such as the U.S.S.R.

When both the world and the oil situation can change rapidly, a large number of scenarios can be envisioned. Selectivity being imperative, however, a brief definition of some aspects is therefore required before discussing alternative departures. The time frame is basically the present decade, using 1985 as shorthand for the middle of the 1980s and 1990 for the period 1989–91. More tentative views will be expressed for the decade up to the year 2000.

European Energy Production

The prospects for European energy production to 1990 have already been surveyed. Given the long lead times of new investment or any new departure, it is impossible to construct any significant new energy facilities by 1985. By 1990, the outlook may be more promising. The nuclear power capacity of Europe is planned to be raised by another 40 percent, at least

Table 7.4 Relative Energy Requirements in Productive Industries in the Early 1970s

Industries have been grouped according to their energy requirements, based on the input-output tables of large European countries. A scale of 1 to 30 has been used, and the position of the industries indicates their relative average energy requirements (per unit of final industrial gross output). Those in group 30 are the most, and those in Group 1 the least, energy demanding *in these terms*. Within the groups, the industries are listed in descending order of energy use.

30 Cement

24 Iron and steel

21 Iron castings

18 Fertilizers
   General chemicals

16 Bricks, refractory goods

14 Aluminum
   Cans, metal boxes

13 Wire
   Other building materials[a]

12 Stone and sand extraction
   Synthetics: rubber, resins, plastics

11 Pottery and glass

10 Textile finishing
   Paper and board

 9 Water supply
   Bolts, nuts, screws
   Dyes, pigments
   Manmade fibers
   Other metal goods

 8 Motor vehicles
   Other vehicles
   Agricultural machinery
   Other chemicals

 7 Other mining and quarrying
   Wheeled tractors
   Other mechanical engineering
   Plastic products   Soap, detergents
   Paint   Pumps
   Domestic electrical appliances
   Rubber
   Cotton spinning and weaving

 6 Construction and mechanical handling equipment
   Electrical machinery
   Machine tools
   Industrial plant and steelwork
   Other nonelectric machinery

   Agriculture
   Forestry and fishing
   Textile machinery
   Other manufactures
   Insulated wires and cables
   Engineers' small tools
   Other manufactures[a]
   Soft drinks
   Shipbuilding and marine engineering
   Other electrical goods
   Toiletries, cosmetics

 5 Other food
   Pharmaceuticals
   Woolen and worsted goods
   Paper packaging materials
   Other paper products
   Construction and building
   Other textiles   Carpets
   Cereals   Sugar
   Office machinery
   Aerospace equipment
   Hosiery, knitwear
   Nonferrous metals[b]

 4 Chocolate and sugar confectionery
   Household textiles
   Leather, fur
   Furniture
   Alcoholic drinks
   Electronics, telecommunication
   Vegetable oils and fats
   Instruments

 3 Footwear
   Grain milling
   Clothing
   Printing, publishing
   Timber

 2 Cutlery   Jewelry

 1 Tobacco

[a] "Other" in all cases means products in the same category (e.g., building materials, manufactures, food) not listed separately.
[b] Excluding aluminum.

in the five years from 1985 to 1990. Beyond that date, the future is blurred indeed.

Many experts consider the decade of the 1990s as decisive for mankind's energy future. Unless some major oil deposits are discovered and begin to be exploited by then, the consensus holds that world oil production may begin to decline sometime during that decade, and new forms and sources of energy that could be developed on a large scale are not likely to be in production until the next century. For the latter part of the 1980s and the 1990s, the options open to Europe are more production and reduced consumption.

On the production side the options are: oil and gas, nuclear, coal, and renewables. Each of them has various degrees of uncertainty.

Neither the North Sea nor other waters around Europe have been fully explored. Indeed, continuing oil or gas finds — admittedly minor ones so far — are being reported, and the already exploited oilfields and gasfields in the North Sea may eventually produce considerably more than presently expected. The high price of oil is, of course, a considerable stimulus to the feasibility of exploitation of some of the smaller deposits already known and those discovered in the future.

The coal option is different in nature. To many consumers, coal is not a convenient fuel in the present general state of coal-burning technology. Large consumers, such as power stations or cement works, may expand their use of coal because they have the means and the techniques to overcome its disadvantages, such as handling and pollution. When the technology is developed and the time is ripe for large-scale gasification and liquefaction, a necessary precondition will be the significant expansion of European coal and lignite mining. This again will present its own problems — mainly ones of manpower — since the possibilities for open-cast mining are limited. Marked technological change (such as remote control and robots) is therefore a condition of further major expansion.

The nuclear option is technically and commercially an obvious one. It is a path on which so far few European countries have started, and it is not free from snags. There is, first, public opposition, of which the most striking examples were the Austrian and other referendums. Second, there are still technical and other problems, such as waste disposal. Breeder and other converter technologies are the obvious choices to widen the potential of the nuclear source, and therefore much depends on the success of the Superphénix, a 1200-megawatt commercial fast-breeder station, the first in Europe, built by the French with German and Italian participation. Still, nuclear power will account for a growing part of total electricity production, though that total is likely to rise at a slower rate than in the past.[3]

Europe's hydroelectricity production will expand but only at a moderate rate, continuing its useful but small contribution to energy supplies.

What about renewable sources? To extract energy from solar power, from waves or wind, from geothermal sources or from "biomass" on a laboratory scale, or even for actual use on a very small scale, is one thing. To do the same on a large scale is quite another. Producing hot water by means of solar collectors in sufficient quantity to satisfy the needs of houses and even larger buildings has been accepted practice around the Mediterranean and elsewhere for some time. Biomass potential may be realized in various communities. New buildings designed with passive solar architecture offer huge energy savings, but in ten or even twenty years they cannot account for more than a fraction of the building stock. Windmills are not new. Geothermal sources help to produce electricity in Italy and elsewhere. The French tidal hydroelectric power station at the estuary of the Rance has been in operation for many years.

These possibilities depend, however, to a very large extent on the geography of Europe. The Mediterranean area may be justified in attaching certain hopes to solar power, as are perhaps the Alpine countries; but the blessings of sunshine are not allocated in similar abundance to the northern parts of Europe. The Rance estuary required building a dam about a mile in length; another estuary, in Britain, where the tide is strong enough to be relied on for electricity generation, is that of the Severn near Bristol, but it is over 12 miles wide! Geography apart, the technology of large-scale use of tidal forces, with all its engineering and other problems, has not yet been solved.

Europe may arrive at a situation where even a small quantity of energy counts. Real relief from the energy constraint, however, will not come from alternative sources unless they supply a certain significant share — say, 10 to 20 percent — of total primary requirements. Most specialists do not expect that this range can be easily reached in this decade or the next. There are studies with more optimistic conclusions that look further ahead.[4] They may read like science fiction today, but many aspects of today's energy scene would have appeared just as incredible fifty years ago — say, in 1932.

Economic Prospects

Quite apart from the energy situation, there are a number of reasons why the high growth rates of the European economies in the golden period of the quarter century following the war and ending in 1973 are generally considered to be unattainable in the future. Postwar reconstruction, the major waves of trade liberalization and European integration — all had their beneficial and long-lasting effects, but they cannot be repeated. The higher share of income to be paid for increasingly (and relatively) more expensive energy is another depressing factor when compared with the past. So is the general endeavor to reduce the high rate of inflation, in the main by deflationary policies.

The 1979–81 oil price increases still have to work themselves through the European economies. According to OECD calculations, the simulated effects of an oil price rise of 10 percent reduce total domestic demand by 0.7 percent, real incomes by 1 percent, and add around 0.5 percent to the domestic inflation rates in OECD countries.[5] Furthermore, they considerably worsen the balance of payments of oil-consuming countries.

Between December 1978 and January 1981, oil prices (Arabian light) rose from $12.70 to $32, or 150 percent. This has greatly contributed to the stagnation and recession in Europe, which started in the second half of 1980 and continued into 1981 and well into 1982. Another result was the undeniable effect of the rising oil prices on inflation. In connection with one of the frequent price increases, the OECD estimated that "about 1.5–2 percentage points of the faster rise can be accounted for by the direct effect of the June [1979] OPEC increase."[6] Repeated further increases had no doubt the same proportional effect.

Assuming that nothing dramatic happens and that the international financial institutions, although under some strain, are able to cope with the recycling of the excess revenues of OPEC countries — the first a bold, and both necessary assumptions — moderate economic growth may be resumed, starting perhaps in 1983. It is unlikely that the average rate of economic advance for the European countries in the period 1980–85 will exceed 2 percent, as a central estimate within a range of 1.5 to 2.5 percent.

Energy Needs

What will be the energy requirements of Europe in the case of this growth path? The general view, based on conventional wisdom, is that for every additional 1 percent economic growth, there will be 0.8 percent additional energy use. But this energy:output coefficient may not be a useful measure or objective. First, it varies greatly by country. Second, it fluctuates from year to year (though it may be somewhat more stable over longer periods) and is arithmetically awkward in years of negative growth.[7] Third, many countries have already reached a coefficient of 0.8, or even lower. In fact, the 0.8 coefficient has been reached — indeed, surpassed — by Western Europe (taken as a unit) in certain periods, as indicated in Column 5 of Table 7.5.

This table indicates the large variations in energy use from year to year. If the coefficient is calculated for periods of five years or more, it probably reflects the changes reasonably well: The sparing use of energy in 1973–78 followed the almost 1:1 relationship in the previous ten years. This, how-

Table 7.5   Western Europe: GNP and Energy Use, 1963–1980

	(1) Output (GDP) $ billion[a]	(2) Output (GDP) Percent[b]	(3) Primary energy Consumption[c]	(4) Primary energy Percent[b]	(5) Energy coefficient[d]	(6) Energy ratio[e]
1963	1084	4.5	984	8.1	1.80	0.91
1964	1150	6.1	1017	3.4	0.56	0.88
1965	1201	4.4	1056	3.8	0.86	0.88
1966	1247	3.8	1085	2.7	0.71	0.87
1967	1288	3.3	1109	2.2	0.67	0.86
1968	1353	5.0	1180	6.4	1.28	0.87
1969	1435	6.1	1270	7.6	1.25	0.89
1970	1510	5.2	1351	6.4	1.23	0.89
1971	1565	3.6	1372	1.6	0.44	0.88
1972	1633	4.3	1440	5.0	1.16	0.88
1973	1729	5.9	1529	6.2	1.05	0.88
1974	1766	2.1	1500	−1.9	(−4.0)	0.85
1975	1745	−1.2	1457	−2.9	(−1.7)	0.83
1976	1825	4.6	1553	6.6	1.43	0.85
1977	1866	2.2	1541	−0.8	(−3.0)	0.83
1978	1920	2.9	1566	1.6	0.55	0.82
1979**	1983	3.3	1622	3.6	1.09**	0.82
1980	2013	1.5	1575	−2.9	(−4.4)	0.78
63–73*		4.8		4.5	0.94	0.88
73–78*		2.2		0.5	0.23	0.84
63–68*		4.5		3.7	0.82	0.88
68–73*		5.0		5.3	1.06	0.88
75–78*		3.2		2.4	0.75	0.83
75–79**		3.2		2.7	0.84	0.83
75–80*		2.9		1.6	0.55	0.82
78–80*		2.4		0.6	0.25	0.81

[a] At 1975 constant prices and exchange rates.
[b] Change over previous year.
[c] Millions of tons of coal equivalent.
[d] Percent growth in energy use for 1 percent GDP growth. (Percent change in energy growth, 4, divided by percent change in economic growth, 2)
[e] Mtce per $1 billion output. (Total energy consumption, 3, divided by GDP, 1)
* The figures in Columns 2 and 4 are annual average compound percent changes.
** Estimate. Because of the unusually cold winter, energy use was extremely high; tentative correction for temperature would reduce the energy coefficient to around 0.75 and the energy ratio to under 0.81; because of the uncertain nature of adjustment, the considerations in the text do *not* take account of the year 1979 nor of the period 1978–80.
Source: *National Accounts of OECD Countries*, OECD, Paris; *World Energy Supplies 1973–78*, United Nations, New York; author's estimates in Table 7.1.

ever, conceals the period of the 1973–74 cutback in oil supplies and the deepest recession since the war. More relevant, therefore, is the 1975–78 period, when the coefficient was 0.75, some 20 percent below the decade before 1973. At least equally valid, however, is the energy:output ratio shown in the last column of the table. This shows the physical units (in millions of tons of coal equivalent — mtce) of energy actually used for each $1 billion worth of output (in 1975 dollars). It is a more stable indicator; its fall after 1973 and its stabilization at a lower level is, therefore, reassuring to some extent. It proves the initial effect of the Europe-wide energy conservation measures, even though it may be considered far from satisfactory.

Thus, if the objective of "reducing the energy coefficient from unity to 0.8" is in any way a guide, it ought to be interpreted as a required reduction of 20 percent in the energy use relative to output. It would mean reducing the 0.75 coefficient of 1975–78 to 0.60 in the coming years. As already discussed, the possibilities of increasing domestic energy supplies are limited. To reduce demand, there is hardly any other option than conservation.

Conservation

Conservation means using energy more efficiently. Table 7.5 earlier indicates that some conservation has already been achieved. Data for 1981 point to a growing contribution from conservation. Higher prices provide a further incentive for conservation; governments can also help by adjusting taxation (e.g., the excise duties on oil products, especially gasoline), although they seem to be proceeding rather gingerly in this direction in view of their fight against inflation. They can also provide tax incentives and help make capital available for conservation investments.

Energy prices to the final users have increased in relative terms, although only to a limited extent. A contrast clearly exists here between the macroeconomic and microeconomic aspects. In 1979, the real price of crude oil imports (that is, the nominal price deflated by the export prices of manufactured goods in world trade) was between three and four times the 1972 price, equaling an increase by a factor of three to four, but the relative price of fuel (money price deflated by the consumer or retail price index) to the final European consumer was only 10 to 50 percent higher generally, depending on the country, the product, and on other factors. The reason for this divergence is clear: Other cost elements in the final product, such as transportation, refining, distribution, and tax, did not rise in line with crude oil price.

Thus, the incentives for saving energy or for investing in conservation have been limited. With this kind of stimulus, users may go some way toward economizing on energy, but not far enough when measured against the possible severity of the future. Given that inflation affects everything else, energy is apparently not yet expensive enough to overcome the users' inertia, stemming from their being locked into their equipment — appliances, houses, cars — and habits.

It does not seem to have been sufficiently recognized by European consumers (and governments) that a gradual transition from imported oil and gas is unavoidable and needs to be encouraged. Hence, more forceful conservation is perhaps the only, and certainly the most immediately efficient, way to diminish the reliance on imported oil. Conservation is therefore a must for Europe. Indeed, this is the main conclusion of a recent analysis prepared for the European Commission by Jean Saint-Geours and a group of independent experts.[8]

Much has been written on conservation in the various European countries. Although it is difficult to draw the dividing line between the scientifically possible and the practically adaptable suggestions, it is clear that significant energy savings can be achieved in all the main consuming sectors. There are many known and proven methods of conservation, from better insulation of dwellings through new or modified industrial technologies to better public transportation. But major changes require major investment; and major investment presupposes relatively rapid economic growth, more rapid than is likely to be achieved in the foreseeable couple of years. This underlines the importance of conservation and the message of the Saint-Geours report and points to a potentially vicious circle: Greater conservation investment is necessary to provide the preconditions for a resumption of economic growth. But it may prove difficult to generate that investment without more rapid growth. Thus, governments may well have to allocate large amounts of capital toward that effort themselves.

Import Prospects: Three Scenarios

Whatever the achievements in the longer term of energy production and conservation, Europe will still need large volumes of energy imports.

The outlook for world energy — and oil — supplies, analyzed in Chapter 2 by Robert Stobaugh, should be converted into prospects for those oil quantities Europe can reasonably reckon with. The Upper Bound case postulates 2.6 percent annual growth in *world* energy supplies. This expected growing quantity, however, should cover non-European requirements as well, which are likely to grow more steeply in some other parts of the world. Available energy supplies for the industrial world are expected to grow at 2 percent.

Of course, it is physically possible that more oil could be forthcoming. Welcome as this would be, we are aiming at surprise-free scenarios, and that should exclude pleasant surprises as well. Thus, a major increase in OPEC production is not among our scenarios, which are as follows:

Scenario A: By 1985, no more than a minor, almost marginal, shortfall in oil supplies (imported from OPEC and other producers) compared with estimated Western European needs after economic growth resumes; this becomes more marked by 1990 and continues to widen.

Scenario B: If the quantities foreseen by OPEC members continue, a sizable shortfall would already be with us in 1985. By 1990 it would assume larger proportions, and the situation would worsen later on.

Scenario C: Given the importance of the turbulent Middle East, it is not unnatural to ask: What is a surprise-free scenario? Suppose some major calamity of political, technical, or, indeed, of any other nature and origin occurs, suddenly reducing OPEC supplies by, say, one half. Thinking of Iran and the many trouble spots in the Middle East, would this be that surprising? We take this as the third scenario. We hope it will not occur but will nevertheless consider it, together with the emergency it would create. Concerning its timing, we place it in the second half of the 1980s.

Major changes in the worldwide availability of oil will naturally affect supplies available for Europe as well as supplies of other fuels. Future United States requirements are surrounded by uncertainty; it remains to be seen to what extent domestic use can be limited and U.S. production of oil and other types of energy maintained or increased. The OPEC countries themselves are rapidly becoming industrial, concentrating on industries that use large amounts of energy. The Third World is also advancing quickly, and so is its energy demand. China's future energy balance is puzzling, but those of the other centrally planned economies is clearer. It seems that the USSR may not be able to increase its oil production sufficiently rapidly to satisfy its domestic demand as well as the needs of Eastern Europe *and* maintain its profitable energy exports to the West. The losers will be the Eastern European countries, whose imports from OPEC will certainly rise. This might be mitigated, however, if optimistic projections of Soviet natural gas development are fulfilled. Thus, whatever the level of OPEC's production, a growing number of consuming countries will require increasing shares of it; thus even increased OPEC exports would not necessarily mean rising availability for Europe.

Europe could manage with tight supplies if they were reasonably assured and continuous. But it seems improbable that Europe's share of OPEC's future exportable production will grow parallel to, and cover, all the needs exceeding domestic and other available supplies. This view concerns the Upper Bound case foreseen in Chapter 2, which expects a modest growth in world energy supplies.

More coal will become available from the major exporting countries —

the United States, Australia, South Africa, and others. Whether these quantities will match the demand depends not only on these countries' ability to raise coal production but also on the rate at which energy importers switch from using oil to coal.

Natural gas may later be marginally short. To some extent it is dependent on the oil trade, as is coal, since when oil becomes scarce, the ensuing scramble for energy affects other basic fuels as well. The possibilities for additional supplies cannot yet be realistically predicted.

Energy Scarcity?

It is not our purpose here to construct a long-term energy forecast for Europe. For purely illustrative purposes, however, some simple estimate nevertheless seems unavoidable, starting with the expected development of European output. A fairly sharply declining energy:output coefficient, going down to 0.6 by the year 2000, has then been applied. (This reduces the energy:output ratio by about 15 percent in twenty years.) The maximum domestic production was assumed, as well as a fairly stiff reduction in oil's share of total energy.

In Scenario A, the shortfall by 1985 is marginal. By 1990 it already reaches a proportion worth noting, 5 percent of all primary energy. By 2000, it is well in excess of 10 percent. Scenario B is stiffer: Already by 1985 the shortfall is about 2 percent for total energy but 10 percent for oil alone, and the oil shortage rises to 30 percent by 2000, raising the shortfall of primary energy requirements (with some contribution from other fuels) to about 17 percent. In Scenario C, the "catastrophic" case, the halving of OPEC exports reduces oil availability to 60–65 percent and, with its effects on other fuels, the shortfall as compared with our "reasonable" estimate of requirements (which remains the same in all cases) is almost one quarter by 1990.

These figures do not mean more than orders of magnitude for further discussion for a number of reasons. An energy "gap" itself may be considered a misnomer; there obviously cannot be any "gap." Either there will be less output with less energy needs, or conservation will go ahead faster than envisaged in the assessment of "reasonable" requirements, or new solutions will change the energy scene radically, leading to entirely novel technologies — or something else. These possibilities should be neither discarded nor belittled. New technology has more than once evolved as the savior in cases of shortages of materials — and, indeed, of energy.[9] Much depends also on the factor so far untouched: prices.

Prices

Oil being the most widely used fuel, its price changes cascade across the board, affecting practically all products and services and in many ways stimulating inflation. By altering the terms of trade, any major increase is always an additional burden on the importing countries' trade balances. Rising prices stimulate conservation or substitution by other fuels, and they also raise the chance of any new source or form of energy becoming economically viable.

We can start in a very simple, orthodox way to approach the question of the likely price development: Whenever demand exceeds supply, the price of any commodity goes up; when demand falls below supply, the price falls. In the case of oil, we have seen both in the last three and a half years, although the increase in price far exceeds the relatively minor slide. Our scenarios exclude any protracted slippage in the price of oil that would imply a persistently serious economic crisis in Western Europe.

With gradually increasing scarcity, prices would increase in real terms. The further we look ahead, the greater the shortage of oil, the more OPEC producers will come close to exhausting their reserves and be more conscious of depletion, and consequently the more they will be inclined to raise their prices.

On the more positive side, the further ahead we think, the more likely it seems that other sources of energy will advance both technologically and economically to replace oil. It is not believed that they can produce energy on a large commercial scale in this century. But once their production problems are solved experimentally, expectation should already affect the prospects for oil and influence oil prices. These expectations could have two effects on producers: either encourage them to push up prices to maximize revenues in a shorter time or encourage them to lower prices to try to undercut the new technologies. The real world, however, may be too uncertain for such calculations to be made.

The Impact of the Oil Situation on the European Economies

Our purpose is certainly not to construct arithmetically foolproof sets of national accounts, modeling the alternative paths of the three scenarios. Rather, our aim is more practical: to give a reasonable idea of the range of possible alternatives. More precise quantified estimates may therefore not be necessary. (Judging by the fate of most earlier long-term forecasts, the likelihood of their proving right is in any case minimal.)

Economic Growth

The year 1973 was a watershed in world economic history. Immediately after 1973, economic growth faltered, dipping into the deepest recession since the war, and it has not recovered its earlier level since. The European economies grew at a rate of 5.1 percent in 1967–73 and at 2.4 percent in 1973–79. The year 1980 was weak; 1981 was a year of negative growth; and the outlook for 1982 suggests more stagnation or, at best, the beginning of a slow recovery.

Reduced growth is not the only indicator of the trouble; high rates of inflation, high levels of unemployment, and oscillations in the operation of the international system were other symptoms. Analysts generally agree that while the sudden jumps in the price of oil are not the only cause, they do carry a considerable part of the blame. Ever more resources were required to secure energy supplies, leaving fewer for other purposes. The terms of trade of European countries suffered from the oil price hike as well as from the rising real prices of many imported industrial materials and food. The effect of the higher energy prices on inflation, which by 1973 was already rapid by previous standards, was of course very significant. Most countries attempted to fight inflation by stricter monetary rule and by general deflationary policies, altogether adding up to high unemployment and unused productive capacities. Several European countries took protectionist measures to ease the pressure on some of their industries. The developing world (outside OPEC) was in similar, and in some cases even greater, trouble. The general malaise reduced the growth rate of world trade.

The second oil price hike, in 1979–80, has had similar effects. Still, conditions in 1980 were not the same as in 1973. The world has changed. It has not yet learned how to cope with oil shocks, but its approach to the resulting problems has changed.

After 1973, the reduced economic growth affected all sectors of demand. Investment suffered most; the growth rate of private consumption was halved. European exports, which were greatly helped by the enormously raised imports of OPEC, fared somewhat better but were still far behind their own record. European imports were more than halved. The rate of fall was smallest in public consumption, the most difficult to tackle. In terms of their share of GDP, foreign trade in both directions rose markedly (exports and imports moved crudely in line), while the shares of private and public consumption increased at the cost of investment. Most marked was, however, the fall in savings. This was a general phenomenon, occurring at the same time though not to the same degree in all the European countries.[10]

Under Scenario A, the physical shortfall will be marginal by 1985; it is unlikely to constrain growth, mainly because the 1979–80 higher oil prices will limit growth in any case. The effect of the prices is already discounted for the near future, with stagnation in 1980–82. Unless there is another sudden increase in real oil prices, the European economies may recover toward the middle of the decade. Scenario B makes energy supplies somewhat scarcer by 1985; it may have some impact on growth, but it is unlikely to be severe. Given, however, the even more limited availability of oil and the consequent pressure on other fuels, prices would rise faster, making it that much more difficult to keep inflation in check and exercising stronger deflationary pressure in general.

Thus, while the price pressure will be growing and will limit general economic advance, no major changes seem to be expected to the middle of the decade. Growth will be sluggish, inflation will remain high, and unemployment will still be with us, probably at a higher scale but not high enough to lead to social upheaval. Relatively more will be spent on energy, and this will stimulate a more active approach to conservation. Gradually, an increasing share of the resources available for investment will be directed toward saving energy in all possible areas: production, transportation, equipment, and even new technologies. There are signs of this already beginning to happen. Real personal incomes will be squeezed by the higher cost of energy, since in the short term people are locked into existing houses and equipment. Although the prices people pay for energy will gradually increase, even in relation to the general rate of consumer price inflation, the difference will not yet become such as to press below certain standards, which will enable consumers to maintain their usual standard of life, albeit lower than in earlier times.

By the end of the decade, all this will change according to each of our scenarios. First, oil — and, to an extent, other fuels — will already be in short supply; all the pressures on growth, inflation, and so forth will be much heavier, and physical shortages will begin to bite (unless, of course, conservation is incomparably more successful than conventionally assumed).

In Scenario A, the shortfall in oil supplies to Europe will be around 8 percent, somewhat more in Scenario B. Pressure on natural gas supplies will be strong, and this source will also stop providing the quantities required. The addition to nuclear capacity is unlikely to fill these gaps by 1990. Coal supplies may still be able to cope with the demand, but European industry and other consumers will not have made the necessary adjustment to coal-burning appliances.

Some measure of physical allocation by government edict cannot be ruled out. This may start with restrictions on private transportation and heating in public buildings, offices, and factories. Industry will be the last to be affected by mandatory measures, but this period of transition will be short lived because the temporary nature of the relatively mild scarcity will soon become obvious.

As the world economy enters the last decade of this century, the scarcity of oil supplies will become gradually more pressing. Although prices will by then be high enough to stimulate conservation on a scale earlier unknown, governments will have to consider more efficient measures both to safeguard supplies to vital sectors and to keep a check on the burden of rising energy import bills.

The international monetary system will be under a permanent and heavy strain; the currencies of many European countries will come under irresistible pressure. In view of the continued high level of imports, many of them will become heavily indebted, and the austerity programs of their governments will help little in solving their economic problems, which in turn may lead to troubles of other kinds. Countries with domestic energy resources in Europe — especially within the EEC — will be pressed to share their supplies with their neighbors.

The situation would, of course, become even more serious under Scenario C, the "catastrophic" scenario. This is a speculative one, assuming OPEC exports are halved, for any reason, before the end of this decade. It would create an emergency situation, and, with the possible exception of the United Kingdom and Norway, no country in Europe could survive without very serious difficulties.

Some may say that such an emergency would do no more than hasten the time when Europe will have to learn how to live with much-reduced energy resources. This may be true, but time is a most important factor: It was the sudden nature of the oil price increases that shook the world; it would be the suddenness of any such emergency that would make it so harmful. Given time, normal (or artificially supported) evolution and accelerated adjustment may provide adequate preparation for the same eventual goal.

European energy supplies under Scenario C would be cut by about a quarter; the most painful, of course, would be the shortfall in oil supplies, which might amount to almost half of the requirements under normal conditions. In such a situation, very powerful intervention would be necessary at all levels and in all sectors. Economic activity would be reduced by between 10 and 20 percent, depending on the preparedness of the various countries, and in those countries without alternative resources the fall might be even more. Governments would concentrate on developing their energy resources, poor as they might be, and on catching up with the research, development, and experimental projects that did not receive sufficient attention in the years of relatively plentiful energy — say, 1975–85 — which will be regarded as "the years of lost opportunity." No more than a handful of activities would be free from some kind of very severe allocation of the remaining supplies of oil and other types of energy — such as military establishments, doctors, ambulances, and the most important public services, such as the water supply. The cuts would be bold and painful, affecting private cars as well as domestic and industrial supplies. The main instrument to keep production going in excess of the allocated small quantities of fuel would be ingenuity.

Structural changes would have to be rapid, but mass unemployment would precede them. Full employment, in the postwar sense, would be unlikely to be achieved again. Hence institutional changes would follow.[11] Those who remained at work would also suffer; they would try to retain the semblance of their earlier way of life, but in the wake of rapid inflation, the choice between spending and saving would become increasingly painful.

Inflation

The oil price increases projected will add to the inflation generated at home. They affect inflation in two ways: directly, as the impact of an all-pervasive factor of production becomes more expensive in real and in money terms; and indirectly, as experience has shown, by reducing output through the complicated mechanism of the terms of trade, resource allocation, and so on, and thereby making it more difficult to absorb a higher share of increasing cost by means of higher production.

Nevertheless, assuming otherwise "normal" conditions, the European economies probably could learn how to live with a certain rate of inflation if the impact from the energy side were gradual. A regular annual real rise in the price of oil, even at a high rate of, say, 6 percent, is much easier to accommodate than sudden jumps, even if it may eventually add up to the same amount by some later year.

The direct effect of an oil price increase may seem modest, but it indirectly cascades through the whole economy. It contributes greatly to those factors that raised the rate of inflation from much smaller beginnings to an annual 8 percent in the ten years ending in 1977, then further, to over 9 percent in 1978, 11 percent in 1979, to over 14 percent in 1980, and then to around 13 percent in 1981. Throughout this period, every single government in Europe has been fighting inflation.

Thus, continued rapid inflation is unlikely to be avoided in any of our scenarios. In Scenario C, however, the price of energy may get entirely out of hand unless it is, along with its physical allocation, strictly set by central authorities — in which case it will also have to be policed.

Employment

The effect of energy developments on employment is perhaps the most delicate and critical aspect of the alternative courses events may take. Unemployment was about twice as high as earlier at the time of the 1975 recession and has been reduced no more than marginally since; in 1981 and early 1982, it was rising rapidly again. In the United Kingdom at the beginning of 1982, it was almost 13 percent, even higher in Belgium, and in many European countries it was not much lower. This is the sad parallel to the lower rate of economic advance.

A second point to remember is the growth of the service industries that have helped to absorb the labor that has been "freed" in other sectors. This is a worldwide phenomenon, occurring equally in Western and Soviet economies. In every year in the ten- to twelve-year period ending in 1976, about 1.25 million persons were added to those working in services in the United States. In the USSR the figure was almost the same; in Japan, 0.7 million; in Western Europe, about 1 million; and in Eastern Europe, about 0.5 million. One cannot help asking how long this flow is likely to continue. On the one hand, services are the most rapidly growing sector of the economy in Europe; on the other hand, rapidly growing sectors always invite more thinking about improving the operation of the organizations involved, which results in an increased rate of innovations. And innovations usually result in saving labor. Thus, it is at least questionable whether services will go on absorbing labor at the earlier rate. The supply of labor will, however, grow in any case, partly for demographic reasons and partly because industrial innovations also usually result in saving labor. Moreover, in some countries, the agricultural labor force may be reduced still further.

If to all this is added the outcome of the depressing effect of the rising oil prices combined with a shortage of energy, the outlook for employment appears very gloomy indeed. It is therefore likely that European governments will have to implement institutional changes. The most probable departures are earlier retirement and a shorter work week. These modifications may create their own social problems but can help to alleviate unemployment, though they are unlikely to solve it. Under such conditions, the "black economy" will flourish and unemployment benefits will present an increasing burden on budgets.

There are, however, some points that may bring a glimmer of hope. In an energy-hungry world, there may be a premium on saving energy at the cost of more labor (insofar as technology permits) since, quite simply, there will be plenty of labor and shrinking quantities of energy. However, energy and labor are not straightforward alternatives in general.

All in all, some optimism is needed to share even the less-than-buoyant views of the distinguished economist Sir Alec Cairncross, who closed his recent public lecture, "Farewell to Full Employment," by saying that it would be wrong

> to leave the impression that we are headed for utter disaster. Things could get a lot worse without any return to the condition of the 1930s. It seems more likely that the world will continue to teeter along from one crisis to another at a disappointingly low level of activity but avoiding a catastrophic fall. The more serious risk is to morale. The public mood may become less tolerant, more impatient, more heated, more receptive of magical and mistaken remedies ... If of necessity we must say farewell for a time to full employment do not let us turn our backs on it or imagine that the world will be a better place without it.[12]

Sectoral and Structural Changes

Structural changes are natural and inevitable. The Industrial Revolution meant the rapid emergence of industry as the main employer of labor and as the main creator of wealth; on a smaller scale, similar developments characterize the more recent rise of the service industries. Technological advances brought to life new industries and services and finished off old, established trades. These changes were brought about by, in general, the relative advantage of the new producer in terms of at least one of the factors of production, interpreted in the widest sense, such as labor, capital, some aspect of management on a national or company level, or a specific endowment with some natural resource. The understandable ambitions of later — or now newly — industrializing countries, usually supported by some considerable advantage in one or more factors, have always meant sharp competition for the already more advanced countries.

Energy-intensive industries were generally built near the source of energy, such as the iron and steel industry and, more recently, the aluminum industry. Elsewhere, other considerations seem to have prevailed: The resources of the forest attracted the pulp paper industries; limestone and clay, the cement and brick industries; the agricultural hinterland, some of the basic food industries, and so forth. If not available locally, energy was brought in. As long as there is no major problem with the physical availability of energy, there will be no pressing need for any major changes, although, of course, as fuel becomes more expensive, any small difference in energy or transport costs will be magnified, bringing growing advantages to industries in energy-producing areas. Other conditions, however, such as infrastructure, skilled manpower, high technology, vicinity of the market, and ingenious management, may provide a certain amount of protection for some European industries, at least for a time.

Among our three scenarios, the main difference is that the likely structural changes may come at different times and with differing speeds. A gradual absorption of any structural change, given time, is of course always easier than solving the problems arising from a very sudden change. In this sense the third, catastrophic, scenario would pose incomparably greater problems than Scenarios A or B. But the problems are not fundamentally different from those experienced by the major textile, shipbuilding, and some other industries in Europe.

The recognition of permanent scarcity will certainly stimulate further innovative activity. What seems desirable and indeed imperative is that any such thinking should not be left until energy becomes the bottleneck but should be activated much earlier, making realistic contingency plans possible.

Without such plans, it will be extremely difficult to allocate the limited quantities of energy. Leaving all other sectors aside for the moment and considering the eighty or so industries listed in Table 7.4, we find that, depending on the country, the top ten or fifteen account for about half of the total industrial energy consumption. It is in these industries that major reductions in energy use can be achieved. On the other hand, all the large users, from cement to steel to plastics and paper, are important to the national economy in one aspect or another, either for strategic reasons or for the supply of the population (e.g., metal boxes or textile finishing). They are likely to find further conservation difficult since, precisely because of the importance of energy in their cost structure, they have always carefully monitored and surveyed their own energy use. Thus, when it comes to a critical shortage, the governments' choice will be very difficult.

How will they do it? The first measure will probably be realistic pricing. This may present a dilemma of another kind, since all the European countries are struggling with inflation. The anti-inflationary stance speaks for keeping energy prices (within limits) low, in view of their pervasive impact. In contrast, the realistic reflection of the pressure of worldwide scarcity suggests higher fuel prices. Although the situation may be different in each country, the latter attitude is likely to prevail gradually, since realistic (i.e., higher) fuel prices can be expected to boost conservation.

There are various estimates of the price elasticity of the demand for energy at the national or lower level. The usual message conveyed by these calculations is that in the short term, the reaction of demand to price changes is low, but in the long run it is high. The question, however, remains: Since all elasticity calculations are based on the past (some of them covering two decades or more), are the results applicable to an energy future that is very different not only in degree but also in kind? No doubt a good deal of caution is advisable here. Simple logic supports the expectation of a marked long-term reaction to rapidly rising energy prices, but it may be false prophecy to expect a solution to scarcity from this angle alone, without considerable support from central authorities in the form of special measures that include significant support for conservation in various, mainly financial, forms. Such measures may help to bridge the present gap between macroeconomic and microeconomic interests: The macro interests favor quick action, while the micro interests find the payoff period of energy-saving investment still too long or have many other projects they consider more urgent or profitable.

This problem, however, points to one area with a promising future: those branches of industry and services that produce energy-saving equipment or otherwise help to save energy. To those we may add similar products and/or services that help environmental protection, which will be increasingly important in line with the change in energy systems toward, for instance, more coal firing or nuclear power or some of the renewable energy sources that are not necessarily friendly to the environment, such as processing biomass. These will be just as important as any branch of industry producing additional energy supplies.

Worldwide energy scarcity will have an impact on the supply and the relative attraction of those materials that at present are the bases of the manufacturing industry. It will also make it more difficult to bridge any shortages that may occur in the supply of now conventional materials. In European history, such shortages were not uncommon and were always solved by some advance in science and technology — but always, without exception, at the cost of higher energy.[13] It is not argued that advances will stop, but in the future they will have to take into account the energy constraint. This will also have an impact on the relative attraction of natural versus manmade materials. The usual raw material for the latter is mostly oil or natural gas and, to a smaller extent, coal. Higher oil prices affect the cost of manufacturing synthetic materials. The increased production of natural products (wool and cotton to replace synthetic fibers, alcohol to replace gasoline) could, under some circumstances, create a competition with food production or at least drive up food prices. It is not far-fetched to imagine a situation where it might be more profitable to process wood into some new form of fuel (apart from fuel wood, such as briquettes or wood powder to be burned in appropriate appliances) than into wood pulp.

Such changes will affect the structure of resource-poor European industry and, in some cases, that of agriculture. They will also point to the need for more conservation of materials that indirectly, in view of the energy embodied in them, result in the conservation of energy as well. There are two ways of achieving this: recycling and prolonging the life of the final product incorporating the material. The technology exists in many areas to reduce in both directions Europe's dependence on imported materials, but the financial incentives to apply them are either insufficient or totally missing. Quite apart from the direct merits of conserving materials and energy, the properly organized recycling of metals, paper, glass, and possibly other materials in quantities exceeding by many times their present level in Europe requires a significant amount of relatively unskilled labor.

The adverse impact of the application of microelectronics and microprocessors on labor requirements has been widely discussed. Their likely influence on energy use has received less attention, but their greatly increased reliability and efficiency will obviously be used to reduce energy needs in many areas, from automobile engines to the control of temperature and fuel in a cement kiln. It is not unreasonable to expect some energy saving from this angle.

It is most difficult to foresee the future of the European automotive industry, especially under Scenario C. The automobile will be caught in a difficult situation. First, it will be gradually more and more expensive to operate a private car; second, there may be less fuel to run it. It will be competing with more essential claimants for scarce supplies of the traditional fuel, and this situation would not be much modified if novel fuels were introduced on a large scale, based probably on agricultural produce or biomass. More efficient design can reduce the fuel requirement of automobiles, but it appears unlikely to secure the survival of the industry in its present size. The dismantling of many public transportation systems will then be regretted, and efforts will be made to revive them in more efficient forms.

The nearer we come to the year 2000 and the harsher the energy scarcity becomes, the clearer it will be that the dwindling quantities of oil in Europe ought to be reserved for those areas where oil products are, at the present state of technology, indispensable. Such a list of areas may be very different by then. Now it consists of aviation fuel, feedstock for the chemical industry, some nonenergy applications (lubricants, solvents, bitumen), road travel, and shipping. In most other applications, oil could be replaced by other types of fuel, at least in principle, although the difficulties of such wholesale replacement will be great. Purists may even question this short list on the grounds of experiments aimed at replacing oil as an automotive fuel by other materials or by pointing to the possibility of using other materials (coal, wood) as chemical feedstock. Speculation along these lines appears useful to indicate the urgent need for concentrated research and development effort in Europe (hand in hand with similar work elsewhere) in order to discover, develop, and build up the energy base of the "post-oil" era. If anything, this ought to be *the* growth industry in Europe. It is also the area where international cooperation, started along the lines laid down within the IEA, ought to be strengthened and its importance repeatedly underlined, in view not only of the long lead times but also of the very large amount of various resources required.

A Way of Life

Differences in history, culture, and climate have shaped the divergent ways of life in the many nations of contemporary Europe, and an energy scarcity will not bring uniformity, creating a real "Mr. or Mrs. Average European." Still, we can make some assumptions for European society as a whole.

There must be a point at which it is no longer possible to continue loading people with personal possessions, but the acquisitive society will probably still be here for the remainder of the century.

People will have more leisure time, either as a result of various work-sharing schemes resulting from an excess of labor (such as a shorter work week), early retirement, or lack of employment. Past experience indicates that people are generally ingenious enough to convert leisure into some do-it-yourself work — including, where possible, "grow-it-yourself" food production.

Extreme political developments can never be entirely written off, but it is to be hoped that their most extreme shapes — the past forms of fascism and communism — can be considered as nonstarters in Western Europe. In the postwar period, however, we have been witnessing other movements aimed at radical change. The hippies and beatniks are on the way out; but the revival of Christianity, which started in the 1960s in the United States, has been spreading over Europe. The spread of such a revival and the resulting changes in attitudes could throw a new light on many problems of today's European society, such as acquisitiveness. However, it could not occur without a genuine miracle, and we have to stay within the realm of a surprise-free future, which excludes major wars and major miracles alike.

The likelihood, therefore, is that society later in this decade and in the next will be rather similar to that of today. There will be differences, of course, but in the historical perspective of Europe they are unlikely to be dramatic, more of degree than of kind. Within this framework, severe energy scarcity could nevertheless turn a difference from a minor into a major one.

Each of our scenarios presents some constraints on economic activity. This will be gradual under Scenarios A and B but come suddenly in C. The gradual tightening in the energy market would eventually bring the people of Europe quite a long way toward the situation that could emerge abruptly in the case of the "catastrophic" turn of Scenario C. Growing unemployment, real incomes lower than they would otherwise be, unmet expectations, and a gradually increasing impact on the life-style will be the main symptoms of energy scarcity felt by the people of Europe. They will have to spend increasing parts of their income first on fuel. With the further rise in energy prices, there will come a point, however, when they will prefer alternatives. They may have to get used to lower temperatures in the winter, and therefore wear warmer clothing. They may use the car less. Their bedtime will be earlier. Outside of work, they will spend more time sharing in household chores and getting involved in hobbies that will gradually turn into more "useful" activities, forced on them by limited real income.

They will be more used to saving energy and materials: People will also move about less because technical development in all areas of communication will make at least a part of the present need for personal meetings unnecessary. They may have more time altogether that can be used for further education or other relatively energy-free activities. Political activity is one of them; its obvious dangers in an unsatisfied society, depressed by energy constraints, should not be underestimated.

At first many Europeans will feel unhappy since most of the changes seem to be reducing their standard of living in a material sense. Some may discover that happiness is not necessarily, and not always, a function of material abundance. Others may take the opposite view. Reactions may be different by country, by age, by profession, and by location: City dwellers and rural people would be differently affected. Thus, any such speculation can be colored by the inclination of the individual reader. Optimists will look forward to an age where severe energy scarcity — in spite of, and allowing for, all the adverse influences on the economy — may result in a more "pastoral" and less materialistic society, emphasizing more the cultural, spiritual, and altruistic sides of life. Pessimists may picture the incalculable consequences of various types of social upheaval and economic collapse rapidly extending into the lives of the average European.

* * *

The two years since the second oil shock have been characterized by an abundant supply of oil. No European country has had any difficulty in securing supplies and building up relatively high stocks. In such a comfortable situation, many of the statements and speculations of this chapter may seem out of place, an unnecessary doomwatching. It is easy to overlook the fact that these were years of economic stagnation, declining industrial activity, and relatively mild winters. In most European countries, energy consumption either fell or rose only a little in this same period. What is too early to know is how much of this "saving" was the result of real conservation and how much simply the effect of the recession and stagnation. It is tempting to be complacent.

The energy market may remain in its present disequilibrium and the resulting oversupply may favor the industrial countries of Europe so long as economic activity does not pick up. We hope that the current slump will be short. It is unlikely that Europe will settle down in the acceptance of practically no economic growth. Otherwise, the social and political consequences could be very difficult. But we should also be aware that economic revival will unavoidably lead to higher energy requirements, which are likely to indicate how delicate the balance is in the energy market.

Once the European economies get back on the track of a reasonably acceptable economic growth rate — and the high level of unemployment is making this an urgent priority in practically every single country — tightening energy supplies may begin to present problems. Still, it can be hoped that solutions will come from three directions: energy conservation, discoveries of new sources of conventional energy in Europe, and entirely novel sources of (possibly renewable) energy. Recurrent tight energy markets are likely to push prices further up, providing additional incentives in all three directions. Beginnings in these directions can already be seen, and if they are reinforced, the present bleak outlook may change. There is plenty of scope left for "technological optimism."

The future need not be bleak, not by any means. Historians report that in the summer of 1454, one year after the fall of Constantinople to the Ottoman Turks, Enea Silvio Piccolomini — later Pope Pius II and considered at the time one of the best informed European statesmen — wrote in a gloomy mood that there was nothing good on the horizon for the future of Europe. Yet, within the next fifty years, Europeans expanded far to the south of the borders of their continent, to Africa, India, and across the Atlantic; less than one hundred years later, the great movements of the Renaissance and the Reformation started to modify fundamental European thinking, opening new horizons, altering the European way of life, and filling it with new content.[14] Though not a guarantee for the future, this recollection leaves no place for despair, either.

8

The Social Contract under Stress in Western Europe

by Jean Saint-Geours

Western Europe created the first industrial revolution with coal. Two centuries later, technological advancement assured the region of a cheap supply of petroleum products from abroad. And from 1950 to 1973, Europe experienced strong economic growth. But that is in the past. The future has to be reconquered.

Energy and Europe: Diversity and Similarities

Over this decade and the next, constraints on energy, reflected in ever higher prices or, in some cases, limited supply, could significantly alter the social, economic, and political conditions of the nations of Western Europe. All of them, with the exception of Ireland, are highly industrial and thus depend on complex, modern energy supply systems. Apart from purely economic changes, these countries will have to reassess their life-styles and relations within Western Europe's political and social systems. This reappraisal will be especially significant because, like the United States, these societies have based their institutions and life-styles on the assumption of unlimited energy supplies at a constantly declining relative cost.

Because the energy problem will not affect each of these countries in the same way, their reactions to this fundamental transformation will vary widely. In fact, the differences over energy among them will tend to increase. Furthermore, although all of these countries are democracies, they still differ on many important social and political issues, including the distribution of power and the relationship of government, business, and the individual citizen. Each nation, in effect, has its own personality, shaped

230

over a long history; so, reactions to the energy problem are bound to be different in Germany, France, Britain, and Italy.

Countries are often treated as single, homogeneous units, lumping together their diverse features. This approach disregards the mix of strategies of the social actors within each country. Still, it is adequate to provide an initial overview of the energy situation and trends of the four major Western European countries, showing how wide the divergencies are throughout Europe.

Great Britain will have no energy problems for the next few decades. Its abundant coal, oil, and gas resources up until the end of this century might even give it an opportunity for an industrial renaissance. Whether Britain will be able to seize this opportunity is its main challenge. There could be a rebirth of initiative that will infuse new life into industry, promote competitive activities, and move the economy forward — provided the country regains its will to work and there is wide enough support for a dynamic blueprint for society. Alternatively, less constrained than it is now by budgetary and balance of payments problems, Britain might be even more attracted to a life-style that rejects the pressures of growth and competition. There is also a danger that Britain's energy advantages may tempt it to revert to an isolationist policy, weakening its international alliances, particularly with the rest of the European Community (EEC).

While Germany's energy position is not as advantageous as that of Great Britain, the social and political factors that will determine its ability to resolve future problems seem much more favorable. In addition to its existing industrial potential, Germany's capacity to adjust to the new constraints of international competition demonstrates a basic agreement among the social classes and a commitment to accept the sacrifices needed to preserve the strength of the Deutschmark. There are signs, too, that the country's traditional sense of discipline, together with the general vigor of the economy, is already resulting in relatively rapid exploitation of the potential for energy conservation.

Germany, however, does have its Achilles' heel, for the options on which its energy policy rests are not secure. The pronounced slowdown in nuclear development has closed one road to reduced dependence on imported oil. Progress in the production and use of coal will take time and may meet environmental opposition. Finally, the country continually needs a large trade surplus in order to pay for oil imports. Moreover, as the drive for national reconstruction following the defeat of 1945 fades into history, it is not certain that the country will be able to retain its industrial advantages, which demand a continuing national effort. The declining size and increasing age of the population could upset the general socioeconomic balance and might cause a reduction in the country's dynamic growth.

Italy's future is even less predictable. Difficulties on the energy front probably will produce further dissension among political and social forces and sharpen conflicts over sharing the burden of austerity. The problem is exacerbated by a constitutionally weak government, which cannot seem to frame and, above all, implement a coherent energy policy, so the possibility of future shortages cannot be ruled out. Finally, the expected increase in unemployment, indirectly caused by the soaring cost of energy, may accentuate the trend toward social disorder.

This somewhat gloomy picture, however, fails to reveal Italy's more positive features. The fact that the establishment carries little weight no doubt has the advantage of affording considerable scope for adaptation and initiative. And entrepreneurial ability and ambition certainly are alive in many sectors of the Italian community, as shown by its export performance at the height of the oil crisis. Last, it should not be forgotten that the structure of the production system — with an underground sector of perhaps 20 percent of GNP — tends to mitigate Italy's weak energy situation.

France is in a rather poor supply position from a natural resource standpoint. This probably accounts both for the long-standing existence of a formal energy policy — to be expected in a country where intervention by the central government is traditional — and for the relatively efficient use of energy. In the face of controversy and opposition, the French government has formulated a particularly clear-cut, determined policy based on the large-scale development of nuclear energy. While a significant majority of the population appears to accept (or to be resigned to) the nuclear alternative, political and social temperatures could well shoot up again as more areas of the country are affected by the program. In the event of another oil crisis, the effects may be magnified by the traditional difficulty the French have in agreeing on rules for sharing the burden of sacrifices. Further potential complicating factors are the lack of social consensus and the weakness of French industry, which limit the possibility to adjust to a changing international division of labor and meet the increasing cost of energy supplies by better export performance.

As this quick look at the four major European powers shows, the situations and constraints determining the outlook in each country vary widely. Thus, it is unlikely that within the nations of Western Europe, the social and political consequences of the energy problem will be similar. Yet all four are industrial countries with democratic regimes and share many basic features: similar political and social forces, with central governments playing a conspicuous economic role; major manufacturing and service industries; trade unions influenced by similar political and social options; and individuals responding to the now familiar incentive of the consumer society. In the future, will the various actors on the European energy scene cooperate or be pitted in conflict?

The energy system must be regarded both as an independent system and as a subset of the larger sociopolitical system. The government, for example, does not exercise its authority in the realm of energy alone, and energy consumption is far from being the largest item of private expenditure. In this way, the energy subsystem meshes with other parts of the social system, and these links involve all sorts of tradeoffs. Cases in point are the relationship of energy purchases to price changes and the overall struggle against inflation or the response of trade unions to energy cost increases, as compared to other demands. In sum, the comparability of the different political and social systems within Europe allows us to identify common aspects of problem-solving, even while noting the key differences, and to anticipate a certain common destiny for Western European nations up to the year 2000.

Recent Trends

Between 1953, when energy problems due to the war and its aftermath had been virtually solved, and 1973, when the price of oil shot up dramatically and supplies were threatened, postwar Europe based its generally sustained economic growth on ever cheaper ample energy resources (see Tables 8.1 and 8.2). Technological innovation, the spread of mass production, higher incomes, wider social security coverage — all were part of a process that helped to create a consumer society, contributing to a homogeneous social structure and laying the foundations for political stability. The considerably increased share of national income compulsorily redistributed through various mechanisms has not brought about an equal degree of increased government intervention in every country. While marked in France and in Great Britain (in spite of attempts to "liberalize"), there is less intervention in Italy and particularly in Germany, even though the

Table 8.1 Economic Growth in OECD and EEC
(Annual growth rate of gross domestic product)

	OECD	EEC (nine countries)
1953–73	4.5%	4.6%
1974–80	2.5	2.2

Source: OECD, *Economic Outlook,* July 1980.

Table 8.2 The Down and Up of Oil Prices
Arabian light oil in constant (1980) dollars per barrel

1953	$ 7.00
1960	$ 5.00
1965	$ 4.20
1970	$ 3.80
1972	$ 4.80
1980	$31.00

Source: French Institute of Petroleum.

public sector there does play a relatively more important role in the economy; investment efforts by the *Lander* (states) are extensive and help to stimulate German industry.

This tendency to harmonize economic and social affairs is contradicted to some extent by the revolutionary changes evident everywhere in lifestyles, interpersonal relations, and in the relationship with such fundamental institutions as government, church and family. The consumer society is a permissive society. Materialism, the accumulation of wealth, scientific and technological advances, the worldwide availability of mass-produced goods, the power of the modern media — all contribute to the liberation of the individual, though not without imposing, in exchange, other constraints. This is not to say that everyone has shared in this material progress. On the contrary, the period 1953–73 was remarkable for the persistence of poverty in some areas and the exclusion of certain groups from the society.

In examining the identifiable and foreseeable effects of the 1973 oil shock on living conditions and the energy situation, it is important to avoid two errors. The first is to regard the energy problem as solely responsible for all social and political change in the last thirty years. For, by the end of the 1960s, it appeared that the political, economic, and financial world order set up in 1945 had begun to deteriorate, while the mass production and consumption society was already being challenged.

The oil problem should not be isolated artificially from the more general context in which it exists, nor should it be used as a scapegoat to avoid other problems. In the middle of the 1960s, the United States, the dominant economic and political world power, refused to reconcile its war effort in Vietnam with its responsibilities to achieve economic and financial balance internally and internationally. In the absence of Western solidarity, the international monetary system set up at Bretton Woods at the end of World War II gradually disintegrated. Some parts of the Third World seized this opportunity to take action on their own, adding this failure of the system to their frequently voiced criticisms of an inequitable world order. The formation of an Arab bloc, together with a cartel controlling the vast majority of oil sales on the international market, was, as we all know, destined to make matters worse.

Over the coming years, the oil question will continue to be closely linked with the vagaries of the international monetary system and with the positions adopted by part of the Third World. The social and political impact of the new energy situation cannot be assessed unless this political and financial context is taken into account. But if energy proves to be lastingly hard to come by, it will be more difficult to solve such other problems as reconstructing an international monetary order, securing full employment, and achieving a social consensus.

The opposition voiced in the 1960s against the mass production and consumption society has been further fueled by the energy crisis. By showing how dependent and vulnerable our housing arrangements, transportation systems, and life-styles are, the energy crisis provided additional ammunition to those who advocate a less affluent society, based on technologies that are less wasteful, less polluting, and less centralized. The rejection of nuclear power plants, the campaign to promote renewable forms of energy, the limitations placed on the private automobile, the indictment of urban sprawl — these criticisms represent a new awareness of the energy problem as well as an overall comment on modern society. While the two trends can be distinguished, it is by no means easy to assess the motivations and behavioral patterns underlying each.

A second error would be to believe that everything has changed since 1973, or even 1968. On the contrary, the vast majority of people in Western Europe have striven to preserve their way of life, habits, and income levels. In France, private car registrations increased from one million to two million in the ten years from 1968 to 1978. Voters throughout Europe put pressure on politicians to reduce unemployment and replicate the growth rates of an earlier era. Wage demands have not diminished. Industrial and engineering firms have stepped up their search for foreign markets in order to balance the higher costs of oil supplies to the national economy. More generally, heightened inflationary pressures show that each socioeconomic category does its best to pass on to others the burden imposed by the higher cost of energy, an important factor affecting industry as well as the family budget.

Initially, one can observe a trend toward saving energy as a result of its higher cost and government-sponsored information and energy conservation programs. So far, however, elimination of only the most obvious waste, rather than any real change in energy use, has predominated. A report released in May 1981 by the Commission of European Communities, *Investment and Jobs in an Energy-Efficient Society*, states: "The results obtained in Europe in 1974 and 1979 were due to better management and greater attention to energy consumption *without major investments having been made* ... The results obtained are largely the result of reduced heating comfort, which occupants of residential buildings accepted in response to the increase in energy prices, and of reduced activity by industries that are particularly high energy consumers."[1]

In fact, the relative price of energy has not increased greatly. In some cases it has declined over the past few years. Also, energy costs often account for only a small share of business or family expenditures. It would take a sudden, dramatic disruption of supplies and/or a very sharp price rise to alter behavior patterns or lead to any fundamental change in lifestyles, such as a major shift to public transportation or relocation. And it would generally take time to replace or renew housing stock, machinery, and vehicles.

Changes in Behavior Patterns and Strategies

The Inadequacy of Our Economic Knowledge

An analysis of the political and social consequences of a future economic situation requires a fairly certain knowledge of its components. While Robert Stobaugh's worst-case scenario logically leads to certain conclusions, the other two possible scenarios — the Upper Bound (2 percent annual growth in availability for the industrial nations as a group) and the Lower Bound (zero growth) — allow plenty of latitude. On the economic side, although price, income, and substitution elasticities have been evaluated in a wide variety of ways, we do not know the long-term possibilities for uncoupling economic growth from energy consumption. But if we can generate sufficient changes in psychological and political attitudes, they appear to be considerable. One recent study for the EEC suggested a possible energy savings of from 20 to 35 percent for transportation, 15 to 35 percent for industry and agriculture, and in the domestic and tertiary sectors, up to 50 percent.[2] Others, however, do not foresee such extensive energy gains in efficiency of use. In addition, assumptions about demand trends in the Third World are necessarily uncertain: A low forecast for those countries greatly reduces the threat of shortages. The Lower Bound scenario, used as one basis for our analysis, assumes that the imbalance between an inadequate energy supply and steadily increasing demand will considerably reduce the economic growth rate and therefore reduce incomes and employment. While this deterioration clearly would have a major impact on the positions of the various social actors, what kind of impact will occur?

In our view, the range of possibilities is wide, at least in a time frame of more than five years. The results will depend a great deal on the progress made in improving energy efficiency and on the relative place of hydrocarbons, both of which, in turn, are a function of technological advances and adjustments in certain life-styles. If enough effort is made through massive investment to promote the efficient use of energy and the development of renewable energy, it would not be overly optimistic to foresee continued economic growth over the longer term, and hence an easing of the pressures.

Here the pace of price increases has a very marked effect. If events take a dramatic turn and sharp, sudden increases occur, then the source of economic growth will be broken, and social and political difficulties, as well as adjustment problems, are likely to result. On the other hand, as pointed out in the report drawn up by the "Group of Wise Men" for the Commission of European Countries:

> Increases in resource costs, especially if progressive, can stimulate a dynamic process — through closer attention to rational decision-making and the introduction of new processes and techniques which re-establish the initial rate of economic growth. Some econometric studies (for example, Hudson and Jorgenson) have suggested that although there might be some loss of production if the price of energy doubles in real terms by the year 2000, it may not be more than the equivalent of one to two years' economic growth provided that the energy price increases are continuous and foreseeable. The loss in production might even be less if stronger dynamic feedbacks are assumed.[3]

The second report to the commission, *Investment and Jobs in an Energy-Efficient Society,* also insists on the importance of economic growth and the need to avoid the pitfalls of short-term policies through a long-term strategy of efficient energy use:

> The investments in an efficient use of energy are particularly welcome in the present circumstances and are predictable in the short-to-medium term, since they reduce pressure on prices, foster activity, make a lasting improvement on the economic situation and eliminate restrictions against a balanced growth. To submit these investments to the restrictions of short-term policy is to help throw the economy into a vicious and depressing circle. For lack of the investments, the relationship between economic growth and consumption of energy will remain strong: It will be necessary to slow down this growth in order to protect monetary stability (rise in prices and deficit in the balance of accounts); investments will be discouraged even more, and we will enter the "vicious circle."[4]

Dynamic adaptation ultimately depends more on a combination of favorable sociopolitical factors than on technological progress itself. There is a considerable gap between passive reaction to an event and the swift and decisive pursuit of a new order through change.

The most useful way of exploring Europe's social and political future in relation to the energy problem is to investigate changes in the relationships among the seven major actors who share the stage: the national government; local government units; energy-consuming industries; producers of new forms of energy, multinational oil companies; producers and distributors of electricity, natural gas, and coal; and individual consumers or organizations representing them.

Development of the Actors' Roles

How will these actors behave, individually and in relation to each other, in dealing with energy? The roles were cast at a time when there were no major energy problems. This was the premise for future action.

The national government began by assuming wider responsibilities in performing its more general functions: to ensure law and order, safeguard national independence, and to provide a solid framework for economic development. The electorate urged such action. Energy was important enough, so basic is it to the entire economy, to warrant such an approach. For the first time, in 1973 Germany adopted a formal energy policy. Public intervention was already a tradition in France, Britain, and even in Italy. In other words, from that point on — and the international political context was a major contributing factor — governments would play a far more active role than before in ensuring a sufficient supply of energy from both domestic and foreign sources. As a result, many other areas of the economy and of the population's day-to-day existence were and would continue to be affected, such as the development of industrial structures, public utility charges, driving habits, and the choice of heating systems. The link with the government's industrial policy, aimed at guiding and supporting the allocation of productive capacity, was to be particularly marked in France. In Germany, the structure of electricity rates charged to households now became the subject of joint action by the federal government and the *Länder*. Even if it sometimes conflicted with the objective of rational energy use, the obligation imposed on the government to protect the environment was not relaxed, for two reasons. First, expansion of the nuclear industry and the renewed use of coal heightened the importance of the environmental issue. Second, the upcoming generation was much more sensitive to the problem than its predecessors.

What impact will the ideology of the party in power have on all these issues? It would seem to be small in Great Britain and Germany, although the conservative parties of these countries are probably more determined than the Socialists to emphasize nuclear power. The difference was likely to be greater in France before the May 1980 elections. The Socialist party officially favored a nuclear moratorium. The CFDT (Confédération Française Démocratique du Travail) had outlined an alternative program that relied heavily on energy conservation, particularly in homes. On the other hand, the policy advocated by the French Communist party was based on nuclear power generation and the extensive use of coal. A few months after the elections, the French government decided to launch five new nuclear plants in the next two years, thus reducing the preceding program by four plants — a good example of political compromise.

In Britain, public opinion generally favors or is indifferent to nuclear power. A survey carried out in December 1979 indicated that 51 percent of those questioned supported the construction of nuclear power plants (although the figure was 65 percent in May 1977). In another poll, only 5 percent of the respondents considered the nuclear problem important.

The situation in Germany has been characterized by uncertainties and changes in opinion during the past few years. A study completed in March 1980 found that 36 percent of those questioned favored the development of nuclear energy, 37 percent opposed the construction of new installations, and 15 percent wanted the existing plants to be shut down. In France, twenty public opinion polls were taken between 1974 and 1980 on the acceptance of nuclear energy. From 1974 to 1977, there was a distinct rejection of nuclear power, reflected in the decline of those supporting it from 76 percent to 37 percent. This trend was then reversed, and in 1980, 56 percent of those questioned supported nuclear power.[5]

On a less general, less political level, local government also will be involved in this broadening of public responsibilities. The change is unlikely to be very great in Germany, where the *Länder* already have a great deal to do with electricity generation and distribution and with district heating. The federal government allows each region to make its energy choices according to its own specific situation, although recommending that the use of coal, heat recovery, and district heating be increased. It is likely that the initiatives of the *Länder* in these areas will involve both individuals and businesses.

In France, on the other hand, the role of the local authorities in energy is currently very limited, but new legislation will make it easier for them to take action, and there is a good chance that they often will; in local politics, energy issues are proving to be an effective platform for political candidates. The general trend toward decentralization (however tentative, at least under the former political majority, but likely to be stronger now that the Socialist party has come into power) can provide these representatives with some room for action. The Southwestern Regional Authority has already taken the initiative of drawing up a plan that takes account of the region's particular situation.

In some regions — for example, where biomass can be readily used — such initiatives may have a certain impact. Similarly, progressive towns will try to introduce district heating, even if it is not easy to satisfy all the economic conditions for profitability. Considerable diversification of energy use will follow. Overall, however, no very great differentiation between the various regions of France is likely to result, and its energy system will remain centralized. It is unlikely, for example, that the political forces in a given region will combine to reject nuclear development when it is accepted elsewhere. However, the regional aspect of the energy problem will be a major subject of discussion and action among local representatives and citizens, with the result that both groups will assume wider responsibilities.

On all these points, initiatives may be stimulated by the new generation. There is no doubt that young people expect a great deal from the new technologies and methods of organization that have been discussed in numerous reports and popularized by the media.

It has yet to be determined whether these youthful aspirations will materialize through some assumption of greater political responsibility. Thus young people, who have a passive outlook and are attracted by the material advantages of modern life, may well be disinclined to participate in local ventures. Their passivity, caused by aversion or a sense of alienation, may exist along with a desire for innovation, particularly in European countries — France or Italy, perhaps — where the gap between words and actions is considerable and is doubtless becoming wider.

Although young people demand a great deal of the business firm as a supplier of income, security, and personal fulfillment, they no longer agree with its premise. Work has been desanctified, and that part of their existence remains isolated from the rest of their lives; the profit motive is often misunderstood, and the manner of distributing profits is a source of bitter conflict. Many young people regard profit as a monopolizing of wealth without understanding that it is a device for both measuring and promoting the health of the firm. The growth of absenteeism in the majority of countries in recent years reflects this attitude. As a consequence, an increasing number of people are becoming individual entrepreneurs, leading to the expansion of an underground economy on the fringes of the system that is subject to social and institutional regulation.

Paradoxically, this phenomenon may engender such flexibility of behavior that, at the level of the individual, society will cope better with the disadvantages of the energy crisis. The occupations we are concerned with are less dependent on plant and equipment, because inventiveness is often considerable and the worker is essentially self-reliant. Nonetheless, it is the larger companies that keep the economy going. Any undermining of their human capital would be an unfortunate development. After all, it is the people running such businesses who have to make the decisions to change and adjust in order to use energy more efficiently, design equipment able to do so, promote renewable forms of energy, and adapt the production process to the new requirements of international competition.

Changes in the distribution of skills do not favor such reform. In most companies — at least in those of a certain size — the gap between a growing upper level of specialists and a large pool of unskilled workers appears to be widening. The first group increasingly aspires to participate in strategic decision-making. The second tends to regard the company only as a way station and a place to earn a living. This is more characteristic of large companies in the service sector than of industrial firms. The growth of data processing, resulting in more sophisticated means of communication and a standardizing of many work assignments, will strongly affect this redistribution of skills.

It is possible that top management may not be best situated to promote the government's economic policy and to motivate its employees. Given generally favorable economic conditions, senior managers will certainly give in to the temptation to turn inward, taking no risks and resorting to protectionism. So far the Western world, as a whole, has largely resisted any such tendency, as demonstrated by the steady increase in international trade, even in the face of the marked slowing of world economic growth. Recently, however, Europe and the United States have taken various protectionist measures, involving restrictions on such imports as iron and steel products, textile fibers, and automobiles.

It may well be that a few additional problems, especially in the context of the Lower Bound energy scenario, will induce a large part of the industrial sector, supported by the unions, to demand and obtain protectionist measures. All projections in this area show that under certain protectionist assumptions, world commerce would stop growing, which would be quite a change from the 8.5 percent annual growth rate between 1963 and 1973. Yet when its own existence is threatened, no individual company or occupational sector is likely to take into account the collective interest of the nation.

We are faced with a fundamental choice between two courses of action. Either most business leaders in Europe, having persuaded wage earners and trade unions of their mission, will use energy problems as a means to reform and revitalize an economy that has been losing momentum for several years, because of higher energy costs as well as slower gains in productivity, slackness of technological progress, and errors of economic policy. Or they will avoid the challenge, will not invest enough in new ventures, and will become anemic, confined to protected sectors. It is not yet possible to determine what course Europe will choose for the 1980s — the dynamics of innovation, even with its disruptive effects, or the stasis of protectionism, which will slowly but surely make the situation worse.

Companies responsible for producing and distributing energy at the national level usually belong to the public or mixed economic sector. This is especially true of France, Britain, and Italy. These production and distribution companies will probably find themselves drawn in two opposite directions.

On the one hand, due to the severity of the energy problem, their role in building the nation's future clearly is gaining in importance. In the realm of planning and financing new plans, as well as facility breakdowns, such as the massive failure of the French electric system in October 1978, the public keeps a watchful eye on the large energy companies and requires them to account for their actions. In Britain, the National Coal Board feels a sense of renewed power and responsibility. In France, the chairman of the Electricity Authority is as familiar a figure and as much in demand as a cabinet minister. This heavier burden of responsibility on the part of public managers seems mainly to derive from the confidence inspired by their competence and interest in the general welfare of society. Yet certain difficulties nevertheless will arise. For instance, the power of these agencies is sometimes regarded as feudal, oppressive, and bureaucratic. Moreover, they are among the most obvious targets of criticism and opposition when major energy policy decisions must be made. Violent demonstrations against the French Electricity Authority are frequent, and its chairman has already been the object of assassination attempts.

On the other hand, a movement to make these national bodies less monolithic is taking shape and will gather momentum in the coming years, especially in France and in Britain, thus affording greater latitude to regional and local government authorities and to personal initiatives. If this goal is to be achieved, the legal framework will have to be modified, making standards more flexible and adjusting the conditions for the purchase of energy. In France, for example, recent legislation makes it easier to harness the power of privately owned waterfalls, and decentralized hydroelectric projects are beginning to see the light of day.

The development of diverse sources of renewable energy — biomass, hydroelectric power, direct solar radiation, geothermal energy — obviously will depend on initiative of the kind not usually taken by giant, single-purpose organizations relying on proven technologies. In addition, the return on such investments may be higher than that imposed by the sponsoring agency. The danger, however, is that the major national energy production agencies — mission oriented, wedded to their own technologies, used to thinking in and dealing with huge projects — may stubbornly resist any diversification of the energy system.

Such opposition might be tempered if those traditionally responsible for producing and distributing energy were made direct partners in the promotion of alternative energy technologies and conservation. In Italy, for example, the Nuclear Energy Commission also deals with renewable forms of energy. In France, the gas and electricity authorities are making special efforts to develop equipment for improving energy efficiency. One cannot overemphasize the significance of this change in approach by organizations, which, until now, have aimed at promoting energy consumption and maximizing sales.

Although the basis for its point of view is unclear, one segment of public opinion blames the energy crisis on the multinational oil companies. Sometimes the multinationals are taken to task for a generally shortsighted attitude, too focused on keeping production costs as low as possible. Alternatively, they are accused of a certain complicity with the owners of energy resources in fabricating a crisis. Or they are denounced for the huge profits that the repeated increases in oil prices supposedly allow.

In short, the international oil companies suffer from a bad press in Europe, except in Germany and the Netherlands. European governments take a rather harsh line toward the multinationals and invariably treat any questions that especially affect the companies — taxation, for example — as politically sensitive issues. Since 1973, all the European governments have strengthened their national oil policies and, more broadly speaking, their general energy policies by making themselves less dependent on the international oil companies. Even Germany, which previously had taken little action, is now moving in this direction.

The multinationals remain a force to be reckoned with on the energy scene, however. They control most of the system linking the oilfields with the retail pump and bear the major burden of finding and developing new reserves, an effort vital to the West. Thus, objectively, the oil companies still will be in a strong position, even if their image and sphere of influence should be permanently weakened.

The multinationals can, and already do, contribute to the development of other forms of energy, such as coal and solar electricity, as well as assisting firms and individuals to use energy more efficiently. Exxon has substantial interests in coal, shale oil, uranium, and synthetic fuels. British Petroleum was involved in producing 40 million tons of coal in 1981, and the goal for Royal Dutch Shell is 28 million tons in 1983. This diversification may well improve the position of the multinationals in the public eye and help them regain their power. However, it is not likely that Europe will develop a position similar to that of the United States, where there is considerable criticism of oil company acquisitions of coal mines, uranium reserves, and solar companies.

It is more likely that in Europe the multinationals will participate in

Table 8.3 Fixed Capital Formation (average annual change)

	1968–73	1973–80
West Germany	6.2%	1.8%
France	8.7	1.1
Italy	3.2	0.5
United Kingdom	2.4	−1.4

Source: OECD, *Main Economic Indicators.*

building up a new energy sector. In addition to the large industrial companies that are gaining a foothold in the area, this new sector will contain young, innovative enterprises and organizations involved in promoting new forms of energy, troubleshooting, and supporting research and development. This emerging force is supported, not only by growing government subsidies, but also by all of the groups that reject continued total dependence on oil and others who oppose any extensive use of nuclear power.

In developing new lines of production to replace existing ones as well as in diversifying, the industrial and service sectors require considerable capital. Over and above these additional demands for investment funds, the temporary lag in capital formation experienced in most European countries since 1974 has delayed routine decisions to replace and modernize plants and equipment (see Table 8.3). Much of the new plant ultimately installed will be designed to use energy more efficiently and introduce new means of energy production.

Some governments, as in Germany and France, are beginning to recognize this interrelationship between investment and energy. In fact, the Energy Commission of the Eighth French Plan has stated explicitly: "An energy policy is an investment policy."[6] The German government has launched a multibillion-dollar program to stimulate investment in energy efficiency.

The report to the commission of the EEC referred to earlier estimates that the level of investment required to increase energy efficiency will fall somewhere between about $90 and $160 billion in the years 1981–90. In order to substitute other traditional sources of energy for oil, an estimated $30 to $90 billion will be needed. Another $380 billion must be added to these amounts to finance the necessary increase in energy supply.

Increased productive investment in the future resulting from efforts in these areas will require additional savings and changes in the allocation of savings. While the drop in new housing requirements in most countries, together with the decline in population, affords some room for maneuvering, this will probably not be enough. Thus, we should expect slower growth in total consumption, which obviously will have an effect on specific goods and services, such as consumer durables, health care, and tourism. Similarly, soaring energy prices may cause a reduction in consumption, accompanied by resistance on the part of individuals to their resulting lower standard of living. For, as already stated, it appears clear that the desire to preserve the characteristics and material advantages of the consumer society will remain strong for many years. Nevertheless, one should not ignore the possibility of significant changes in patterns of consumption, such as an increased emphasis on leisure activities and communication conveniences.

This objective may well be attained under the optimistic scenario, which includes two strong elements of flexibility: a reduction in the amount of energy needed for each unit of GNP and the possibility of developing alternative forms of energy. Under the pessimistic scenario, however, it does not seem feasible. The worst-case scenario, as we shall see, raises other problems.

What mechanisms will national communities employ in relocating their resources? Clearly, this is a political issue touching both on the relations between individuals and their governments and on the perception of social justice by the different socioeconomic groups. The danger is that without sufficient consensus and financial expertise, some countries may have to confront increased social tensions or resort to another dangerous dose of inflation. In any event, governments will be held responsible, and once a certain threshold has been reached, one cannot exclude the possibility of subversive action or other forms of economic and social anarchy.[7]

Two factors, however, justify some cautious optimism. First, the market for many consumer durables may be approaching saturation.[8] Second, since individuals will have to pay to improve household energy efficiency or to convert their homes and offices to alternative energy sources, they will be able to judge for themselves whether these investments are worthwhile. It is essential, therefore, that individuals have access to information and assistance concerning energy improvements, from both technical and financial standpoints.

It is the behavior of the network formed by these individual producers, consumers, and citizens that is hardest to predict, since it is really a composite, whose varying tendencies are not easy to assess. In France, considerable research over the past ten years has been devoted to examining and plotting the course of leading sociocultural trends in order to identify new social demands, changes in customs and ideas, personal relationships, and attitudes toward authority as well as to reveal the points of social and political friction. A desire to forecast future development, of course, underlies all such surveys. The gradual emergence of an energy problem has added

Table 8.4   Value Priorities in Western Europe

	Age 65 and over	Ages 25–29
"Materialist"	Between 37% (U.K.) and 62% (West Germany)	Between 18% (Belgium) and 27% (U.K.)
"Post-Materialist"	Between 6% (Switzerland) and 1% (Germany)	Between 23% (Belgium) and 11% (U.K.)

Source: OECD *Interfutures,* p. 104.

another element to the many factors already at work. As will be shown later, this new element seems to engender contradictory responses by stiffening the opposition of some to the consumer society while reinforcing the commitment of others who fear its demise.

The observations made by various researchers over a decade are largely in agreement. Bernard Cathelat has worked out a detailed typology of group psychological trends and measured changes in behavioral and verbal responses to the basic concerns of existence, and he has interpreted these trends in terms of life-style.[9] His findings indicate a gradual return in France to an "inner-directed" attitude — hedonistic, calculating, and cautious — as opposed to the "outer-directed" attitude that prevailed between 1950 and 1970. Such a trend, on the whole, would seem to favor slowing the pace of the consumer society and using energy more efficiently. But it also risks stemming the necessary flow of investment and tilting toward protectionism. The trend does not appear likely to result in sweeping changes in general living conditions, which the French do not seem to want in any event.

Other research points to strong social currents that, although not yet affecting the majority of the people, do affect some 50 percent of the teenage population. These currents appear to have gained steadily in momentum over the past ten years or more (see Table 8.4).[10] As will be shown, the need to conserve energy and the danger of energy dependence are likely to reinforce these movements, four of which are highly relevant: a desire for self-expression; the search for roots in a social unit, such as a neighborhood, community, workshop, or linguistic or cultural minority; sensualism, which emphasizes emotions, feelings, and intuition, while questioning rational thought, judgments, and intellect; and last, the rejection of established authority.

Such attitudes, reflected in new types of personal and social relationships, will use the energy crisis as a pretext for preferring leisure to work and pursuing emotional or sensory gratification (for example, through various kinds of entertainment). In addition, these attitudes favor producing goods by oneself or within a small community, where individual initiative is still possible rather than subject to constraints imposed by some large entity relying on hard technology. Communication on all levels is paramount, and people will demand communication services, environmental protection, and welfare benefits (health and education), rather than material goods. This outlook places much greater emphasis on the environmental aspects of energy development.

It would be surprising if such a movement could eventually sweep the majority along with it to change the face of society. More likely, the movement will encounter a still more basic faith in the work ethic, mass production, the maximizing of income, and the earning of a return on investment. The notion will gradually emerge, whether explicitly or implicitly, that we can reconcile increased automation and computerization, the conservation of energy, and shorter work hours with increased opportunities for consumption and leisure activities.

This conclusion accords with the apparently contradictory results of a number of public opinion polls, particularly where the respondents were young.[11] A high percentage of those questioned were concerned, first and foremost, with economic security. In work, the main concern was self-interest. Economic growth and an annual increase in real income are political and social objectives that people will not renounce easily; for example, not even to relieve the continuing anxiety aroused by the planned reliance on nuclear power. Finally — and this appears fundamental — a clear majority believed that technological progress will permit mankind to develop alternative forms of energy (including nuclear energy) as well as to overcome current obstacles to economic growth.

The Interface of Strategies: Conflict and Cooperation

Thus the main actors in the new crisis have begun to adapt their individual functions and interests to the new conditions. Nevertheless, in a complex society that is undergoing change in one of its basic building blocks — the availability of energy — the behavior and strategy of one actor are not independent of those of others. In fact, just the opposite occurs: The interrelationships, conflicts, and cooperation among strategies result in their modification and the creation of new interaction.

This process appears most clearly from the point of view of energy policy. As Leon Lindberg points out in *The Energy Syndrome,* modern energy policy shares the essential characteristics of any complex political problem.[12] In the first place, because of the increasing number and complexity of interdependencies among the different sectors of modern society and the many value systems operating in each sector, policy decisions impinge on an even wider spectrum of values. Thus it becomes increasingly difficult to achieve a consensus among the interested actors. In this way, the range of the impact of an energy policy tends to widen, and more transactions and compromises are required. Price-setting for petroleum products is a good example. In the majority of the European countries, this process has evolved out of a subtle, problematic, and changing compromise among the government (concerned with financial policy and economic balance), oil companies, electricity producers, industrial companies, and consumer representatives.

Second, the conditions under which energy policy is implemented have become increasingly uncertain. In theory, this ought to cause the actors in a given country *either* to take greater risks *or* to provide for greater margins of safety, depending upon their respective positions and attitudes. In fact, the trend probably will be toward taking greater risks for a number of reasons. One is the difficulty of harmonizing widely divergent interests. Another is the virtual impossibility of imposing authoritarian solutions, except during a complete breakdown or emergency. A third is the skepticism of many toward government warnings and orders.

The Upper Bound scenario does not seem to create any inordinate obstructions to adjustment. The inadequacy of supply can be overcome if a substantial effort is undertaken to rationalize the use of energy and to develop new energy sources. Although the various interests at stake do not agree, ways to bring them together can be found, and the economy may even benefit from these dynamic forces. The main pressures would probably arise from the social effects of higher prices and the reallocation of production resources required by the restructuring of national economies.

The crisis of the worst-case scenario paradoxically seems to afford greater opportunities to solve the energy problem. This perilous situation clearly calls for a national mobilization and the adoption, at least initially, of a system like that of a wartime economy. The government would be justified in exercising special powers. Scarce resources would be allocated in an authoritarian manner. Yet a public consensus would probably arise spontaneously, and exceptional measures taken to obtain basic supplies for the public welfare would not provoke much resistance.

In the circumstances described by the Lower Bound scenario — perhaps the most likely one — conflicting strategies would probably create the greatest tensions. Every actor in the energy crisis finds reasons to shift the burden to others, but the constraints are so extreme that, on the whole, the effort does not seem worth it or can only proceed in fits and starts. The major areas of conflict or cooperation among the strategies of our actors include:

- the level and manner of setting utility and fuel oil rates to industrial and residential customers, as well as the prices of transportation fuels;
- the distribution of the increased production costs, caused by rising energy prices, between producers and consumers and among different income groups;
- the appropriate designation and exercise of authority to produce and distribute energy;
- the choice of producers to develop different energy sources and the pace of such development;
- the selection of methods to finance the development of traditional energy sources, new forms of energy, and improvements in energy efficiency;
- the identification of desirable ways to improve energy efficiency, whether through standards, regulations, incentives, information, or exhortation.

Our analysis shows five key points where the energy actors meet either in conflict or cooperation: (1) responding to continually rising prices; (2) reducing energy dependence through conservation, especially through government intervention to lower consumption; (3) reducing energy dependence by increasing domestic production and diversifying supplies from outside (through government intervention); (4) managing an officially declared shortage following a crisis; (5) restructuring the domestic economy.

Price Increases

Insofar as governments have the power to control prices — and most of them do, at the very least, through tax policies — they will vacillate between a purely laissez-faire policy and a more interventionist approach, designed to mitigate the social effects of higher prices. The level of energy prices compared to prices of other goods and services directly influences household consumption and the competitive position of major industrial sectors.

A laissez-faire policy leaves the task of distributing the higher costs to the market and vigorously promotes the efficient use of energy. Companies that are big energy users run the risk of a loss in competitiveness if the prices their competitors have to pay for energy have not risen at the same rate, and they might even see their products forced out of the market if other, more competitive substitutes exist. Such a hands-off policy is particularly hard on low-income groups, as they are least able to switch quickly to lower-cost substitutes. They are, in effect, prisoners of a life-style wedded to the availability of cheap energy through the location of housing, dependence on electricity, and use of the automobile. For these reasons, a policy of adjustment aims at softening the shocks, thereby minimizing the inflationary pressures in the short term, affording industries and individuals time to adapt and reducing social opposition.

It is more difficult to choose between the two policies (or more realistically, to combine some features of each to evolve a third, intermediate policy) in the Lower Bound scenario. By contrast, in the Upper Bound case, it is fairly easy to envision a policy of progressively more realistic prices accompanied by compensating social benefits. In the no-growth scenario (where energy resources do not increase), a policy of realistic prices is even more essential.

But prices are likely to rise sharply and quickly, imposing heavy costs that are difficult to offset satisfactorily. If the effects of these price hikes are not too severe, one could presume that the European governments would try to mediate among the conflicting pressures by including or accepting a progressive increase in prices while granting the lowest income groups compensation in order to redistribute some of the burden more fairly. Industry would be required to adjust and, if necessary, would be given temporary financial assistance to make an adjustment. By these means, the government could prevent an "artificial" economy from springing up.

The issue is, however, complex, and not only from the point of view of international competition and social justice. On the one hand, the costs of developing the various sources of energy are not the same. On the other hand, it may be to a country's advantage to give preference to a particular energy source over others in order to reduce its dependence. Moreover, anti-inflationary policy poses a significant general constraint. For all of these reasons, the energy source that has the widest, most varied application in Europe — electricity — will not pay for itself in the long run, and current users are in a relatively favored position.

Reducing Energy Dependence through Conservation

The initiative in this area is in the hands of the national government, which can impose a general conservation policy in spite of opposition by industries and individuals. This does not, however, guarantee full compliance for reasons that have been analyzed innumerable times since 1973, such as insufficient information, inertia, inadequate financial resources, reluctance to change life-styles, institutional obstacles.

A major source of conflict may arise between the national governments and the oil-importing multinationals; rightly or wrongly, the companies may be accused of acting against the national interest. Another major possible conflict involves the relations of the national government with consumers or their representatives (local authorities, associations).

The first type of conflict is unlikely to occur or, if it does, will be extremely limited. The multinationals are in a weak position in this respect. Most of them, moreover, are now adopting efficient energy use as a selling point. And the second area of conflict can be avoided if consumers are made aware of the need for energy conservation, hence the importance of information. This is the crucial point: Inadequate energy savings will result, above all, from ignorance, habit, and apathy.

A commonality of interest may help to generate the necessary awareness, first, by reinforcing cooperation among European governments, needed to reach fuller integration. Second, there is a convergence of interest between consumers and alternative energy industries. Finally, consumers and the organizations closest to them (local governments, associations) share common interests, for two reasons: Local governments may be regarded as the champions of local interests in dealings with the central government; obviously a city will be more concerned with maintaining employment in its area than with reducing energy dependence at the national level. And the demands of a conservation policy often can be met through such actions at the local level as district heating, reorganizing transit systems, and developing local energy production systems.

These interrelationships suggest three important phenomena: at the social level, the trend toward centralization resulting from the need for an overall energy conservation policy; the importance of information management and distribution; and the stronger links between citizens and social and political organizations locally, which in the long run may counter the trend toward centralization. In addition, considering the dynamics of the general situation, we could predict that the pressures created in the Lower Bound scenario will encourage, even require, substantial efforts to save energy. But the general context could prove less conducive to conservation. For example, an economy weakened by the lack of energy resources or soaring costs might be unable to invest enough funds to conserve energy; or, again, severe social tensions might challenge government authority.

Increased Energy Supply

Increased domestic production of energy may result indirectly from conservation, but it also may be a goal of national and local government policy. Here, possible conflicts between the national government and the multinationals will occur over the development of domestic production and alternative energy sources, diversification of external sources, and the relationship of the companies to foreign governments, whether of oil-consuming or oil-producing nations.

Conflicts certainly will arise in Europe regarding tradeoffs between reduced energy dependence and protecting the environment. One camp, championing the national welfare, will consist of the national government and producers of conventional forms of energy (coal, nuclear, etc.). The other, concerned with protecting its own interests, will include those local authorities most affected by the new projects. The positions of individual consumers also may diverge over the relative importance of the two questions.[13]

The national government and domestic producers of conventional or alternative energy will continue to cooperate on supply, especially because many such producers belong to the public sector, even in Germany and especially in France, where the managers of national industries and the senior civil servants to whom they report have been trained in the same school and consider themselves members of the same elite. Such links account for the great pressure to develop non-oil-based forms of energy, whether conventional or alternative. Such moves, however, will be hampered by the conflict between the government and the private sector over financing investments, competition for capital in financial markets, or disagreement between energy-producing and energy-consuming industries over the rate of reinvestment of the profits that will develop. Already in France and Italy, the national electricity producers are tapping a considerable share of the available capital resources.

The dynamics of this kind of competition tend to favor centralization, as evidenced, for example, by the Danish experience in developing nuclear power and offshore energy resources. The strong central government intervenes to ensure that the national interest prevails over local concerns and that potentially dangerous technologies are developed safely. The scarcer energy supplies become — as under the Lower Bound scenario — the stronger this centralizing trend.

Managing an Official Shortage

As has already been suggested, a shortage may in fact be far less a source of conflict since it would require dealing with a crisis caused by external forces. Although it would be fairly easy to persuade the various energy actors within each country to form a united front, whether such unity also would develop among the European countries is open to question.

To date, there is little evidence of a coherent European energy policy, although some progress has been made. There are agreed-upon goals, demonstrated especially in 1980 in decisions of the European Community and then at the seven-nation Venice summit, which seek to decouple energy consumption and economic growth, place increasing reliance on nuclear energy and coal, and emphasize the role of price to induce conservation. In addition, the European Community has adopted certain statutory and budgetary measures in the field of energy. For example, in 1978 it committed about $55 million over four years to finance demonstration projects for energy conservation. Of loans totaling about $1.6 billion per year (through the NIC, European Investment Bank, and Coal and Steel Community), $200 million is designated for projects concerning the demand for energy. The statutory measures cover the performance of heat generators and the production of hot water and provide for labeling household appliances with their energy efficiency ratings. Finally, all the members of the EEC have agreed to the same basic obligations concerning the stockpiling of petroleum products and emergency measures.[14]

Nevertheless, selfish and divisive forces are bound to surface because of the diverse interests at stake and conflicting interpretations of the crisis. Governments will vacillate between mandatory measures, such as rationing, with its accompanying red tape and potential for fraud, and an appeal to the community spirit, which may well go unheeded. One way of predicting future behavior is to examine the responses of various countries to restraints imposed in the past. Germany leads in solid consensus and discipline; France, in organization and also in cheating, although limited. Italy is outstanding for its lack of formal organization and its corresponding ability to "muddle through" by resorting to a system of mutual assistance within small groups. Great Britain is characterized by its unpredictable behavior; can it still unite in the face of national peril, as in World War II, or have hedonism, class conflict, and lack of discipline now become the dominant national traits?

In every country, if the shortage drags on, the danger of political and social upheaval will grow; national consensus will deteriorate, and challenges to the legitimacy of the existing government may rise. Hoarding behavior may send real prices (at least those of petroleum products) skyrocketing, in spite of efforts to contain them. There is also bound to be a sharp increase in unemployment.

In *France without Oil*, H. Aujac and J. de Rouville describe the situation during a shortage: "With industrial production slowed by energy restrictions, companies lay off non-essential workers. Moreover, despite extensive aid to businesses by banks, the number of bankruptcies rise sharply. The French are stunned; scarcely a few months ago distinguished economists were seriously debating whether French society had entered the post-industrial era; now the same society has become paralyzed."

Aujac and de Rouville have divided the population into three age groups; the over-60 group, accustomed to adversity; the 30–60 group, which finds it hard to accept and adjust to setbacks; and the under-30 group, which is rediscovering the ideals of 1968. The last group sees "a future far removed from the consumer society, one where the dominant values would no longer be based on Mammon, and where the car, TV, refrigerator and second home would be replaced, among other things, by greater goodwill among people, smoother human relations, a return to the simplicity and restfulness of nature, and rejection of centralized systems of social control, which are viewed as an intolerable impingement on individual freedom."[15]

This point of view is important, since it reflects both opposition to the consumer society, which, as mentioned earlier, existed well before the energy crisis, and the reactions of certain social groups to the energy crisis itself. The likelihood of such a scenario, however, is open to question for every European country. It is doubtful whether the student protests of 1968 can be ascribed to a rejection of the material advantages of the consumer society, since the former rebels have returned to this materialistic life-style. But, above all, if there is a serious energy shortage, many people will be in an extremely difficult position. The possibility of returning to the simplicity and serenity of nature will not exist for most of the population; they may even regard the idea as a bad joke. Unless there is a change in general living conditions over the long term, society is likely to be organized around extensive government intervention aimed at controlling energy prices, allocating resources, compensating for social inequities, and promoting the more efficient use of energy. The basic conditions of energy use will change to meet the long-term objectives, and most measures taken will assume an economic and social structure based on mass production and consumption.

In this event, the national energy producers and distributors would become an instrument of the government rather than a countervailing power, since their freedom of action would be severely limited. Local governments, on the other hand, would enjoy positive relations with the companies and residents within their boundaries, since they would probably be involved in protecting local employment and stimulating innovation and research that could help problems caused by the energy shortage.

Adjusting and Restructuring the Economy

While only three energy actors — the national government, energy-consuming firms, and individuals — are involved in this area, the relationship is an absolutely vital one. Adjusting and restructuring the economy in the developed countries is motivated by industry's need to adapt to the new economic conditions and by the government's determination to promote energy-saving technologies and activities. In Germany, where the movement began as far back as the early 1970s, the process is likely to gain momentum. In France, despite a slow start, restructuring of the economy should progress, since government and big business have gradually reached a basic agreement on the subject.

The main conflict will be between workers, anxious to safeguard the stability of their employment, and management, concerned with maintaining the company's profitability in the face of domestic and foreign competition. The position of the trade unions will also be an important factor in all the European countries. The chances are that some of the unions, whether they like it or not, will adopt the views of government and management. One should not underestimate the seriousness of the conflict, however. That is, even if national unity is achieved, it can hardly be expected to last, once the first shock of a shortage is over. At that point the conflict is apt to take on political overtones, with a revival of working-class and middle-class parties.

On the other hand, if the economy is reorganized to promote the most energy-efficient technologies, this may help lower unemployment, especially if governments support the process by substituting a tax on energy consumption for the value-added tax.

In such a conflict-ridden situation, an easy way out would be for governments to limit the need to restructure the economy by closing the borders to foreign competition through protectionism. Such action might be taken, partly in response to pressure from certain sectors and unions, and also as a result of government initiative in an attempt to improve the social and economic balance.

Plausible Future Events

The choice of the word "plausible" to qualify the social and political effects of the new energy situation is an act of restraint. The links between the energy subsystem and the overall sociopolitical system are complex and uncertain. Nor is the energy problem the only factor of crisis or change in our society. Is it not possible, after all, to consider the energy problem as one aspect of the North-South confrontation and the breakdown in the international monetary system? The resolution of the energy problem cannot be dissociated from efforts to change life-styles in order to minimize the effects of the crisis, nor from the technological advances that will help overcome the problem. Nor should it be forgotten that prospects for resolving it will vary according to the economic growth rate, which, in turn, depends on the availability of energy and, above all, on the joint efforts and determination of people and on the shrewdness of economic policy. Whether all these factors can be combined satisfactorily is impossible to foresee.

Our energy future revolves, in its sociopolitical aspects, around three crucial issues: the soundness of the social contract, the distribution of power, and the adoption of new life-styles. While all three are already subjects of debate, it appears that their strongest impacts will occur successively. We are already embroiled in the first question; the second will evolve in the medium term, when the effect of political decisions becomes clear; and the last belongs to a more distant future because of the time it takes to replace inventories, change behavior, and relocate facilities.

The Soundness of the Social Contract

The increase in energy prices, especially sharp in the worst-case scenario, will mean that needs now regarded as essential, and so far satisfied on reasonably fair terms, can no longer be fulfilled for the lower income groups at all or only by sacrificing other, equally important needs. These needs are constrained as much by social organization (such as the distance between home and place of employment, public transportation, the type of heating facilities in multiple-unit residential buildings) as by people's habits and preferences. Even a brief increase in unemployment following a period of industrial adaptation, necessitated by the energy crisis and the new conditions of international competition, will further accentuate the sense of impending doom and injustice that builds social tensions. The rise in unemployment throughout Europe points to an increase in such tensions (see Table 8.5).

In at least three European countries — Italy, France, Great Britain — a large segment of the population considered the social system to be unjust long before the energy crisis. In these countries the disparity of incomes and living standards remains extreme, and the number of people denied the benefits of the consumer society is large.

Table 8.5 Unemployment (% of workforce)

	1973	January 1981	January 1982
West Germany	1.0	5.0	7.5
France	1.8	7.4	9.0
Italy	4.9	8.7	9.9
Great Britain	2.5	9.3	11.8
EEC	2.5	7.7	9.7

Source: European Community.

In short, while there are undoubtedly "winners" in modern life, there are also "losers." The enduring nature of the energy problem ensures that this opposition between winners and losers will gain center stage. Organized social groups will oppose energy prices based on the marginal cost of new supplies and will seek economic assistance and subsidies, demanding action from government authorities. All these forces may converge toward increasingly centralized decision-making and more government intervention. As the situation worsens, heavy pressure will be brought to bear on the central government to exercise greater control over strategic economic activities and to socialize many aspects of the problem. Each social class will try to shift the burden to others, perhaps forcing the government to adopt wage and price controls, regardless of their doubtful effectiveness. Should some outside event suddenly cause an acute shortage, such a major threat might restore the social consensus. But for this to occur, governments must affirm their authority, and must do so immediately.

This pessimistic vision of the weakening of the social contract might be tempered, however, for various reasons. An external event could eliminate any ideological content in the conflict among social groups and ease the pressure. In communities linked by functional ties — industries, large urban areas — strong cooperation may actually develop to counterbalance generally divisive forces. This will often be a question of leadership, the need for which has diminished during the period of prosperity in the West. Finally, although most social groups will strive to maintain their positions or regain lost ground, an increasing number of individuals may tire of the struggle and cut back their energy spending in order to relieve the family budget.

Toward a New Division of Authority

Since the government is on the opposite side from all the other energy actors, there is no way of knowing which actors will lose authority if that of the government increases — or conversely, which will benefit from a reduction in the power of the government. The pattern will vary among the European countries, since it is precisely in terms of the distribution of political and social forces that these countries differ the most.

In certain areas of economic and social life, no European country presumably will avoid increased central government authority. The major energy options (coal, nuclear, solar, and imported or indigenous oil and gas), various consumption standards for residential heating and certain industrial processes, and fundamental pricing principles are all cases in point. In all the countries, the government and leading domestic producers will develop a close association, since the national government's decision-making role necessarily will increase.

In the longer term, multinationals will also suffer a loss of autonomy. Throughout Europe a "nationalization" of the energy problem may occur, settling many issues bilaterally between European states and oil-producing countries. Furthermore, the profit margins of the major oil groups are under close scrutiny, with governments tending to increase taxes on the income of these companies. This assuages public opinion and helps to redistribute the additional burden of price hikes.

In any event, whether as users of energy, producers of new forms of energy, or as manufacturers of energy-saving materials, industrial companies seem likely to enjoy a strengthening of their inherent authority, because responsibility for decisive improvement rests with them and the seriousness of the situation tends to limit the intensity of employees' demands. The companies have an advantage over the other actors, subject, however, to two conditions: an adequate response by company managers to employees' demands, particularly those of senior employees to participate in strategic decision-making and management; and government action to prevent any significant increase in social injustice, should the energy situation worsen. Thus, the position of the company itself, if bolstered by restraint on the part of the unions, may be strengthened.

Finally, the authority of local governments will most likely increase, particularly in France and Great Britain, and as a result, so will the power of individuals at the local level. This movement may be expected to arise as steps are taken to decentralize energy management by the use of biomass and cogeneration of electricity as well as through developing urban centers, managing public buildings, shaping a public transit policy, and so on.[16] This may provide both a technically sound and humane approach for overcoming the problems and severe constraints of the Lower Bound scenario.

The regional or local community provides, furthermore, an ideal medium for the development of a communal life-style in response to constraints imposed by the energy problem. Private, nonprofit groups will seek to minimize energy expenditures and an overly individualistic life-style through organization and mutual assistance in such areas as cooking, child care, and transportation. It is also at the regional or local level that environmental movements can have the greatest impact and where they converge with and complement the development of conservation associations and projects.

The direction and scale of such transfers of authority remain uncertain, since they depend on major psychological, social, and political attitudes largely transcending the realm of energy. In short, there are three fundamental choices: shaping society through market forces or government intervention; managing the political and administrative systems primarily at the national level or by regional and local authorities; and developing a greater community spirit or persisting with individualism.

Market Forces versus Government Intervention

Political parties are firmly divided on this issue. The Communists and the Socialists with Marxist leanings strongly favor government intervention, and they are seconded by the Labour party in Britain, which advocates nationalization. The Social and Christian Democrats, on the whole, prefer reliance on free market forces but allow for certain adjustments on social issues. The Conservatives in Great Britain and France seek to reestablish the free interaction of supply and demand while following a fairly active industrial policy.

Future choices, therefore, will depend a great deal on the political stripe of the parties in power. Nearly all have learned lessons from the economic experiments of recent years. For instance, in Great Britain, the experiments of wage and price controls by the Labour governments were not very successful, and France under Giscard d'Estaing completely abandoned a very complex and bureaucratic system of controls. Thus, it is unlikely that in the future any government would impose a system of overall price control, even in the event of a critical energy shortage. Western economies are highly interdependent, as shown by the increasing complementarity of their industries and the significant volume of trade among them, and no government can follow a policy of self-sufficiency without severing these links.

One may therefore expect the governments in all the Western European countries to control economic and social developments arising from the new energy situation through specific measures: imposing standards for use, setting schedules of rates, and, most important, aiding investment in energy research and development, production of new energy forms, and improved energy use. Each government, moreover, may be expected to take special steps to counter unemployment directly if it exceeds a certain level.

The task of modifying energy consumption, in part by substituting products of relatively lower energy intensity wherever possible, falls to the market. Under the Lower Bound scenario, however, it is likely that governments will try to promote more rapid adjustment by imposing standards to phase out certain energy-intensive products while providing financial assistance.

Central Government versus Local and Regional Authorities

It is reasonable to assume that governments will intervene broadly but delegate to provinces and towns the initiative and power to act. As a result, private individuals will be more closely involved in energy management, and local requirements and characteristics will be taken into account.

As argued earlier, major political and social movements toward decentralization can balance the economic and technical trends toward further concentration of power in the central government. Whatever the advantages of decentralization, in some cases it risks excessive fragmentation within a country. In Europe, most people clearly believe that the individual nation no longer has its former prestige and authority, which evolved out of an ideology and a history of reunification and recurrent wars. Some maintain that this is a welcome development, for the individual can now regain a personality less bellicose toward other countries. Unfortunately, the world is not in tune with such a trend. To begin with, the European Community, consistently unable to integrate, is not a real political force. Moreover, the dangers inherent in the economic and political situation of Third World countries and their relationships with developed countries are increasing.

For reasons that have little to do with the energy problem, governments recognize the risk of divisiveness in too generous a transfer of power to the provinces. In Great Britain the authorities are curbing movements toward autonomy, sometimes exacerbated, as in Scotland, by the discovery of energy resources. Italy has gone a long way toward giving power to the provinces. While this major reform may have enabled the country to stay on its feet at a time when its political structure threatened to collapse, such decentralization seems to make it more difficult to adopt a clear plan of action (especially regarding choice of energy sources), which is needed to overcome the energy problem.

Finally, in France, unified by the national government's power to arbitrate differences and muster support, the fear of centrifugal forces remains strong. Despite arguments in favor of transferring authority, the national government only grudgingly metes out power to local authorities and firmly squashes any fringe movements for autonomy. In line with the Jacobin tradition, the more that the external dangers require cohesion in the interest of national security, the more certain that no further decentralization will be tolerated.

Community versus Individualism

The rise of the "mass" society with its materialistic values undeniably has fostered individualism at the expense of communitarianism. Major forms of institutionalized solidarity at the national level (such as social security) are no substitute for the warm, personal relationships that exist within smaller social units. The ready availability of energy and the high degree of flexibility that electricity and petroleum products afford have contributed greatly to individualism and the creation of small, autonomous family units. The energy problem thus may trigger a return to more communal values and behavior patterns.

Some of the more extreme individualistic practices, in heating and transportation, for instance, probably will begin to disappear. But it would be better for the government to encourage such a process, since habits and attitudes change slowly. The factors that seem likely to work in favor of a radical change in values and attitudes include a disillusionment with the materialistic society, which dispenses welfare in a bureaucratic, impersonal way; a revival of religious feeling and community spirit; and the development of a more regional life-style. But for the next ten years, at least, the impact of these three factors is likely to be confined within the fairly narrow limits set out in this paper.

Our society will remain motivated by the goal of sustained economic growth and a steady rise in incomes that generates greater material satisfaction. Furthermore, the falling birth rate and the aging of the population will tend to narrow further the outlook of the family unit and to exacerbate egocentric attitudes while the proportion of active, productive people diminishes. Last, the microcomputer revolution occurring both in homes and business will contribute significantly toward dispensing the means of production and consumption, thereby fostering individualism. This technology will encourage working at home (by means of terminals, display screens, and microprocessors) and enjoying leisure at home, using video equipment. Because such changes will help to save energy, they will increasingly gain favor.

A Different Society?

A study group convened by the Commission of the European Communities investigated the new characteristics of socioeconomic development and published a report, *A Blueprint for Europe*.[17] It notes that in Europe, "the political demands of the citizens are directed not only toward obtaining a more balanced distribution of economic resources and a more efficient provision of social services, but also toward more direct involvement by citizens in the management of economic relationships, social services, and cultural activities." Such demands derive mainly from a challenge to the principles and practices of industrial society (even though it may already be largely obsolete); the more sensible use of energy is only one of the aspects involved. It is possible, however, that these circumstances will become the principle factor in the future, a sort of model of social change based on a "third system of social organization, which will take its place alongside the market system and the administrative system."

This third system could take various forms. One could envision the development of autonomous cooperatives and the expansion of a home-based economy. Families, associated in interfamily groups, could produce an increasing number of things with materials and equipment supplied by the commercial sector of the economy. There would be a revival of craftsmanship in a modern domestic form — a collective do-it-yourself movement — which could help solve the problem of certain personal and domestic services (public transportation, maintenance, repairs, etc.), which the commercial sector is less and less able to supply efficiently and at a reasonable cost.[18]

Participation in the management of social and cultural activities may relieve the pressure on energy resources by raising the information content and lowering the energy content. Examples include town planning and experiments in educational, social, and health services that already have been implemented by cities in Italy, Germany, and France.

As suggested by Aujac and de Rouville, a serious and lasting energy shortage would encourage such trends to develop. Theorists of postindustrial society believe, however, that the ability to organize society in autonomous groups depends on "untrammeled economic growth and productivity," and that an extensive self-managed sector thus will probably play a supporting, rather than a key, role in promoting economic and social change. And unless society returns to extreme frugality, the kind of barter that forms the basis of self-sufficient activities is bound to remain limited. It is the market economy that will have to supply most of the material goods required by society, even if services are provided free.

In fact, the authors of *A Blueprint for Europe* anticipate the gradual emergence of an autonomous sector of society, based on two assumptions: the "shift away from individual consumption in favor of collective activities and consumption" less wasteful of material resources; and the possibility of "assuring everyone's material welfare through a sparer system that is more *socially egalitarian*" (emphasis added). This, in a nutshell, describes the political and social problems as well as the uncertainties posed by such significant change.

Technological progress, mainly in data processing and telecommunications and largely spurred by the energy problem, will afford greater individual scope for autonomy in education, work, and leisure and at the same time relieve energy constraints.

It appears possible, in the years ahead, to harmonize the relationships among the home, place of work, and the community. Achieving such a transformation in life-style presupposes a far-sighted approach by individuals, groups, and governments in technological research, land use and urban planning, and especially in the management of time. Certainly no official decree can effect such changes. Things will happen spontaneously. The home and office or factory will be brought together through the subdivision of production units, the extension of work in the home, the promotion of public transportation, shorter working hours, less specialization, and more do-it-yourself activities.

Such spontaneous developments, however, will not be enough. Some overall planning, based on a minimal consensus of the people affected, will have to be implemented. In a future marked by energy problems, we must rely on collective understanding and resolve even more than on individual awareness. In democracies, it is the task of elected representatives to work diligently toward this end.

Finally, the main question is whether the political regimes of Western Europe are capable of handling a difficult or serious situation over the next twenty years. Once again, it should be noted that energy is only one facet of the larger problem confronting Western society: the threat of international disruption as a result of population explosion in the Third World. Clearly, if energy problems persist or worsen, many other problems will be harder to solve, including resuming sustained economic growth, controlling inflation, and achieving international financial equilibrium. The need of the developed world to use energy more efficiently may perhaps prove to be the best means of defusing world economic and political tensions. Because the efforts required to bring about change do offer an opportunity for innovation and will stimulate economic development, governments may find in them a way to marshal the support of their citizens. In addition, since most of the public neither holds the government responsible for the energy problem nor imbues it with much ideological content, the danger of political instability diminishes.

The Upper Bound scenario takes into account the time that will be needed for Europe to respond to the energy crisis and overcome its difficulties. The potential for social discontent, arising from the manner of allocating the necessary sacrifices, characterizes the Lower Bound scenario and will require increased government power and intervention. Politics might then evolve toward somewhat socialist experiments.

The crisis scenario provides an opportunity for decisive action, which all governments, even in Italy, are capable of taking. A majority of the population would be prepared to support such a wide scope of government action. Here, however, the unpredictable and sheer luck play an important part; success or failure will depend on circumstances, the caliber and authority of those in power, and the way information is handled. Everything may hang on a statesman's illness, a disquieting scandal, or on an emotional reaction by the community at large.

These contingencies should not prevent us from acting insofar as we can. The efficient use of energy, through careful and thoughtful conservation, must be a major goal in the interest of national welfare and humanity. A massive effort toward conservation can best promote the economic, political, and social balance of Western Europe, however the energy question may unfold in the future.

9

The Global Poor

by Althea L. Duersten
and Arpad von Lazar

The world of energy offers the challenge of opportunities and the anxiety of scarcities. The period after 1973, of readjustment in international energy trade, brought about changes affecting both the industrialized and the developing world. Our objective here is twofold: to delineate the impact of the international energy crisis on those less-developed countries that are not major oil producers and exporters, and to discuss the choices these countries face in adjusting to a world of higher-cost energy. Our main focus is on the oil-importing developing countries (which we refer to as the OIDCs), although we will also discuss the problems of the countries with limited production of oil and gas.

Our universe is the world of the global poor. Three of the four billion people of the world live in countries where population growth rates have continued to exceed those of the industrialized world. A large part of this population resides in rural areas, and agriculture remains a very important factor in terms of both gross national product and employment. Industrial development has proceeded to varying degrees, often creating enclaves of modernity in economic structure and social development. The rate of migration from rural to urban areas is high, resulting in overcrowded cities and mounting social and political tensions.

The pattern of energy consumption in developing countries is that of societies in transition. This is a world in which the use of traditional, non-commercial fuels — wood, agricultural and animal residues and waste — as well as animal and human labor is still widespread. But it is also a world where commercial energy resources — oil, gas, coal, and electric power — are increasingly important. Oil and petroleum-based products supply the major share of commercial energy in the OIDCs. This growth reflects the development of industrial and urban sectors as well as the modernization of the agricultural sector.

Within this world, the pattern of impact and adjustment to the world of higher energy prices has varied widely.

From a global perspective, the OIDCs have remarkably little impact on the world oil market. Fifty-nine of them import less than 10,000 barrels per day each; thirteen, less than 20,000 barrels per day; another thirteen, less than 60,000 barrels per day; the current U.S. import requirement is around 6 million barrels per day. The total net imports of all the OIDCs in 1980 were 4.5 million barrels per day. In terms of world oil consumption, these countries are modest players in the world market. But it is the impact of expensive energy on their domestic economies that represents the real concern in the 1980s. The cost of imported energy will mean that the OIDCs will have less to spend on domestic programs; they will have to sacrifice attempts to help the poor and disadvantaged in their own societies. Their chances for real growth in the 1980s will be threatened and the social cost of economic stagnation, extremely high. Expectations may prove as difficult to deflate as an overheated economy.

The differences among the almost ninety developing countries are great. Yet we can clarify the problems and distinctions among them by dividing them into three categories as a better way to highlight their different policy options.

The first group consists of marginal oil- and/or gas-producing countries that have the potential either for a limited export capacity, for domestic self-sufficiency, or even for domestic energy deficits, depending largely on policies followed at home. Egypt and Peru are examples of this group. But the political and social conditions facing policy-makers in the developing world make planning for energy use and demand management use difficult. The pressures of increased urbanization, the relative decline in the importance of the agricultural sector, and the growth in the industrial sector in these countries will place large demands on domestic oil and gas resources. Moreover, in societies in which consumer subsidies for energy products — such as cooking fuel, fertilizers, and electric power — have become established political traditions, change will be difficult and in some cases politically impossible.

The second group includes countries with partly industrialized economies and high levels of dependence on imported petroleum supplies. Patterns of industrial expansion, the lack of domestic oil and gas, and the structure of demand in the industrial sector all limit policy flexibility. The cost of the imported energy required to serve the industrial sector has led, in some cases, to a continued and growing need to borrow both private and public international capital to meet growing oil import bills. Consumption patterns reflect, not only the importance of the industrial sector, but also the growth of commercial energy demand in both the residential and transport sectors, owing to increases in per capita income as well as social choices embodied in the changing life-styles patterned on those of the industrialized world. Disparities in income, inequities between the well-integrated modern industrial sector and the more traditional agricultural sector, and the existence of the urban poor within the modern sector all place energy policy within clear-cut and limiting political constraints. Solutions require demand management and the development of alternatives to oil. Turkey and Brazil are countries typical of this group.

A third group of countries faces a double energy crisis. On the one hand, increasingly expensive oil import requirements have resulted in severe balance of payments problems. On the other, noncommercial or traditional sources of energy, including firewood and animal and plant wastes, still constitute a large share of the demand and are in increasingly scarce supply. Many of these countries are also primary commodity exporters and have little trade flexibility in the short term. Jamaica and Kenya are two such countries that face both particularly severe macroeconomic difficulties as a result of the cost of imported energy and a "traditional" energy crisis at home. Again, the adjustment process could entail unpopular domestic policies, which would exacerbate political and social instability.

A consideration of the energy future of the OIDCs raises several questions. Will they need to shift away from previously preferred development strategies and courses of action? Do we have a new dependency problem at hand, where the constraints of energy scarcity erode government control over both the use of resources and the definition of broader development policies? Will governments be able to cope with the cross pressures of reduced growth rates and accelerated demands for goods and services? What are the consequences of these new development constraints in terms of social change and political stability? We will discuss these questions within the context of our three broadly defined categories.

International Economic and Financial Considerations

In a world of alarming statistics, the impact on the OIDCs of the continued escalation of the price of oil looms large. The cost of their net oil imports has risen tenfold in constant terms in the past decade. The 1979–80 oil price increases left the OIDCs with a net final import bill of $74 billion, or 5.3 percent of GNP, compared with 2.8 percent in 1978. It has become increasingly difficult for these countries to finance the higher cost of oil imports, and their deficit on the current account has risen sharply, from $27 billion in 1978 to $61 billion in 1980, and is estimated to have remained at about the same level in 1981.[1]

While the developing countries will share in the global adjustment to higher energy prices, both the burden and the impact on these countries will be different from those of the developed world. Indeed, that adjustment process will be more difficult for the OIDCs for several reasons. First, energy conservation, however desirable, will be difficult to implement. Energy consumption levels are already low, and further economic growth will increase the demand for commercial energy as the economy becomes more industrial and urban and shifts occur to more energy-intensive agricultural production. Energy savings can be achieved particularly in industry and transportation, but require such politically difficult demand restraint measures as the removal of domestic price subsidies. Also, the investment required to improve energy efficiency may simply be too costly.

Thus, both the growth in demand and the shift in composition of the demand toward commercial sources have imposed a substantial financial burden on the OIDCs, one they are ill equipped to manage. In 1970, the current account deficit of OIDCs totaled 2.3 percent of GNP; by 1980, current account deficits increased to 3.9 percent of GNP.

Projections for the current account situation of the OIDCs in the coming decade are naturally tentative and depend on assumptions about the price of oil, rates of economic growth in both developed and developing countries, and economic policies undertaken to facilitate the adjustment process. Nonetheless, it is clear that current account deficits are likely to rise in nominal terms and probably in real terms, representing a monumental financing problem.

That burden will continue to be particularly severe in the low-income OIDCs. In 1974–78, the average annual GNP growth rate of the OIDCs was cut by one fourth; per capita GNP growth rates fell from 3.1 to 2.7

Table 9.1 Commercial Primary Energy in the OIDCs
(mbdoe)

	1970		1980		1990*	
	Production	Consumption	Production	Consumption	Production	Consumption
Oil	1.2	4.2	1.5	7.3	2.8	11.2
Natural Gas	0.3	0.3	0.5	−0.7	1.6	1.6
Solid Fuels	2.3	2.4	3.5	3.7	5.6	6.4
Primary Electricity	0.9	0.9	2.0	−2.0	5.1	5.1

* Projected.
Source: World Bank, *World Development Report 1981*.

Table 9.2   Current Account Balances of OIDCs, 1970–90
(billion 1978 $)

Country group	1970	1975	1978	1980	1985 High	1990 High	1985 Low	1990 Low
Low income	− 3.5	− 7.0	− 5.1	− 8.1	−12	−15	− 8	− 9
Middle income	−15.0	−42.8	−20.4	−44.1	−37	−45	−33	−34

Source: World Bank, *World Development Report 1981*.

percent. For low-income OIDCs, the impact was even more marked; average annual per capita GNP growth was more than halved, from 1.8 percent in the 1960s to 0.8 percent in the 1970s.[2] While greatly expanded borrowing (together with official capital flows for the low-income OIDCs) allowed the OIDCs to cover their past current account deficits, the continued need to finance current account deficits through external borrowing raises two questions about debt servicing problems in the years ahead: How will the OIDCs manage existing debt burdens? And from what sources and on what terms will they be able to borrow in the future? These problems will be exacerbated in those countries in which a large portion of the new debt will reflect borrowing in international financial markets on terms less favorable than those associated with traditional development financing.

While many OIDCs have in the past had easy access to large private capital flows from the market, efforts to borrow further may be more difficult in the 1980s. Private banks will be increasingly constrained by so-called country lending limits and concern over portfolio exposure. In some cases, a country's recent economic performance may not have been strong, and thus prospects for continued high levels of borrowing will not be favorable. Moreover, private commercial lenders will be scrutinizing not only a country's economic performance but also its political stability during a period of difficult economic adjustment.

The overall economic prospects of the OIDCs, as well as their ability both to service existing debts and to finance future capital requirements, will depend heavily upon export performance. But efforts to stimulate the growth of exports will be strongly affected by economic growth rates and policies in the developed countries, which will influence both demand and market access for goods produced in the developing world. For those OIDCs with a substantial industrial base, their continued rapid growth in manufactured exports will be critical to their ability to pay growing oil bills. Government policies designed to promote exports have already be-

Table 9.3   Selected Countries: Energy Imports,
Total Exports, and Debt Service, 1980

	Oil Imports as a Percentage of Exports	Debt Service as a Percentage of Exports
Marginal Oil and Gas Producers		
Egypt	0	18.9
Peru	0	31.9
Semi-Industrialized OIDCs		
Brazil	50	34.5*
Turkey	100	14.3
Low-Income OIDCs		
Jamaica	48	12.8
Kenya	21	8.9

* Reflects public debt only.
Source: World Bank.

come so crucial that access to the developed nations' markets has replaced national sovereignty over domestic resources as a prime consideration of the continuing North-South dialogue. The developing countries are actively pressing for greater access to the markets of the industrial world. But such efforts at promoting exports run directly against the growing pressure toward increased protectionism in industrial countries, which are also struggling with an overall economic slowdown. For manufactured exports — particularly footwear, steel and electrical consumer goods, textiles, and clothing — protectionist politics in developed countries could put a serious brake on the export performance of the developing countries. In addition, a lower rate of economic growth in the industrial countries will also have a marked impact on the export earnings of those OIDCs that export mainly primary commodities, such as mineral ores, cotton, jute, and wood for industrial use.

The low-income OIDCs generally cannot borrow heavily from the private international banks that have contributed so heavily to the financing of current account deficits in middle-income OIDCs. Furthermore, the low-income OIDCs have not been in a position to accumulate financial reserves from which they might otherwise finance further increases in their current account deficits. As a result, they have less flexibility in dealing with the impact of increased oil prices and have to depend more on official public sources of capital to meet their needs.

Since the 1970s, the low-income countries have experienced a larger deterioration in their trade balances than the developing countries as a whole. Since many of these countries are primary product exporters, the slow growth in the industrial countries' demand, particularly for agricultural products, depresses their export performance. Faced with substantial current account deficits as well as with limited access to private international capital flows, these low-income OIDCs will remain dependent on international aid. However, the budgetary constraints imposed by economic recession in the industrial world, and an increased concern for adjustment at home, dim the prospects for increases in concessional funds.

For those low-income countries in which increased oil import bills must be financed out of domestic resources, a substantial diversion of funds from development investments and expenditures may be the only option. Increased oil prices coupled with international inflation have already forced many governments to adopt fiscal, monetary, and income policies aimed at restraining aggregate demand. As necessary as these policies may be, their political and social costs are expressed by their conflict with government objectives for alleviating poverty. Public investment programs targeted to low-income groups, designed to create employment and improve living standards, will have to be curtailed sharply due to the lack of public funds. Thus, the real cost of higher oil prices can be measured in the heavy toll on development.

The impact of the increased oil prices on the OIDCs has international financial and domestic policy implications. The price increases of 1979 alone resulted in an estimated increase of about $35 billion in the oil bills of the OIDCs in 1980. Thus, developing countries will require large capital inflows in order to finance energy imports. For middle-income, semi-industrialized OIDCs, the bulk of this financing will have to be raised in the private international capital markets. Given the exposure limits of individual banks as well as credit considerations, countries with relatively weaker economic prospects or troublesome political problems will find it difficult to raise adequate new private capital.

Managing the existing debt could prove problematic. Following the 1973 price increases in oil, there was a marked increase in the number of countries encountering debt servicing problems. During 1974–78, as the most immediate adjustment occurred, the number of countries with arrears on current payments or conducting or seeking multilateral debt renegotiations increased from three to eighteen. At the end of 1978, the debt of these countries accounted for about 12 percent of the total outstanding external debt of developing countries. In the majority of these cases, deterioration in the balance of payments, in part brought about by a decline in the growth rate of exports, was responsible for debt servicing problems. Thus, the indicator to watch is export performance; while the total indebtedness in these countries did not differ greatly from those in the other developing countries, a relatively weak export performance led to far greater difficulties in servicing existing debts.

In the 1980s, debt rescheduling will be more likely than large-scale default. The buildup of past debt has been concentrated in better-performing developing countries, like Brazil and Mexico; and the industrial countries cannot afford to let their major debtors disintegrate into economic crisis and political instability.

On the developing country side, however, servicing debt may force hard domestic choices, particularly when stabilization programs supported by public international capital are involved. While the removal of price subsidies for food and fuel, for example, may make good economic sense and may be comforting to financial institutions, the consequent food riots, general strikes, and political protests reflect a less happy reality at home. And the tightening of fiscal reins can slow economic expansion. Unemployment is a concern in all developing countries, and a slackening in the opportunities to enter the work force can radicalize large groups of people for whom ideology is not necessarily a foregone conclusion.

It is exactly this circumstance of thwarted aspirations, of slowed or blocked social mobility, and of a constriction of personal life choices that suggests a twilight world of ambiguities for these countries in the 1980s. The true horror of poverty is that it demonstratively reduces the individual's time horizon — everything is here and now, all opportunities lost are lost forever. In countries with large income inequalities between the ruling few and the rural and urban masses, the slowing of economic expansion inevitable in an energy crunch endangers and often totally cancels out government efforts to reduce the painful effects of marginality through social policies.

The initial crunch is felt in the mushrooming megacities of the developing world. Reduced employment opportunities, less money for social services, and less upper- and middle-class willingness to pay for social peace through financing social reforms — all these hit first and hardest at those who are at the lowest and most marginal level of society: recent rural migrants, the unskilled, and the young, who are the potential entrants into the work force. The energy crisis already has postponed the integration of these social groups into the modern industrial or service sector.

In addition, the poor, because they are poor and on the margin of the monetary economy, do tend to spend a larger proportion of their income directly on energy than middle- or upper-class families. Especially in the city, where many of the makeshift options of rural life are not available, this means a life of sustained deprivation. An energy crisis in the OIDCs will solidify a state of permanent misery for those on the lower part of the income ladder. Not only does the economic pie not grow bigger, but there is also less and less chance for recutting and redistributing the same pie for an ever-growing number of aspirants. Such circumstances will have the most serious consequences for the political order of the developing world.

Energy Policy Options

Oil has become expensive and is likely to become even more so. The OIDCs have a particular need to develop effective energy policies that maintain oil imports at financially manageable levels and, to the degree possible, stimulate the development of domestic energy, including oil, gas, hydroelectric, coal, and renewable resources. The need is particularly acute because the energy requirements of the developing countries are likely to increase at much faster rates than those in the developed countries. Economic growth and modernization in the developing countries will be accompanied by a structural change toward an increasing share of energy claimed by the industrial and urban sectors, both of which are generally associated with increased commercial energy use. Increasing per capita income levels will also encourage the substitution of commercial for noncommercial energy sources, thus adding to the demand for oil-based products. Finally, the potential for developing countries to practice conservation is limited, particularly as the use of oil products for private transport, residential heating, and commercial cooling — areas targeted for savings in the developed world — is minimal.

Marginal Oil and Gas Producers

For countries that presently have, or possess the potential for, oil and gas production, an important priority is an increase in investment in exploration and development of indigenous oil and gas resources, which may require revising existing legislation and incentives in order to attract private capital. If national oil companies are to take the lead, improvements in their financial performance and technical strength will be necessary.

An important policy choice will be faced concerning the disposition of production. It may be feasible for some countries to attempt to meet internal energy demands through substituting fuel of lower value, thus releasing oil for export. While an export potential also exists for natural gas and coal, the scale is more limited due to the difficulties in transportation to markets and the large investment required for transport and processing facilities. Attempts to conserve domestic production for export will also depend on patterns of internal demand. Prices play a critical role in the evolution of demand. Subsidizing domestic oil product prices over the longer term, particularly for OIDC producers with a limited export potential, is an expensive policy; subsidized prices encourage a pattern of consumption and industrialization based upon unrealistic energy costs, which will cause more painful longer-term adjustment once domestic oil and gas reserves have been exhausted.

The benefits for developing countries in stimulating either domestic self-sufficiency or an export potential when possible are obviously great. The cases of Peru and Egypt are typical of the countries in this group. Peru's resource base is diverse, with estimated known oil reserves of one billion barrels, economically viable hydroelectric sites, coal reserves, and small natural gas reserves. However, crude oil is the dominant energy source in Peru, supplying approximately 80 percent of its commercial energy requirements. While the potential exists to develop alternative energy sources, the gestation period for these investments is long, and oil will still provide the major part of Peru's requirements through the rest of the century.

On the basis of current trends and current known reserves, Peru should be able to maintain a small overall exportable oil surplus, at least through 1985. However, further exports will depend upon the level of exploratory work and resulting deposits discovered in the Amazon, the rate of growth of alternate energy source investments, including hydroelectric power, and demand management policies for the domestic market. There will be problems because the transportation sector is likely to increase its share in total oil consumption, and the use of oil-based electric power generating capacity will be necessary to compensate for any shortfalls in the growth of hydroelectric and coal-fired power stations.

On the production side, high priority must be given to the stimulation of investment in the exploration and development of domestic oil and gas reserves. If developing countries wish to obtain private capital, attractive and internationally competitive contractual terms will be necessary. Efforts to attract private companies can be complemented by measures to strengthen and improve the efficiency of national oil companies involved in joint-venture or wholly owned operations. As public sector enterprises, national oil companies are often hampered by personnel policies that are unduly rigid and related to civil service regulations rather than to the needs of business enterprise and by domestic taxation policies that make internal generation of needed funds for investment impossible.

Another important element of policy is the development of domestic alternatives to imported oil. This type of substitution effort is particularly important, given the overall pattern of demand and product use. For developing countries as a whole, electricity consumption is growing at a much higher rate than that of developed countries, reflecting in part the low base of electricity use in the developing countries. Furthermore, two thirds of the electric energy consumed in developing countries is used by industry; increasing industrialization and urbanization will result in continued high growth rates for electric power demand. Countries, like Peru, that have both hydroelectric and coal potential, as well as the production capacity for small quantities of oil, are in a relatively favorable position. They can substitute coal and hydroelectric power for oil in electric power generation and for some industrial uses, thus consuming indigenous, not easily exportable resources and freeing a high-value surplus of oil for export. But the magnitude of the capital investment required for new hydroelectric power generation projects and for fuel substitution efforts will be a serious constraint on the efforts to develop substitutes, particularly in low-income developing countries.

An important part of government policy, given the magnitude of the financing problem, is the preservation of the financial integrity of national power companies so that they can raise needed funds to invest in new power projects and ensure financial self-sufficiency. This will require that rate structures keep pace with rising marginal supply costs. Admittedly, this will be a difficult policy for governments both to support and to implement. Consumer resistance to rate increases is at least as vociferous and political in the developing countries as in the developed world. And in countries with high domestic inflation, rate increases may run counter to government efforts to bring down domestic inflation to more acceptable levels.

Domestic prices for petroleum products in many developing countries are heavily subsidized, and subsidies have become an undesirable yet tenacious political feature. In Peru, for example, after the 1973 oil price increase, the government attempted to protect the domestic economy from the impact of international oil price increases; thus, prices for petroleum products remained unchanged from 1969 through 1975. As a result, a rapid increase in the consumption of these fuels, particularly of gasoline, occurred. Thereafter, the government began to increase prices to reflect international prices and to take inflation into account. During 1980, for example, gasoline prices were increased by about 25 percent. By January 1981, gasoline was $1.03 per gallon, which was a significant step achieved despite considerable consumer pressure to maintain subsidized prices. This new pricing has indeed affected demand. Present domestic subsidies may prove to be an expensive policy, ensuring future consumption problems. Domestic demand in countries like Peru is met from domestic production and so reduces the exportable surplus. A subsidized domestic price structure encourages the growth of domestic consumption. Low prices also reduce the financial viability of national oil companies, which badly need to be profitable in order to raise funds for their own programs of exploration and production.

Developing countries with natural gas potential, such as Egypt, face somewhat different options. In many of these countries, associated gas deposits have been flared — that is, simply burned where produced — because of the lack of internal demand for natural gas and the lack of an infrastructure for gathering, processing, transporting and distributing it. Development on nonassociated gas fields has been slowed by the same factors. This waste is particularly serious. Once flared, this gas is gone forever.

The case of Egypt illustrates the range of issues facing a developing country with both gas and oil potential. Egypt presently produces about 670,000 barrels of oil per day and around 100 million cubic feet per day of gas from three major fields. Government policy aims to increase investment in the exploitation of natural gas in order to make a qualitative change in the pattern of domestic consumption, increasingly replacing oil by natural gas in the national economy.

Achieving this type of demand shift requires several steps. First, efforts must be made to increase the economy's potential to absorb natural gas in a way that uses to best advantage a scarce and valuable resource. Once decisions have been made on the use of gas, substantial investments in processing, transportation, and distribution facilities will be required. Some countries may also be able to upgrade natural gas from being primarily a boiler fuel to higher uses, such as for fertilizer, feedstocks, and domestic fuel. The planning and provision of the complementary investment packages are of high priority in gas-producing developing countries.

Semi-Industrialized OIDCs

A second group of developing countries are both middle-income and semi-industrialized oil importers. Here the main policy issues are based on two needs: in the immediate term, to manage the financial impact of a continued high reliance on oil imports, and over the longer term, to develop viable alternatives to oil consumption within the constraints of existing industrial patterns and structures. While facing a severe short-term financial crisis, these countries generally have a stronger capacity to plan and implement needed policy and investment packages for the energy sector.

The cases of Turkey and Brazil are good examples. Turkey, which also depends greatly on imported oil, has fared far less well than Brazil in adjusting to the world of higher oil prices since 1974, particularly because of its poor performance as an exporter. For a country of Turkey's size and per capita income, the unusually low level of exports relative to GNP prevented it from earning enough foreign exchange to finance imports, particularly oil. Borrowing surged in the international capital markets until the economic crisis at home was translated into a negative judgment on the country's credit and new sources of private capital virtually disappeared.

Brazil, in contrast, enjoyed high rates of economic growth between 1967 and 1973, led by a 25 percent annual growth rate in manufactured exports. However, Brazil imports over 80 percent of its oil and as a consequence of the price increases, a formerly manageable import bill has risen to staggering levels. The 1973 price increase alone caused the annual Brazilian oil bill to rise from $711 million to $2.8 billion. Between 1978 and 1982, Brazil's oil bill doubled, from about $5 billion to $10 billion, or to about one half of total export earnings. Nonetheless, the Brazilian economy proved remarkably resilient in withstanding the impact of these ever-growing bills. This was due to a growth in manufactured exports, liquidity in international financial markets, and a continued faith in Brazil's overall creditworthiness.

But over the next decade, Brazil will need to develop domestic energy resources on a scale large enough to reduce the drain on the economy caused by oil imports. As part of that effort, the government has taken steps to increase exploration for domestic oil and gas deposits and to take advantage of its advanced technological base and resource endowment to explore and develop new technologies that could develop substitutes for petroleum products, including an ambitious alcohol fuels development program. Brazil's objective is to find alternatives for 45 percent of its projected gasoline consumption by 1985. Brazilian consumers have already accepted a blend of gasoline and alcohol, the use of which rose from 1 percent in 1975 to 19 percent in 1980. Alcohol can provide a premium liquid fuel alternative to petroleum products, but the economic viability of alcohol is limited to countries, like Brazil, that have a technologically sophisticated industrial base, sufficient agricultural land and surplus agricultural production that can be devoted to necessary raw materials (sugar cane, cassava, sweet sorghum), extensive experience in industrial plant management, and large domestic markets and relatively low labor costs.[3] The capital investment is estimated at $8.9 billion for the 1981–85 program, but in countries like Brazil, the economics are likely to be attractive and, where the transportation sector is a large consumer of petroleum products, the potential savings large.

Given the possible impact of substitution efforts and the relatively large industrial base of middle-income OIDCs, options in other sectors — particularly electric power generation — should be explored. The industrial sector is a main user of electric power as well as an important source of export earnings and employment. Many middle-income OIDCs, like Brazil, whose industrial sectors are large consumers of electrical power, can: increase fuel substitution through hydroelectric and coal-fired generating plants; strengthen existing transmission and distribution systems in order to reduce energy and capacity losses and gain substantial energy savings;

and invest in connecting grids that can result in important energy savings. Finally, savings can also be obtained in the industrial sector itself through improvements in housekeeping, maintenance, and operating practices and as a result of investment in retrofitting and process changes designed to reduce the energy intensity of production.

No discussion of the industrial, middle-income OIDCs' energy situation is complete without considering the perplexities of nuclear options. The most powerful and painfully evident characteristics of nuclear power in a developing country are its enormous capital cost, a very high dependence on imported costly technology, and the considerable size of economically feasible installations versus the size of networks and the demand to be serviced. Add to these the political vagaries of fuel supply and restrictions on fuel reprocessing.

In Brazil's case, nuclear power is planned to satisfy the high growth and industrial-urban network of Rio de Janeiro and São Paulo. In 1975, Brazil ordered eight 13-megawatt nuclear power stations from West Germany in addition to setting up joint projects for uranium mining and processing, uranium enrichment, and the reprocessing of spent fuel.

Brazil's entry into a nuclear future has encountered considerable birth pangs. Technology exchange problems persistently plagued the Brazilian proponents of the program, and a lack of satisfactory uranium enrichment arrangements meant that Brazil had to rely on outside suppliers. In this sense the nuclear option further aggravated its external dependency, which already was bothersome enough in terms of oil import requirements. The foreign policy disagreements that persisted during the Carter administration only added to the prevailing view among Brazilians that although there is now a clear-cut commitment to include nuclear power as an energy policy option, it would have to be on a more modest scale.

This kind of foreign policy entanglement confuses nuclear policy even more for the developing world. The twilight zone between the peaceful and military uses of nuclear materials remains the battleground of nationalistic pride and perceived economic need. Some industrial countries want to slow and control the spread of nuclear military potential; this is certain to complicate relations between the OIDCs and the developed world.[4] The OIDCs, in pursuing the nuclear option, may be importing not only a new technology but a new and potentially acute foreign policy problem.

Traditional Fuels: A "Second" Energy Crisis in Low-Income Developing Countries

For a number of developing countries, the energy crisis is twofold, affecting not only the cost of imported energy but also the availability of traditional energy sources. Traditional or noncommercial energy sources such as firewood or dung still make up a large percentage of the total energy supply. But these resources are becoming increasingly scarce. Efforts to meet the rising domestic demand have resulted in severe deforestation in some countries as well as the diversion of needed animal wastes from fertilizer to fuel uses. Countries like Jamaica, Nepal, and the Sahelian countries of West Africa are thus faced not only with rising oil import bills, but also with dwindling supplies of noncommercial, traditional energy.

Jamaica presents a particularly stark case; with almost no indigenous conventional energy resources, it imports approximately 99 percent of its commercial energy needs. What makes the situation even more difficult is that a single sector, the bauxite-alumina industry, is both the country's leading export earner and imported energy consumer, using almost 50 percent of total oil imports. Conservation is thus difficult to accomplish without cutting into needed export earnings. The government of Jamaica has therefore adopted a dual policy of passing on the increased cost of energy to consumers and at the same time pushing for the development of local renewable energy sources. Studies are under way on the feasibility of small hydroelectric projects, the extent and feasibility of peat resources, and the possible uses of wind and solar power.

Kenya faces a more specific dual crisis. Some 55 percent of its total energy consumption is accounted for by firewood, an additional 15 percent by charcoal; oil accounts for 25 percent. Thus it is caught in a two-way dilemma: oil imports are increasingly expensive, and firewood supplies are not keeping up with the demand. Energy substitutes require both capital and time to develop. Kenya is attempting to accelerate reforestation, and to develop its hydroelectric and geothermal potential, and to explore for petroleum, but in the short term little can be done to reduce the demand for petroleum in essential sectors (transportation, manufacturing, and electric power generation) without slowing economic growth. And reforestation is an investment in the long-term future.

Future economic growth in these countries will stimulate their demand for petroleum products, and difficult macroeconomic problems will be faced in financing necessary oil imports. Renewable energy sources can thus be important in meeting the "second" energy crisis.

Renewable energy sources vary widely in scale and technological sophistication, commercial availability, and current and future competitiveness with conventional sources. Some can be expected to substitute for petroleum products in the modern sector. Others will be more useful in meeting traditional fuel needs, such as heating and cooking, particularly in rural and low-income urban areas.[5]

At the present time biomass — principally firewood and agricultural and animal wastes — is still the main source of energy for rural households and, to a lesser degree for low-income urban groups, resulting in near ecological disaster. Overuse, including deforestation, has brought about serious soil erosion and declining agricultural fertility. Rural families spend increasingly long hours searching for firewood on denuded hills. The urban poor spend larger and larger percentages of their income on firewood and other traditional fuel. Thus, there is an urgent need to increase firewood production through reforestation in addition to increasing the efficiency of traditional fuels. Reforestation technology is relatively simple, through firewood production either on large-scale plantations, in village woodlots, or in small landholder plots.[6] Similarly, there are technologies to increase fuel efficiency through chipping, pelletizing, and charcoaling; fuel-efficient stoves could also be beneficial. However, while the technology is fairly simple, there are problems of wide-scale acceptance, diffusion, and distribution. A second promising area is the application of updated or new technologies, such as small hydroelectric plants, windmills, and solar applications for heating and cooking.[7] But these technologies require considerable capital, and constraints also exist in moving from the laboratory prototype to the market shelf. But an effort must be made to develop these technologies on a scale wide enough to reduce unit costs and to provide for a sustainable delivery system. This could include, for example, local manufacture and distribution of locally adapted technologies by private entrepreneurs in developing countries. Such an approach would also help overcome problems of maintenance and spare parts, which can be critical obstacles to the use of relatively sophisticated imported technologies.

A third technology that could meet rural energy needs over the next twenty years is the production of biogas from the anaerobic decomposition of animal, plant, and human wastes by a methane generator or digestor. The biogas can then be used directly in cooking, thereby substituting for firewood in rural areas or for petroleum products (bottled gas) in urban areas. Biogas generation has been tried in both India and China with reasonable success. The main constraint to its more widespread adoption, at least in the Indian experience, appears to be the high initial cost of the digestor unit as well as inadequate training and technical support.

But the real problem in expanding the use of these technologies may be more perceptual than technological. No one likes to be a second-class world citizen. Anything less than what is fashionable and innovative in the United States appears backward to societies in transition, which are striving to look ahead, not back. And the need to accelerate development to break the cycles of underdevelopment and poverty provides an especially compelling reason to focus on high technology solutions. The essential magic of high technology rests, after all, in its promise to effect quick (and perhaps automatic) conversions on a large scale. Reaching the rural and urban poor quickly and efficiently is both a "populist-radical" promise and a pragmatically helpful way to govern. In the wake of the victorious Sandinista revolution, a young member of the Nicaraguan government spoke about "one giant nuclear power plant" for his country to supply the rural poor with electricity, as though it were the first priority for energy policy. Of course, economically — and politically — one of the last things Nicaragua needs is to concentrate its scarce development capital on a nuclear power plant. Yet, hemmed in politically among a distrustful United States, an uneasy Mexico, and politically fermenting small Central American neighbors, the Nicaragua of the Sandinistas has to deliver quick, decisive, and highly visible results, especially for those who made and supported revolutionary change: the poor and marginal of society. The dream of high technology is clearly a Faustian bargain; in this case it is wishful thinking and not a realistic or reasonable policy alternative.

Energy Policy Constraints

The world of policy options rests on the broad horizon of the possible. But a short step back from that horizon offers a more sober vision of a world of choices, limited by the very real constraints of national politics and economic conditions and by the wisdom with which individual countries and their leaders have approached their energy dilemmas.

Financial Constraints

The majority of the OIDCs will have difficulties attracting the financing they need to develop their energy resources. Despite worldwide increases in oil and gas exploration, most of the effort has not taken place in the OIDCs. The 1980 investment plans of the major oil companies indicated that most of the exploration expenditure was planned for the United States, Europe, and Canada, with 75 percent of the growth in drilling activity in the United States. In 1981, the exploration effort increased by less than 5 percent over 1980 levels outside the United States and Canada.[8] The public international financial resources, such as multilateral development banks and bilateral aid programs, represent only a fraction of the capital needed for energy development and particularly to open the often environmentally difficult and high-risk oilfields for exploration and potential production. Most experts generally agree that the chances for massive onshore oilfield discoveries in the OIDCs are marginal, while substantial offshore discoveries are probable. To stop the rapid decline of the productive capacity of currently functioning but rapidly "aging" fields requires a large-scale, quick infusion of capital in secondary and tertiary recovery techniques. Overall, for oil and gas alone, it is estimated that $40 to $60 billion (in 1980 dollars) will be required over the decade to reach an admittedly ambitious target of 4.5 mbd OIDC oil production in 1990, from a current level of 1.7 mbd.[9] The readiness of the private sector to commit available capital is a key variable. Private companies are hesitant to undertake operations in high-risk offshore exploration and production and in sophisticated secondary and tertiary recovery processes where the size of potential discoveries, while important in the context of the individual country's needs, would not be attractive from a company's perspective.

The perception of the private sector in this case is crucial. It is not only the availability of capital that places many of the oil multinationals in a potential lead role, but also their control of complex technology and manpower. Political (as compared to technical) risk-taking has *not* been the hallmark of the last few years' exploratory activities, although oil companies, especially crude-short multinationals, have actively pursued even marginal prospects in relatively politically safe investment areas.

Still, extensive exploratory and production activity remains a high-risk proposition for the private sector in most OIDCs. The primary risk, of course, is political and relates to the stability both of governments and of the contracts entered into. The threat of nationalization is no longer the key. The multinationals have learned — and quite effectively — to live with and benefit from the new circumstances of national dominance and ownership of resources. At stake is the potential for change in contractual terms that could reduce a private company's benefits once the initial investment and potential discoveries have been made.

The balance between the host government's desire for control of resource management decisions and the private sector's desire to obtain a secure and profitable supply of crude for international sales is a delicate one. In some cases, the political, supply, and market risks may appear more acceptable if the cooperative arrangements are agreed upon, including participation by international lending agencies. The private sector wants to ensure the stability of reasonable returns over time. The host government wants to attract needed capital for oil and gas exploration and the technical assistance that the international agencies can provide.

While national oil companies are politically important, in the crucial realm of exploration it is the private energy companies that will continue to carry the burden of searching for and exploiting new oil and gas riches. Recognizing that this pattern will probably persist into the late 1980s, private foreign involvement in exploration and production in developing countries will remain a critical necessity.[10]

Resource Constraints

Even with additional financing, simple resource realities will limit future options.[11] While prospects for additional oil and gas finds in the OIDCs are fair to good, no single major reserve discovery on the scale of that in Mexico seems likely. But although, in terms of the world market, the potential for major oil discoveries is modest, in relation to domestic consumption of the OIDCs, potential discoveries could be important. Gas prospects in the Northern Argentine basin, offshore Thailand, Malaysia, Brunei, and Bangladesh are promising, but these are difficult areas, requiring costly and complex exploration and production. Ironically, much of the new OIDC production during the coming decade will be in gas, not oil. The transportation of gas to international markets poses such considerable technical and financial constraints that much of this production will be for domestic consumption, at least until the technology and pricing considerations better approximate the risk and financial outlays by developers, whether national or private.

For oil exploration, in many cases, there is little if any commercial interest in developing marginal fields that offer low yields at great expense. The problem is that without the private sector, except for a very few fortunate geological "cakewalk" cases, many developing countries lack the institutional capacity to explore for and exploit oil and gas reserves.[12] National oil companies are often weak, lacking technical expertise and financial resources, and the technical weaknesses are often exacerbated by poor financial performance, due in part to low domestic prices.

Technological Constraints

In an era of absolute shortages, every little bit helps. If that is so, the appeals of "small is beautiful" and alternate energy sources and appropriate technologies are self-evident.[13] New sources such as photovoltaics represent new and costly technological applications.[14] Although few experts doubt that the twenty-first century will see the substantial use of these energy sources in some form, the initial enthusiasm for them overestimated their industrial applicability and underestimated the costs of technology and a management network development. The simpler renewable technologies, like firewood and alcohol, represent a more realistic option for this decade.

Planning and Policy: The Eternal Dilemma

Before 1973, energy planning received little attention. As the main actor in the energy world, the oil industry did a fair job in matching supply with demand, and the developing countries generally assumed that they could follow the developed world in counting on oil.

What little pressure did exist to match short-term problem-solving with long-term development trends produced planning that was long on supply projections and short on demand patterns and the options for conservation. In addition to the dominance of the private sector in an era of supply abundance, the public sector's energy institutions were persistently populated by former oil and power company technicians who, with few exceptions, represented the supply side of the planning equation.

In the current world of energy scarcity and imbalance, effective OIDC energy planning is constrained by factors including the very perception of the problem, the lack of manageable policy implementation procedures, and a near total lack of the necessary human resource base.[15] A minimal amount of "homework" in accumulating information and basic data on the energy circumstances of a country is essential for a coherent energy planning process. A domestic resource inventory, the identification of main supply scenarios, the structuring of projected demand profiles, relating these factors to the country's overall development planning objectives — these should be, in simple terms, the focus of a national energy assessment. Most OIDCs do not have accurate data on their energy resources. Even fewer are equipped to link supply scenarios to the performance of the various sectors of the economy. And there is a reluctance to link domestic pricing and market mechanisms to changes in the international energy markets.[16]

Although some promising first steps have been taken in some OIDCs to identify domestic energy resources, these initial efforts have failed to make a significant dent in determining the limits of energy supply and demand and linking them to various development options. Similarly, the political aspects of using domestic resources represent an often neglected constraint of considerable dimensions. We have already referred to the dilemmas inherent in weighing the options of the accelerated employment of potential oil and gas reserves for domestic and/or export use.

If lack of a data base, adequate information on domestic resources, and an assessment of sectoral connections are weak links in OIDC energy planning, then the shortage of skilled manpower in the energy sector and in planning institutions is nothing less than catastrophic. The energy manpower needs of the OIDCs are on two levels, both unfortunately in scarce supply. The first is less glamorous: What is needed are programs to train and retrain technicians in modern but locally applicable nonrenewable and renewable technologies. Such training would have to encompass planning, local management, project evaluation, and review processes in order to provide the basis for a practical energy extension effort.

The second level of scarcity involves the highly skilled experts conversant in planning, management, and policy skills. The current technocratic domination of the energy decision-making sphere offers little hope of bridging the gap between the "rational" worlds of supply and demand equations and the more politically bound spheres of real world planning. There is a definite need for training planners to link better the ideal work of forecasting to the imperfect world of policy implementation.

The disarray of institutional structures and their functioning is also critical. OIDC institutions in either energy-planning or decision-making areas are products of a quick "fix and conversion" process that was considered necessary in the immediate period after 1973. As a result, these institutions have emerged either as converted and expanded structures of past regulatory control agencies, such as hydroelectric power authorities and coal boards, or as brand-new creations having no authority but using technical skills. The importance of the energy issue argues for political centralization in decision-making, even though such an approach would leave the process vulnerable to the vagaries of short-term political change.

Of course, planning is not the miraculous key to problems rising from institutional weakness, skill shortages, or plain confusion in policy-making and implementation. One might ask whether planning institutions and skilled personnel are a step toward solving the energy policy predicaments of the OIDCs. Or are we merely substituting a state of ill-managed despair for one that has been characterized by a Latin American politician as "dynamic stagnation" — great institutional complexity, elaborate roles and functions, furious activity, and, alas, zero output! The only realistic answer is that planning is a modest step that could help national energy policy problem-solving. Alone, it certainly cannot and will not do the job.

The financial, resource, technological, and planning constraints all contribute to the difficulties currently faced by the OIDCs. They add up to a sizable fare, especially in the context of economic constraints, the demands for quick political solutions, and the gratification of immediate sectoral and class needs. In the short term, the management of financial issues will be both costly and at times politically painful, yet unavoidable. Remedies for planning constraints will require time and patience in building institutions and substantial investment in the development of human resources.

* * *

The 1980s and 1990s will continue to be a difficult and unpredictable period in world energy, and particularly in oil markets. While the OIDCs are not major actors in terms of their share of world trade, their demand for energy, particularly for oil, will grow at a faster rate than that of the developed countries. Thus as Robert Stobaugh has detailed, the average annual growth rates for energy availability could range from 0 to 2 percent for the OECD countries and from 2 to 4 percent for the developing countries outside OPEC under the Lower and Upper Bound scenarios to the year 2000. Future economic growth rates will, therefore, be particularly vulnerable to fluctuations in oil prices and supply.

Under conditions of critical supply shortages or even market tightness and in the absence of formal supply allocations by OPEC, the OIDCs, with their relatively limited financial and institutional capacity, would be particularly hard hit. Indeed, there is some evidence that during the 1979 Iranian supply crisis, many developing countries had difficulty obtaining sufficient crude supplies or could only obtain supplies in the relatively more expensive spot markets. The issue of access to supply is, therefore, a crucial one for the OIDCs. Political rhetoric aside, the prospects for bartering between oil-importing and oil-exporting developing countries are not that bright. We cannot overlook the fact that many of the oil-exporting countries are themselves developing countries with substantial domestic resource and technological needs; the industrial countries may therefore have more success than the OIDCs in arranging barters that can be based on high-priority technology packages or market access and trade arrangements. On the margin, the OIDCs may have recourse to special supply arrangements with individual OPEC members or, through a decision by OPEC as an institution, be able to work out special arrangements. Thus, for example, Venezuela brought forward plans to guarantee oil supplies to nine Central American and Caribbean countries at special rates. Mexico, not a member of OPEC, joined Venezuela in this arrangement, stressing its commitment to help poor Central American countries. In addition, some recent discussions within OPEC have addressed the possibility of establishing a system for supplying oil to the OIDCs, which would allocate supplies on a priority basis during periods of overall market shortage.

The OIDCs will continue to struggle with high oil import bills and the financial consequences of a dependence on oil. Some may find it increasingly difficult to finance oil imports through private borrowing. And the poorest OIDCs will remain dependent on concessional aid flows to meet their financial needs. Once again, the possible future role of OPEC will be critical; some members have called for the establishment of a joint fund to help developing nations pay for their oil imports. Whether the industrial countries, particularly while facing their own difficult economic adjustment problems, would be either willing or politically able to participate financially in such a scheme is uncertain.

The OIDCs will also continue to be faced with the need to manage the "second" energy crisis, the increasing scarcity of traditional fuels. While the development of renewable resources may in time meet some of these domestic energy needs, many technologies are not out of the experimental stage. The newer technologies, such as synthetic fuels (methanol, coal gasification and liquefaction) will require large amounts of capital and private sector industrial know-how to develop and will not provide a viable alternative for the OIDCs for decades, at the very least.

International confrontations over supply and demand problems are most likely to take place between the major actors, the oil-exporting developing countries and the industrial world; and while the economic future of the OIDCs will continue to hang very much in the balance, the OIDCs as a political force will remain very much on the sidelines.

Yet a significant imponderable for the future is OPEC and the course it may take in its relationship with the OIDCs. The history of OPEC–Third World relations is one of considerable promise and expectation as well as much confusion and some bitterness. OPEC countries feel that they could be the most effective bargaining spearhead for the Third World countries in the confrontation with the developed world, and they have increasingly accepted the burden of responsibility in helping energy-poor countries to ease the strain of increasing oil import bills.[17] Accordingly, at the May 1980 Taif OPEC Ministerial Meeting, OPEC's relations with the Third World occupied center stage. OPEC's long-term strategy-drafting committee basically accepted a Venezuelan proposal to convert OPEC's Special Fund into an international development agency, with an initial capital of $20 billion.

But the question remains: Can and will OPEC bail out the OIDCs? As stated earlier, OPEC, or at least some of its members, believe less and less in "handouts"; that is, subsidies and grants. OPEC's objective is to find and develop an energy future for the OIDCs that is less dependent on imported oil and more geared to using domestic resources.

This all sounds good. But is OPEC prepared and ready to assist in the development of such an "energy future"? It is, after all, no more an organization than the total of the priorities and preferences of its membership, and short-term national goals have a detrimental effect on its global efforts and their effectiveness. For example, the confrontation between Iraq and Iran has effectively paralyzed the development assistance efforts of Iraq, one of the emerging aspirants for a leadership role within OPEC. The developing countries do represent a serious concern for OPEC, but they also represent a political headache and liability. When more fundamental issues confront the organization, the poor of the world and their concerns tend to slip back to second place. The fact is that in terms of the oil supply and demand equation, the Third World is not very important.

This realization is at the very heart of some of the bitterness that OIDC political leaders have voiced about OPEC and the direct and indirect price they have to pay for imported oil. Some oil-importing countries in Africa, especially Kenya, have been very outspoken in their attacks on OPEC and at times on the Arab member countries of the organization. As the *Nairobi Times* stated in June 1979: "Whole economies of several Third World countries are about to be put out of business. The very security of nations, to say nothing about the welfare of millions of Third World people, will be at stake. OPEC nations appear to be oblivious to the harm they are causing." For the most part, however, the OIDCs are silent. When the non-aligned nations met in India in February 1981, their communiqué blamed the economic problems of the OIDCs on the policies of the advanced industrial countries, and despite the impact of the 1979–80 oil price increases, they exonerated the oil producers.[18] "We are trapped in the solidarity of the decolonized," said the former foreign secretary of a leading Third World nation. "In private, we are very angry about oil prices and what they are doing to us. In public, we are very careful, for we hope to get something from the oil producers, and we don't want to offend them."

Some of these commentators and critics tend to recognize that the OPEC members closest to their own development predicament are the most sympathetic. Yet, Algeria, Nigeria, and often Venezuela have been persistent price hawks over the past eight years exactly because of perceived domestic pressures to generate revenues through oil sales that would enable them to accelerate their own domestic development. Algeria and Nigeria are both in desperate need to reform their agricultural sector to feed a rapidly growing population, which wants to know nothing of the shabbiness of villages, craves the lights of the city, wants to eat a diet of imported food, and seeks to emulate a Western, high-energy-consuming life-style. It is highly doubtful that citizens of such countries are eager to sacrifice their own aspirations to those less fortunate countries. Thus, the problem is that the most logical supporters of OIDC development within OPEC are also the ones who themselves have the greatest current domestic need for an accelerated flow of riches.

The "oil weapon" will be played out in terms of producer-exporter and industrialized country confrontations, not within the context of either party's relationship to the Third World. For better or worse, these countries will remain on the sidelines during the course of the confrontations over pricing and production levels. But OPEC is concerned over the longer-term energy future of the OIDCs and has stressed the need to assist them in the current adjustment process, both through concessional assistance to provide for balance of payments financing and through the development of indigenous energy resources. On paper, OPEC's aid record is impressive; OPEC countries have been providing roughly $4 to $5 billion per year in official development assistance through a variety of national and international institutions, particularly the OPEC Fund for International Development.

But it is not only financial assistance that is needed for the OIDCs, but also technology, training and institutional development, project management, and financing. In these areas, OPEC remains a helpless giant, more able to deliver short-term financial help than longer-term technical solutions.

All the above raises the specter of a new type of external dependency developing, based on energy deficiencies. The oil bill will further deepen the foreign assistance dependency of the poorest and energy-deficient developing countries. In the industrializing developing countries like Turkey and Brazil, the new link is between international financial lending institutions, with OPEC and non-OPEC oil exporters as the major holders of all trumps: As a São Paulo industrialist stated, "Without big banks there are no big bucks for oil, and without OPEC there is no oil for industrialization — to repay the loans to the big banks!"

The domestic implications of the energy crisis also present a formidable challenge for the OIDCs. The financial burden of rising oil import bills will have a critical impact on domestic public policies. Domestic growth rates may be threatened, equity objectives jeopardized, and the adjustment process slow and painful. In a world of shrinking resources available for public investment programs, the pressing social or welfare programs will be the first candidates for curtailment and elimination. This may exacerbate social tensions in the OIDCs, particularly in countries with large disparities in income distribution and inequitable access to services. The warning is clear: The tenuous social contract that holds together the fragile body of societies in transition may well be strained beyond limits by a world of uncertain and expensive energy.

10

Burdens of Debt and the New Protectionism

by Hans-Eckart Scharrer

The international constraints on energy that became evident in 1973 are bringing a number of highly significant changes in the international economy: lower rates of growth of world production, changes in world demand and industrial structures, changes in relative prices, and a redistribution of world income.[1] These changes are inevitably exerting a strong influence on the pattern of international trade, payments, and investment. Further constraints in the years ahead will intensify these trends, altering even more the world economy and raising serious questions about its stability. Problems of debt and a revival of protectionism may well become central questions of world politics.

According to our two energy scenarios, in the period to 1990 (and on to 2000) the world supply of energy would be rising by an annual 2.6 percent in the Upper Bound case and by only 0.8 percent in the more pessimistic Lower Bound case. World petroleum production (except in Comecom countries) is projected to expand by 0.9 percent per year in the Upper Bound scenario and to contract by 1.0 percent in the Lower Bound case.[2] This might limit the annual growth of production in the world economy to an average 4.5 percent at best and 2.5 percent at worst.[3]

How, then, would the total increase in production be distributed internationally? The "best" performance to be expected over the decade would be an average 5.5 to 6 percent GNP growth for the developing countries without oil and an average 3.5 to 4 percent for industrial economies. On the basis of the Lower Bound scenario, GNP growth rates would be in the range of 3.5 to 4 percent and 1.5 to 2 percent, respectively. Economic growth, then, could well be even slower than in the years since 1973, a period in which growth was generally considered too low and unemployment too high.[4]

For the industrial countries, average growth rates of about 2.5 percent are the minimum needed to prevent absolute employment levels from falling (assuming that labor productivity will continue to grow by that amount). But a number of countries already have a high rate of unemployment; moreover, in the 1980s the working population is again rising. Thus, growth rates well above the minimum are needed to attain full employment. Indeed, that is what all countries are aiming at. Under these circumstances, any energy constraints would place the economies of the industrial countries under a great deal of strain.

For the oil-importing developing countries (OIDCs), average GNP growth rates of 3.5 to 4 percent, such as would emerge under the Lower Bound scenario, would clearly result in a major setback to development. With the population growing at 2 to 2.5 percent, real per capita incomes would be rising by no more than 1 to 1.5 percent on the average. And it is likely that even this small increase would be eroded by deteriorating terms of trade vis-à-vis the oil producers and, perhaps, the industrial countries.[5] Thus, for the great majority of countries and peoples, the 1980s could well turn out to be a decade not of development but of major disappointment and indeed despair.

The growth scenarios of the World Bank and the United Nations offer little, if any, consolation. The World Bank projects average growth rates of GNP at 3 to 3.5 percent for the industrial countries and from 4 to 5.5 percent for the OIDCs, with great regional disparities and variations among countries at different stages of economic development. This implies, for example, a zero or even negative growth rate of per capita income for the African countries.[6]

Even this modest achievement depends upon several requirements that cannot be guaranteed: There must be a sustained level of investment sufficient both to stimulate demand and employment in the short run in a general climate of recession and to contribute critically to medium-term economic growth and adjustment.

Other critical requirements include the development of ways to channel enough capital to OIDCs to permit them to maintain their previous import levels in real terms and to finance the investment required to adjust to the new international economic situation; the maintenance of a liberal international trade system; and a world economy that does not suffer additional economic and political shocks.

There are disturbing signs that the World Bank's low case is already developing. Aid for low-income countries is unlikely to meet the modest requirements of the high case, with some middle-income countries even now experiencing both debt and political difficulties. Without strong international action in the 1980s, the World Bank warns, the low case is the likelier outcome, and a number of factors, including serious political instability, major problems in capital markets, or a breakdown of world economic cooperation, could bring about a much worse outcome.[7]

If we accept the projections for oil and energy supplies as being fairly realistic — and in any case not overly pessimistic — and if we further consider the limitations that the energy supply will impose on economic growth, then it is also realistic to assume that oil consumers will compete for the scarce resource — with what consequence? Oil prices, as well as prices of other fuels, will periodically go up, relative to the prices of other goods. How far is an open question. If we take as a yardstick the former price target of the most important oil-exporting country, Saudi Arabia — the driving force behind OPEC's efforts to establish a mechanism of "crawling" oil price increases — then the price should rise by about 3 percent per year in real terms.[8] However, Saudi Arabia now seems more interested in level prices for the next few years in order to protect oil's share of the world energy market and to allow the international economy (in which Saudi Arabia now has a large stake) time for economic recovery. The Upper and Lower Bounds of this book suggest annual real increases between 1980 and 2000 of 2 to 4.5 percent, although the actual moves are likely to follow the pattern of "jagged peaks and sloping plateaus."

Where does all this lead with respect to international trade and payments? Let us begin with what has happened since the first oil price increase.

The Future Pattern of World Trade

The Experience of the First Oil Price Rise

The quadrupling of the oil price in 1973–74 was a watershed in the development of the world economy:

The annual rate of growth of world production dropped sharply from 5.4 percent in the period 1961–73 to only 3.6 percent in 1974–79. As far as the developed countries are concerned, their average growth rate was almost cut in half — from 5.0 to 2.7 percent; while the OIDCs faced only a slight drop in growth — from 5.4 to 5.2 percent.[9]

At the same time, the era of full employment in the developed market economies ended. The unemployment rate in the OECD jumped from an average 3.0 percent during 1964–73 to 4.9 percent in 1974–79. In 1979, five years after the first oil shock and on the brink of a new wave of oil price increases, unemployment was still at a level of 5.1 percent.[10]

Inflation, which had already started to accelerate in 1972, received a major push from the oil price increase. The attempts of organized labor to shift — through high wage increases — the burden of higher oil prices onto business, and the reaction of business on the price front, led to a rise in consumer prices at an average rate of 11 percent a year over the period 1974–79. Compare this to the 4 percent price rise in the sixties and early seventies.[11] Expansionary fiscal policies and accommodating monetary policies in many countries compounded the inflationary pressure, with results varying according to differing national policies.[12]

The current account surplus of the oil-producing countries increased from $7.5 billion in 1973 to almost $60 billion in 1974 (see Table 10.1). At the same time, the OECD countries' current account position deteriorated by $36 billion, to a negative $26 billion, and the developing countries' by $18.5 billion, also to a negative $26 billion. Only four years later, however, the OPEC current account surplus had virtually disappeared and the industrial countries had restored their 1973 surplus of $10 billion. On the other hand, the $23 billion deficit of the developing countries in 1978 was at about the same level as in 1974.[13]

The different exposures and national policy responses to the internal and external shocks caused by the oil price increase also produced large fluctuations and alterations in exchange rates.

In line with these developments, the rate of expansion of world trade slackened from an average 8.5 percent during 1963–73 to a mere 4.5 percent in 1973–79. Trade thus grew only moderately faster than production, which meant that the process of international integration and division of labor slowed down. As in the entire postwar period, total trade growth depended heavily on the expansion of trade in manufactured products. More than before, mineral resources lagged behind overall trade expansion (see Table 10.2).

Until 1972, trade volume was expanding faster, and in the sixties much faster, than export prices. This pattern has since been reversed, which can be explained mainly by the rise in oil prices. Whereas the share of crude petroleum in the total volume of exports declined from 5.5 to 4.1 percent between 1970 and 1978, its share in trade value increased to 12 percent (see Table 10.3). The recent round of oil price increases has accentuated this upward trend. The increase in the oil share took place at the expense of other raw materials as well as industrial goods.

As a result of the shifts in the terms of trade to the benefit of OPEC countries, their share in the total value of world exports increased from 7.5 percent in 1973 to 11 percent in 1978. Their share in the world import of commodities did not increase to the same extent, rising "only" from 3.5 to 7.5 percent in the same period.[14] It should be borne in mind, however, that OPEC countries are large importers of services. For goods and services

Table 10.1 Current Account Balances of Selected Countries and Country Groups: 1973–1982
(billions of dollars)

	1973	1974	1975	1976	1977	1978	1979	1980	1981[a]	1982[b]
United States	7.1	4.9	18.3	4.4	−14.1	−14.1	1.4	3.7	8.8	3
Canada	0.1	− 1.5	− 4.7	− 3.9	− 4.0	− 4.3	− 4.2	− 1.6	− 7.5	−10.8
Japan	− 0.1	− 4.7	− 0.7	3.7	10.9	16.5	− 8.8	−10.7	5.5	17
France	− 1.0	− 6.2	− 0.2	− 5.7	− 3.1	3.3	1.2	− 7.4	− 6.5	− 6.8
Germany	4.6	10.3	4.0	3.9	4.1	9.2	− 5.3	−16.4	− 8.5	− 1.8
Italy	− 2.7	− 8.0	− 0.8	− 2.8	2.5	6.2	5.5	− 9.6	− 9.5	− 5
United Kingdom	− 2.4	− 7.7	− 3.4	− 1.6	− 0.1	1.8	− 1.8	7.5	14.3	2.3
OECD	10.5	−26	0	−18	−24	10	−33	−73	−35	−27
OPEC	7.5	59.5	27	36	29	4	62	110	60	35[c]
Non-oil LDCs	− 7.5	−26	−30	−17	−12	−23	−38	−60	−68	−71
Other countries	− 3.5	− 9.5	−18	−13	− 8	− 9	− 4	− 1	− 5	− 8
Statistical discrepancy	7.0	− 2	−20	−12	−16	−18	−13	−24	−48	−69

[a] Estimate.
[b] Projection.
[c] Range of 15–35.
Source: *OECD Economic Outlook*, various issues.

Table 10.2 Growth of World Exports and Production, 1963–1980
(annual average percentage rate of change in volume)

	1963–73	1973–79	1974	1975	1976	1977	1978	1979	1980
World commodity output	6	3½	2½	−1	7	4½	4	3½	1
World exports									
Total	8½	4½	3½	−3	11	4½	5½	6	1½
Agricultural products	4	4½	−3½	5	9½	2	9	7	1
Minerals*	7	½	−2½	−7½	4½	2	1½	4	−7
Manufactures	11	5	8	−4½	13	5	5	5½	3½

* Including fuels and nonferrous metals.
Source: GATT, *International Trade 1980–81,* Geneva 1981, p. 2.

combined, their export share in 1978 equaled their import share in world trade. And even for commodity trade proper, the expansion of their imports was impressive — 38 percent per year during the period 1973–78.
The OIDCs' share in world exports was unaffected by events in the oil

Table 10.3 Primary Commodities and Manufactures in World Export: 1970–1979*
(shares of trade value)

Year	All primary commodities	Primary commodities except crude oil	Crude oil	Manufactures
1970	37.7	32.2	5.5	62.3
1973	38.4	31.6	6.8	61.6
1974	44.2	30.6	13.6	55.8
1975	41.7	28.1	13.6	58.3
1978	38.2	26.2	12.0	61.8
1979	41.1	28.4	12.7	58.9
1980	44	28.6	15.4	56.0

*Without state trading countries.
Sources: United Nations, *Yearbook of International Trade Statistics,* various years; UN, *Monthly Bulletin of Statistics* 34, no. 6 (June 1980), p. xx; IMF, *International Financial Statistics* 33, no. 7 (July 1980), p. 48; HWWA Commodity Price Index; HWWA calculations and estimates.

Table 10.4   Import Growth of Oil-Exporting Countries 1974–1980
(percentage changes of dollar imports)

	1974	1975	1976	1977	1978	1979	1980	Ø 73–80
Capital surplus countries ("low absorbers")[a]								
Kuwait*	46.2	53.5	39.2	45.7	− 4.9	12.9	35.7	31.0
Saudi Arabia*	44.6	47.4	106.3	68.6	39.4	19.8	26.9	48.2
Iran*	60.6	90.7	24.6	20.3	25.8	−54.2	33.4	19.7
United Arab Emirates*	112.5	54.6	24.2	51.7	3.8	29.8	29.8	40.5
Iraq*	161.3	77.8	− 7.2	14.9	− 6.0	130.5	39.0	47.1
Libya*	60.4	28.2	− 9.3	17.7	21.7	72.5	25.5	28.5
Qatar*	39.3	51.1	103.5	47.1	− 3.4	20.4	− 8.0	31.3
Others ("high absorbers")[a]								
Indonesia*	40.7	24.2	19.0	9.8	7.4	8.0	71.3	24.1
Mexico	58.9	8.6	− 8.3	− 9.1	37.8	59.9	61.6	26.3
Nigeria*	49.0	117.5	36.2	34.2	16.2	−27.7	74.6	36.2
Algeria*	76.3	41.6	−11.3	33.4	21.9	− 2.3	25.7	23.7
Venezuela*	50.4	40.5	21.3	39.4	17.7	− 9.3	3.9	21.7
TOTAL, oil-exporting countries[b]	62.3	58.2	23.6	34.0	18.6	0.7 (13.9)[c]	33.8	31.5

* OPEC members.
[a] Ranked in the order of their balance on current account in 1978 (excluding transfers).
[b] Without Gabon, Trinidad and Tobago, Bahrain, Ecuador, and Mexico.
[c] Without Iran.
Source: IMF, Direction of Trade Yearbook 1981; author's own computations.

markets. With 12.5 percent, it was roughly the same in 1978 as it was in 1968 and 1973. Indeed, the developing countries were able to step up their exports more than the industrial economies did. This success, which was achieved in the face of a slowing down of world economic activity, was due mainly to their being able to raise the share of manufactured products in their total export basket from 34 percent in 1973 to 40 percent in 1980.[15] Particularly noticeable was the growth of their exports of engineering products.

Yet we need some qualification here. Most OIDCs did not manage to adjust to the changes in world demand. Trade dynamism was concentrated within a rather narrow group of about ten newly industrializing countries (NICs), whose share of world manufacturing production increased between 1963 and 1973 from 5.4 to 7.6 percent and continued to grow after the oil shock to reach 9.3 percent in 1977.[16] Most of these countries, with the possible exceptions of Brazil and Mexico, pursued an outward-looking growth strategy, led by exports. Accordingly, their share in world exports of manufactures, which had grown from 2.6 to 6.3 percent in 1963–73, expanded further to over 7 percent, whereas the share of the other developing countries declined from an already low level of 2.3 percent in 1973 to 1.6 percent in 1976. The most important market for the newly industrializing countries was the OECD area.[17]

These countries also provided an important outlet for the export of capital goods from the developed areas in the recession years after the oil shock. In fact, they were able to decouple themselves from the downturn that hit the industrial economies and instead managed to continue to achieve satisfactory, even though slightly lower, rates of economic expansion. The reasons for this can be found in their investment strategy: They judged that the recession in the industrial countries would be short; therefore, they used the high liquidity in international financial markets to obtain credit at favorable terms to finance their current account deficits and their investment programs.[18]

Indeed, it appears that not only the newly industrializing economies but other oil-importing countries, too, were both prepared and able to finance sizable current account deficits. Thus, although trade expansion slowed down due to the world recession, availability of finance was generally not a constraint to trade growth. This meant, on the other hand, that both the foreign indebtedness and debt service obligations of the developing world — and of the middle-income new industrializers in particular — increased during the period after 1973 (see Table 10.5). A major portion of the total credit expansion was provided by private banks in the form of syndicated loans at commercial terms.

Trade after the Second Oil Shock

After the initial crisis of 1973–74, a process of readjustment certainly began in the world economy. Some of it was solidly grounded, some of it was shaky, some of it was very costly. But it was taking place, although in many cases, the initial impetus was being overcome by complacency. In 1979, half a decade later, oil prices again exploded. Today, in the early 1980s, we can see the effects of the second price increase on the world economy. What is clear is that the second shock has considerably complicated the process of adjustment.

The industrial countries on both sides of the Atlantic — the most im-

Table 10.5 Public and Private Debt of the Developing Countries
(billions of dollars)

	1970	1975	1978	1979	1980[a]
Disbursed debt outstanding at year end	62.4	168.1	316.9	369.2	416.0
Official sources	34.9	74.2	122.2	136.2	154.0
Private sources[b]	27.6	93.9	194.7	233.0	262.0
Debt service	7.8	24.2	51.9	67.2	76.0
Official sources	2.5	5.5	9.2	11.5	14.0
Private sources[b]	5.3	18.7	42.7	55.7	62.0

[a] Estimated.
[b] Includes some lending by official sources that is not guaranteed by a public body in the borrowing country.
Note: Details may not add up to totals because of rounding.
Source: World Bank, *Annual Report 1981* (Washington, D.C.: 1981), p. 24.

portant markets for exports — are far from full employment and price stability, however those two central goals might be defined. They are in the midst both of stagnation and of a painful process of restructuring their economies in response to the change in energy prices and increasing competition from Japan and the newly industrializing nations. In the first part of 1982, their rate of unemployment was running at over 9 percent and their average inflation rate was still over 10 percent. Meanwhile, within two years, 1979 and 1980, the OPEC countries again acquired a heavy current surplus, some $110 billion in 1980. This surplus declined to $60 billion in 1981. It was balanced by the aggregate deficits of the industrial countries of some $75 billion in 1980 and $35 billion in 1981, and of the OIDCs of about $60 billion in 1980 and $68 billion in 1981 (see Table 10.1).

The surge in oil prices combined with the current account deficit of the OIDCs that followed has again, as after 1973, created heavy unemployment, adjustment, and financial burden for the world economy. From a global perspective, the initial situation may appear better this time, for a greater portion of the total swing into deficit was borne by the developed nations — almost 80 percent of the total turnaround of $105 billion compared to 70 percent of $52 billion then. Moreover, two "strong" economies, Japan's and Germany's, were initially subjected to more than half of the total swing. This is quite different from the situation in 1973–74, when the burden fell mainly on Italy, the United Kingdom, and France, with Japan carrying only 9 percent of the swing and Germany actually doubling its surplus. Yet the situation is inherently unstable since no country is actually prepared to tolerate a deficit for an extended period, and there is a strong disposition to shift deficits to other countries.

The level of international indebtedness is seen as unreasonably high by some borrowers and even more so by private financial institutions and banking authorities. Unlike the situation after 1973, there is now little scope for much further expansion of bank lending. Given this state of affairs, it is somewhat hazardous to apply mechanically the trade:output ratio of the post-1973 period to forecast trade development in the 1980s. Nevertheless, when this is done, we arrive at growth rates of world exports of 5.5 to 6 percent at best and 3 to 3.5 percent at worst. Realization of the best case would thus bring some improvement over the 1973–79 period (4.5 percent) but would still be well behind the rate of 8.5 percent achieved in 1963–73. Trade performance under this scenario would be very much in line with the high case of the World Bank, which assumes a growth of exports of 5.2 percent per year during 1980–85 and of 5.7 percent during 1985–90.

But various objections may be raised against this moderately optimistic outlook:[19]

In industrial countries, governments faced with low rates of economic growth and mounting problems of unemployment may increasingly take recourse to protectionist measures to shield their industries (and workers) from competition from abroad. The objective of reducing current account deficits will work in the same direction. The theoretical support of such policies is already in evidence in several countries.[20] In fact, because of different economic and social structures, the OECD countries will be affected unevenly by oil scarcity and rising oil prices. Some countries will be better placed to cope with the burden of adjustment than others. In particular, those having difficulty will experience growing protectionist pressures. However, even those that are adapting better will not be inclined toward further trade liberalization, and on the whole, protectionism is likely to rise and trade expansion to be curbed.

Significant questions hang over the future imports of the OPEC countries. The Iranian revolution and the seizure of the mosque in Mecca have demonstrated to the ruling elites the risks of overambitious growth and development strategies. High rates of inflation, rising disparities of incomes, the social uprooting of major parts of the population from the traditional sectors of the economy (Bedouins, farmers, craftsmen), the influx of foreign workers, technicians, and advisers, and, in consequence, the decay of traditional value systems — all these are the price to be paid for a forced development strategy. In addition, the effort to modernize a country in great haste entails high economic costs, given the existing technical and employment bottlenecks. Inadequate ports and transport facilities, a lack of warehouses, deficiencies in planning and coordination — all drive import costs up and give rise to major waste. These social and economic "misdevelopments" are a burden to the population, undercutting to some degree the positive effects of public spending. They also pose a threat to the power of the ruling elites, and not only in the more feudal regimes. It is fair to assume that the political leaders have learned some lessons and will try to control the economic development with the consequence that the growth of imports will be slowed.

The OIDCs are faced with large and persistent current account deficits. This time, the level of external debt accumulated in the 1970s may limit their ability to increase imports at the "historical" rate. This applies in particular to the new industrializers. Whereas, after the first oil shock, they generally had been able to maintain their demand for capital goods, quite a few of them are now among the problem countries in terms of international indebtedness.

On the whole, these factors would suggest that the rate of trade growth will lag behind the projections. On the other hand, some factors may be expected to work in a more favorable direction. First, the absorptive capacity of oil-producing countries in the 1980s should not be underestimated. Up to now, there have been no real signs that the oil-exporting countries have actually scaled down their infrastructural and industrial planning for the years to come. For the population-rich "high absorbers," this is indeed unlikely to happen since they are faced with all the bottlenecks hampering development typical of the Third World at large. They therefore need massive transfers of real resources from the industrial world. The "low absorbers," on the other hand, could become "high absorbers" in the future. The rapid rate of growth in income increases their demand for foodstuffs, consumer durables, and foreign services, including travel. In addition, these countries will have to continue and even expand their imports of capital goods and of technological and managerial know-how from the industrial and newly industrializing countries in order to overcome their almost complete dependence on petroleum exports by creating an efficient manufacturing industry. Priority will be given to projects that use the available oil and gas as material or energy resources, such as refineries, petrochemical complexes, fertilizer plants and steel mills.

For the infrastructure, too, the challenges are enormous. Transportation and communications, electricity and water supply, education and health — these are areas where development is still in an early phase, despite major efforts undertaken in the past, and will require large imports of goods and services. However, industrialization may in the course of time lead to a replacement of imports by domestic products, and the oil-producing countries will eventually sell their own oil and energy-intensive products in international markets.

In sum, the annual growth of imports of the OPEC countries can be expected to be in the range of 20 to 30 percent in nominal terms; that is, at rates comparable to those realized at the end of the 1970s. A recurrence of the over-expansion of imports experienced in the first few years after the 1973–74 oil price increase is, however, as unlikely as the reverse, a major decline in import growth.

In the face of energy constraints, the industrial nations and the OIDCs will face major adjustments, which can be summarized as follows: the adoption of energy-saving production techniques, energy conservation by private households, substitution for products with a high oil or energy content by others using less energy, development of new energy resources.

Moreover, the effects of the adjustments in terms of trade on the level and growth of private disposable income will bring about alterations in domestic and international consumption patterns. All of these changes in the pattern of demand and supply will require an acceleration of private and public investment worldwide and corresponding flows of plant and equipment.

In terms of volume, gas, coal, and machinery and equipment both for energy-saving investment and for the production of energy will take up a larger share of world trade. Manufactured commodities with a high oil and gas content and energy-intensive products will decrease their share. In terms of value, however, higher shares of petroleum and petroleum products, gas, and, to a lesser degree, coal in total trade may be expected. Manufactured products should maintain their share, whereas the proportion of raw materials (other than energy) is likely to fall. Energy producers, including "new" suppliers, will increase their share of world trade, the newly industrializing will remain roughly the same, and the shares of the industrialized economies and the OIDCs will both decline. The oil-exporting countries will increase their share at the expense of both the industrial economies and the OIDCs.[21]

Under the impact of the jumps in oil prices, the system of international trade is also undergoing changes. First, protectionism is on the rise, generally through nontariff barriers, including "voluntary" restraints. The target countries are mainly newly industrializing countries and "aggressive" (or merely successful) industrial economies.

Second, an increasing portion of all trade is likely to be government dependent or "managed" trade in the forms of directly and indirectly subsidized exports and of state-to-state trading of oil, gas, and other commodities. The latter would imply a creeping relapse into bilateralism, often with visible or invisible political strings attached. Both managed trade and increasing trade barriers will detract further from an efficient international allocation of resources. Thus — and this is a critical point — such policies are equivalent to a renunciation of potential economic and income growth.

Third, trade conflicts could arise from divergent oil and energy policies, especially price and tax policies, which could give producers an artificial advantage against their foreign competitors. The controversy in 1980 about U.S. exports of synthetic fibers to the United Kingdom is an example of this type of conflict.

Can the liberal international trade system survive these pressures? This is a major question for the decade, for the liberal trading system may well not survive unless the governments and business enterprises in the industrial world undertake dynamic, forward-looking adjustment strategies rather than try to conserve traditional production, consumption, and income patterns in a rapidly changing international environment. The OPEC countries also must make a positive contribution through their oil production, oil price, and import policies. An objective of the major industrial economies must be to strengthen the moderates in the OPEC camp by a constructive policy approach that considers their legitimate interests, particularly with respect to the investment of their proceeds from oil exports.

The Future Pattern of International Payments

It is the system of international payments that knits the world economy together. The dramatic rise in oil prices and the resulting shifts in international trade and national economies have sent shock waves through that system. Thus, it is particularly important to understand how the international payments system will evolve in the new environment.

The global pattern of world payments in the 1980s will be determined, first of all, by the OPEC countries' pricing policy and their aggregate current account surplus. It is reasonable to assume that the OPEC surplus — and the corresponding deficit of the oil-consuming world — will be sustained in the years to come, although at much lower levels then anticipated immediately after the price hike[22] (see Table 10.1). A number of problems and questions that had an easy answer in the 1970s may well come up again, but in a much more difficult version in the rest of the 1980s. There are four main issues: the distribution of the current account deficit among oil-importing countries, the financing of current account deficits, the deployment of OPEC's surplus funds, and the role and instruments of adjustment policies.

The Pattern of Current Account Deficits

The aggregate current surplus of the OPEC countries is matched by a corresponding deficit in the rest of the world. How this deficit is distributed is important mainly for two reasons:

First, the entire deficit expresses a primary deflationary gap resulting from the transfer of income — and hence purchasing power — from countries with a high propensity to spend to economies with a high propensity to save. The distribution of the deficit determines each country's share in the total loss of world demand, production, and employment.[23] Second, to finance a current account deficit, foreign capital (private and government loans and credits, equity finance) has to be attracted. This may compel a government to change its domestic policy in order to improve its international credit but, in the process, to give up priorities otherwise deemed essential. In addition, generating and transferring debt service payments will put a burden on future incomes and balance of payments.

These fears and preoccupations are not, by any means, always well founded. Yet they are a political reality and may indeed influence political action in most countries and hence the development of international economic and monetary relations in the years to come.

Since the oil-producing countries will be likely to expand their demand for high-quality investment and consumer goods, industrial countries and, increasingly, the newly industrializing economies will be in a better position to balance their accounts than the large group of other OIDCs. Therefore, in the course of time, a disproportionate share of the aggregate deficit is likely to fall on the latter, always assuming that it can somehow be financed in the first place. This would be in line with the experience after the first oil shock.

How will the deficits be distributed between the advanced countries and the new industrializers? In addition to OPEC import demand, much will depend on the import policies of the industrial countries. NIC products generally compete with those of "traditional" industries in the advanced economies, industries that are declining and facing worldwide overcapacities. Thus protectionist pressures are likely to inhibit the rapid growth of imports from the new industrializers. It is therefore safe to assume that the new industrializers will not be able to compensate for their higher oil bill to any major extent through improvements in their trade position vis-à-vis that of the developed countries.

Initially, the overall deficit of the industrial countries was more evenly spread among them than after the first oil shock. No country had a substantial current account surplus, and even "structural" surplus countries such as Germany and Japan were running large deficits (Japan also had a large deficit in 1974).

Over time, however, greater economic and monetary differentiation can be expected. Not all developed economies are equally well equipped to cope with the challenge of rising real prices of oil and energy. Great differences in rates of inflation and unemployment, as well as in the ability to adjust production to the changes in prices and demand, will make for a great variety of reactions, as the chapters on Western Europe, Japan, and the United States demonstrate. Apart from a major redistribution of current account deficits, we would therefore also expect larger fluctuations and continuous movements of exchange rates.

Such changes help to bring about external and internal adjustment. It is therefore important that the monetary authorities should not interfere with exchange rate trends. This applies especially to the European Monetary System, where clinging to rigid rates in the face of divergent inflation and expansion rates is bound to add to oil-related problems.

Uncertainty about future exchange rates is likely to promote the diversification of assets of public and private financial interest. After the move to floating exchange rates, an increasing proportion of exchange reserves shifted into currencies other than that of the U.S. dollar. By the end of 1978, roughly one third of the foreign exchange reserves of countries other than the six reserve centers were invested in these different currencies.[24] Recent trends on the exchange markets related to interest rates and current account balances have significantly reversed this shift. The future of the diversification process will, of course, depend to a large degree on the strength of the dollar vis-à-vis its substitutes and on the relative yields and marketability of dollar and nondollar assets. To channel this process smoothly will be a major task for national and international monetary authorities. The provision of instruments denominated in special drawing rights by the IMF could well contribute to more stability in international exchange relations. Closer cooperation between authorities of the reserve centers and those wishing to diversify their official reserves is also needed.[25]

Financing Current Account Deficits

The aggregate deficit of the oil-producing countries is at any moment automatically and fully financed by the OPEC capital surplus, which is the counterpart of their surplus on current account. It is also true that the actual deficit of any individual country is always financed; this is an accounting identity. This identity does not, by any means however, indicate a stable state. The method of financing may simply postpone the trouble to a not too distant future; for example, by expensive and short-term borrowing. Or the actual volume of imports and the resulting current account deficit may be far below the desired volume; that is, the volume necessary to maintain a certain minimum standard of economic activity.

The industrial economies can draw on a variety of sources to finance their deficits. Indeed, it is difficult to imagine that any of them would face a serious financing problem in the years to come. Their international credit is supported by a number of factors: Their net external indebtedness is generally low, with few exceptions. Their ability to generate and transfer future debt service payments is considered high and is less likely to be impaired by trade restrictions than is the case for less developed countries. Their membership in "clubs" like the OECD and the European Community gives them some guarantee of the economic and financial support of economically strong partners should the need arise. The political stability of the Western economies, combined with their legal system and tradition, tends to minimize the political risk for any lender. Moreover, the experience of Italy and the United Kingdom has shown that critical financial problems such as those that came up after the first oil price increase can be overcome rather quickly.

Developed countries are therefore the ideal candidates not only for bank financing but also for direct credits and long-term investment from capital-surplus oil countries. In addition they have access (as does every country) to the financing facilities of the IMF (which, in the case of the Group of Ten, are strengthened by the collective SDR 6.5 billion pledge under the General Arrangements to Borrow).[26] EEC countries benefit also from the $30 billion short- and medium-term financing facilities established under the European Monetary System and from ad hoc instruments such as the Ortoli facility.

In order to assess the present and foreseeable financial position of the OIDCs, it is useful to study the experience of the period following the first oil shock.

Of the total deficits accumulated over the years 1974–79, about half was financed by borrowing from banks (see Table 10.6). Altogether, about 70 percent was provided by official and private transfers, direct investment, and official development aid.[27] The contribution of reserve-related credit facilities, which includes the use of IMF credit, was significantly less than the amount provided by the private bond market. Indeed, an abundant supply of private and government finance enabled the OIDCs not only to offset their current account deficits but to augment their reserves substantially.

As a result, total foreign indebtedness of non-OPEC developing countries almost tripled, from about $100 billion at the end of 1973 to $272 billion at the end of 1978.[28] At the same time, their debt service payments increased rapidly, from $14 billion to $44 billion in this period, as a consequence of their greater reliance on more expensive, shorter-term private credits rather than on concessionary official finance.

But not all developing countries were able to draw on the various sources of finance, as the World Bank pointed out:

Table 10.6 Non-Oil-Developing Countries: Current Account Positions[a] and Aspects of Their Financing, 1971–1980

(in billions of dollars)

	1973	1974	1975	1976	1977	1978	1979	1980
Sum of current account deficits	−16	−38	−48	−38	−36	−42	−61	−83
Sum of current account surpluses	4	1	1	5	7	5	3	1
Accumulation of reserves (−)	−10	−2	2	−13	−13	−15	−12	−2
Total financing requirement	−22	−39	−45	−46	−42	−52	−70	−84
Nondebt-creating flows (net)[b]	10	13	12	12	15	15	22	21
Long-term borrowing from official sources (net)	6	10	12	11	13	14	15	21
Use of reserve related facilities (net)[c]	—	2	2	4	−1	−1	—	3
Other borrowing (net)[d]	6	14	19	19	15	24	33	39
Memorandum items								
Borrowing in bond markets, net of repayment	1	1	1	2	3	4	3	2
Borrowing from banks, net of repayment[e]	10	15	15	21	14	25	41	52

[a] Balance on goods, services, and private transfers.
[b] Official transfers, SDR allocations, gold monetization, valuation adjustments, and direct investment flows (net).
[c] Use of Fund credit and short-term borrowing by monetary authorities from other monetary authorities.
[d] Including errors and omissions.
[e] BIS series, which includes all private and public borrowing from banks.

Source: IMF, *International Capital Markets — Recent Developments and Short-Term Prospects, 1981*, Occasional Paper No. 7 (Washington, D.C.: August 1981), p. 48.

Most middle-income countries (and one low-income country, Indonesia) resorted to borrowing from private sources. But most low-income countries had to rely on Official Development Assistance (ODA) — concessional loans and grants. [As a result] the middle-income countries as a group accounted for 83 percent of the total debt of developing countries at the end of 1978. The portion of debt from private sources was even more highly concentrated; 90 percent was owed by middle-income countries. The middle-income countries stand in sharp contrast with the low-income countries, whose borrowing grew much more slowly because of their limited creditworthiness and the limited availability of official lending.[29]

After the second oil shock, the banks were as liquid as after the first, and as long as the OPEC surplus prevails, this situation is unlikely to change. But will they be prepared to continue their lending to developing countries at the same pace as before? In 1979 and 1980, this has indeed been the case (see Table 10.6). As a result, the developing countries' indebtedness to banks increased by almost 60 percent, to $237 billion, within these two years. With greatly increased credit exposure mainly to middle-income LDCs, risk is considered to be higher. First, although major borrowing countries have truly used the loan proceeds for investment purposes, this could not prevent their debt service ratios from deteriorating. With these higher ratios, their vulnerability to external shocks has become greater. This has caused concern, especially when measured against the probability of future oil price increases, continued OPEC surpluses, and protectionist tendencies in the industrial countries, the major markets for the products of debtor countries. Second, several countries are already experiencing problems with the repayment of their old bank credits, and some have had to ask for a formal or informal rescheduling of their debt.[30] Third, the events in Iran, as well as in Nicaragua and El Salvador, have highlighted the political risk involved in lending to developing countries. Poland's debt problems have further accentuated concern about creditworthiness in country lending.

With spreads and fees at a low, the return on new loans is often inadequate in view of the risks. A number of medium-sized banks have already retreated from the market. The major international banks are therefore compelled to keep larger quotas in their portfolios, so their lending capacity is exhausted sooner. At the same time, the need to maintain, for prudential reasons, certain capital-asset ratios and the increasing control exercised by supervisory authorities over lending policies are bound to limit the banks' ability to expand their LDC lending at the same rate as before, which in the years 1973–80 was at an annual rate of 25 percent. American banks, which in the first years after the 1973–74 oil price increase were the major driving force in lending to less-developed countries, have for some time slowed down this activity.[31] European and Japanese banks have taken their place, although it is difficult to tell how long they will be prepared and able to maintain their commitment. But it seems likely that the rate of lending to developing countries could be reduced.

There is obviously a case for strengthening the role of the multilateral financing institutions — the IMF, the World Bank, regional development banks — and for increasing both the OECD's official assistance and direct financing by OPEC countries. The amounts provided so far by OPEC financing have been quite substantial, both in absolute terms and relative to the GNP of the capital-surplus countries and their current account balance. However, financial assistance has been lopsided geographically, a major portion having flown to Arab countries.

Since the second oil price hike, there seems to be a growing awareness among OPEC countries of the problems facing the Third World at large. There are some encouraging signs that the oil producers, both inside OPEC and out, will assume a greater role in the recycling process; to some degree this policy has already been implemented.[32]

On the other hand, OPEC's awareness of the risks greatly limits its willingness to expand its direct lending to the developing world. It prefers, in the words of the chairman of the Union des Banques Arabes et Françaises, to "extend credits to the Third World via the international banking system, for two reasons. First, debtor countries behave much more cautiously when it comes to servicing commercial credits than with regard to their obligations to foreign governments. Second, the banking community is both better suited and more business minded as far as risk assessment is concerned."[33] Moreover, there is the unstated expectation that the industrial countries through their monetary authorities — as lenders of last resort for the commercial banks — will de facto guarantee the OPEC countries' deposits. This guarantee is lacking in the case of direct lending to developing countries.

For all these reasons, official development assistance by industrial countries is unlikely to be raised significantly above the traditional level of 0.35 percent of GNP. According to World Bank projections, the aggregate annual amount (including assistance to multilateral institutions) is assumed to rise from $24.6 billion in 1980 to $30.8 billion in 1985, when measured at constant prices, and to $44.1 billion when valued at current prices. This increase would cover about one third of the developing countries' additional expenses from oil imports. Quite apart from their own economic difficulties and preoccupations, the developed economies appear unprepared to assume a greater role in easing the external constraints faced by the OIDCs as a result of the oil price increase. They argue that the problem has been created by OPEC and that it is up to OPEC to solve it. They also resent the tendency of the non-oil developing countries to blame the industrial countries for their economic problems.

But this reasoning is incomplete. Indeed, "we cannot see recycling as simply OPEC's responsibility," observed Richard Cooper, a former U.S. undersecretary of state for economic affairs. "Industrial economies must work harder to stem inflation, so that oil surplus countries can feel they can earn a real rate of return on their investments in the industrial countries. Developing countries will have to improve their own investment and borrowing climate, so as to attract funds not just from the North, but from the South as well."[34]

Moreover, both the OPEC and OECD countries have an obvious interest in the political and economic stability of the developing countries, be it only to prevent a collapse of the international financial system. On this basis of mutual interest and responsibilities, some sort of agreement to share the burden of larger financial (and real) transfers to the OIDCs, and more particularly to the poorer ones, should theoretically be possible. Yet there is some doubt as to how such an agreement would actually contribute, in the shorter term, toward solving the pressing problems of the poorer countries.

Rather than rely solely on unilateral or joint action by the OECD and OPEC countries, the multilateral financial institutions, in particular the IMF and the World Bank, will have to play a more important role in financing balance of payments deficits than they did after 1973–74, not only because of the limitations in commercial bank lending and bilateral development assistance, but also as a way to induce deficit countries to adopt adjustment measures more timely and actively than they did in the seventies.

In the spring of 1980, the World Bank embarked on a program of nonproject "structural adjustment lending," designed to help recipients "reduce their current account deficit to more manageable proportions over the medium term by supporting programs of adjustment that encompass specific policy, industrial, and other changes."[35] This could be supplemented by commercial bank lending under cofinancing arrangements.

There are also proposals to revise the bank's Articles of Agreement, which limit the total amount of outstanding loans to its own capital and reserves. By allowing a higher gearings ratio, lending could be expanded with the existing capital base. Plans have been shelved for the much-discussed creation of a separately financed Energy Affiliate to the World Bank to assist the developing countries in developing their energy resources.[36]

The IMF, too, is changing. There is now general recognition that both the nature of balance of payments deficits and the adjustment policies required are different under the global conditions of a high, sustained OPEC surplus from what they traditionally have been. The response of the fund has been in four areas: an increase in lending capacity, a lengthening of maturities, interest subsidies to the poorer member countries and a more supply-oriented approach to adjustment.[37]

Although the IMF's most pressing financial problems seem to have been solved, in 1980 the fund almost tripled its loan commitments over 1979 and 1978, to a total of SDR 9.5 billion.[38] If this trend should continue, the fund will periodically have to seek additional resources from its wealthier countries and perhaps from the international financial markets.

For the future, it is intolerable that the fund's ability to fulfill its proper tasks should depend, in the face of highly liquid international financial markets, on the preparedness of a few countries to replenish its resources. A more flexible approach that makes full use of the available opportunities for financing is therefore needed.

What is encouraging is that this time the fund has refrained from establishing new facilities with a low degree of conditionality, as was the case with the oil facility after the first oil price increase. This is all the more remarkable since the developing countries, through the Group of Twenty-four, have time and again "reaffirmed their view that in the present difficult situation additional balance of payments support should be provided with the minimum of conditionality."[39] Such an approach, if adopted, could only lead to disaster. True, the precise contents of conditionality need to take account of the fundamental changes that have occurred in the world economy. Indeed, given the external cause of the oil-importing countries' current account deficits, the traditional recipes of a demand-oriented adjustment policy need to be supplemented by more medium- and long-term supply-oriented policies aimed at strengthening the productive base of deficit countries. However, to infer from the "exogenous" OPEC surplus that financing should substitute for adjustment would be a costly misunderstanding for individual countries as well as for the international economic and financial system at large. Rather, as the managing director of the IMF said, "Financing and adjustment must go hand in hand," with financing needed to bridge the time between now and a situation in the future when the oil-exporting countries, too, will have adjusted to their greatly increased revenues.[40]

More could be done to support the international financial institutions in promoting adjustment. Among other things, the industrial and OPEC countries should consider transferring, voluntarily, at least part of their allocations of Special Drawing Rights irrevocably to the World Bank and the IMF to support medium-term adjustment programs for the OIDCs. It is indeed paradoxical that those countries that are best equipped to cope with the adjustment burden have received three quarters of the SDR 12 billion of newly created special drawing rights allocated in the three-year period 1979–81, whereas only one quarter has benefited the really needy. This is not to argue for a formal link between SDR creation and development financing, but for a temporary measure of international sharing under specific circumstances. Besides strengthening the financial base of the IMF and the World Bank and providing the final beneficiaries with conditional balance of payments support, a valve would be opened with respect to the industrial countries' incompatible efforts to balance their current accounts.

It should now be clear that shifting a greater portion of the oil-importing countries' aggregate deficit to the less developed countries is obviously bound to compound the debt problems of the latter and consequently the vulnerability of the financial system at large. This is true even if the transfer is done through international institutions and coupled with adjustment programs. Given that special drawing rights are created as a sort of "manna money," without any quid pro quo on the part of the recipients, assistance could be extended in the form of conditional grants, with repayment obligations only in case the beneficiary fails to comply with the policy prescriptions.

With the OPEC surplus expected to prevail for quite some time, there seems to be no easy solution to financing the deficits of oil-importing countries, and in particular of the OIDCs. But, without a solution, the liberal system of world trade and the international financial system may no longer be able to function.

The Investment Policy of the OPEC Surplus Countries

In the years to come, a major preoccupation of the OPEC countries, and more particularly of the countries with less capacity for foreign goods and services, will be to find secure and profitable investment outlets for their surplus funds.

Table 10.7 shows the quantitative, regional, and time dimension of the problem in retrospect:[41]

• The external assets (net) of the OPEC countries increased from $7 billion at the end of 1973 to $160 billion in 1978 and then more than doubled, to $387 billion, at the end of 1981.

• Only seven countries — Saudi Arabia, Kuwait, Iraq, the United Arab Emirates, Iran, Libya, and Qatar — were faced with high and growing financial surpluses. At the end of 1981, their net external assets were equal to the aggregate assets of all the OPEC countries combined, because of the net indebtedness of other OPEC members. Qatar, with its $16 billion, remains markedly behind the other countries in this category.

Table 10.7 Net Foreign Assets of OPEC Countries, 1973–1980
(billions of dollars)

	Net Assets December 1973	Net Assets December 1978	Net Assets December 1980	Net Assets December 1981
Saudi Arabia	4.2	67.3	118.2	161.6
Kuwait	3.7	35.7	67.1	76.2
Iraq	1.0	17.2	50.3	31.8
U.A.E.	0.5	15.0	35.0	38.6
Iran	0.7	26.5	31.8	28.2*
Libya	3.2	11.3	24.7	33.4
Qatar	0.9	6.0	10.8	16.1
Nigeria	−0.6	− 0.8	16.6	11.6
Gabon	−0.3	0.1	1.5	0.7
Indonesia	−4.2	− 6.9	0.8	− 1.3
Venezuela	1.0	0.3	0.7	7.7
Ecuador	−0.1	− 0.8	− 1.7	− 3.3
Algeria	−2.8	−11.4	−12.7	−13.9
OPEC TOTAL	7.2	159.5	343.1	387.4

* Includes substantial assets whose ownership is in dispute.

Source: Susan Bluff, "OPEC's $350 Billion Balance Sheet," *Euromoney* (September 1980); Bluff, "OPEC Surplus," Bankers Trust Company.

Even before the Iran-Iraq war, it seemed likely that Iran would produce oil at rates significantly below those at the time of the shah, although also cutting back on the extensive development program. The Iran-Iraq war has subsequently forced those two countries to draw down sharply on their foreign assets, and there have been large capital flows from other Arab states to Iraq, as well. When the war ends, both Iran and Iraq will require very large sums for reconstruction, which will put pressure on them to increase oil exports and to draw down any foreign assets that might be built up.

Therefore, just five countries — Saudi Arabia, Kuwait, the UAE, Libya, and (with a question mark) Iraq — are not only important oil exporters but, because of the volume of their financial assets and flows, are of importance to the international financial and monetary system as well. In addition, because of their low demand for foreign goods compared to earnings, these countries are best able to scale down their oil production without hurting their own growth. Prominent among them is Saudi Arabia, which, with its $162 billion at the end of 1981, alone accounts for more than 40 percent of the total net asset position of all the OPEC countries combined.

All the remaining oil exporters, except for Gabon, had a current deficit over the period 1974–78. The surplus of revenues over expenditures and the resulting improvement of the net external position achieved in Nigeria, Indonesia, and Venezuela after the second oil price rise proved short-lived in the face of the sharp decline in demand for OPEC oil. Ecuador and Algeria have continued to increase their net indebtedness to the rest of the world.

The different demand patterns of the OPEC countries are important for their investment objectives and strategies.[42] The first objective is always to maintain a country's international liquidity under conditions of fluctuating foreign exchange revenues and expenditures. This is the typical purpose for keeping foreign exchange reserves. Investment periods are short, usually three months; investment instruments are treasury bills and term deposits with foreign and "Euro"-commercial banks. For the "high absorbers," this is generally the only objective.

"Low absorbers" will, as a rule, devote only part of their total financial surpluses to monetary reserves. The rest, perhaps the greater part, will be invested with a view to spending the money in the future to import goods and services and/or to secure a continuous stream of income in the face of depleting oil reserves. The actual investment strategy depends on the development strategy pursued. Priority is always given to the development of the domestic economy. "Several states, particularly Saudi Arabia, consider the accumulated foreign assets as a kind of deferred expenditure, to be spent developing their country," the deputy governor of the Saudi Arabian Monetary Authority has explained. For these assets, "the time frame is essentially medium term. With this fairly defined time period, the preference will be towards fixed income instruments."[43]

On the other hand, "those assets designated, either implicitly or explicitly, as an endowment fund to counter the gradual depletion of the oil reserves, are obviously long-term. Several of the Gulf states have invested amounts in equities, as a means of preserving the real value of their assets over time. Fixed income securities of longer maturity are used, with liquidity less important than yield. Other non-financial assets, such as property or precious metals, have also been considered as areas of investment."[44]

The actual investment pattern conforms to this view. Whereas the statistical evidence is inconclusive in terms of precise figures, it appears that in the years between the oil price increases, i.e., 1975–78, at least 40 percent — and probably more — of OPEC's net cash surplus was invested in the medium and long term. Only in 1974, 1979, and 1980, years in which the cash surplus multiplied due to price increases, was the proportion lower, reflecting financial adjustment problems of the oil-exporting countries.

It should be emphasized that for the OPEC capital-surplus countries, piling up foreign assets is not an objective in itself. Constant current account surpluses and increasing stocks of foreign assets are rather tolerated for economic and political reasons, if and as long as the disproportion between the expected returns on investment in oil (i.e., leaving the oil in the ground) and in financial assets does not become too great. The deputy governor of the Saudi Arabian Monetary Authority expressed the matter thus:

> The crucial factor is the investors' view of the relative performance of financial assets vis-à-vis oil. Once OPEC countries have accumulated sufficient assets to satisfy their perceived needs for official foreign reserves, the balance represents, in terms of their domestic economies, the over-production of oil. It is this trade-off between accumulating financial assets, and leaving oil in the ground, that is central to the whole issue of OPEC surpluses. The more unfavourable the relative return between these two options, because of inflation and currency instability, the more difficult the decision to maintain production levels, let alone to increase them.[45]

This view is mainly derived from the fact that oil is a finite, depletable resource that in the course of time is likely to become increasingly precious. As observed in Chapter 2, the evaluation is based not only on purely economic considerations; that is, not only on a comparison of returns on oil reserves and financial assets.[46] Representatives of OPEC surplus countries are also stressing the notion of their responsibility to future generations, whom they feel are entitled to receive assets comparable to the ones that the present generation has inherited in the form of oil. From this consideration there follows, inter alia, a strong risk aversion, typical for such investors as insurance companies and pension funds, in industrial countries. Financial resources are generally invested in first-class assets only.[47]

This explains why OPEC is not prepared to finance a larger part of the developing countries' current account deficits by direct credits. From this risk aversion there also follows a certain inclination to diversify financial investments not only by instruments but also by currencies. Whereas the strong dollar orientation of OPEC's foreign trade and payments transactions has an important bearing on the composition of their currency deposits, currency diversification is considered important, not least because of the need to maintain the real value of financial assets.[48]

What consequences for the organization of the international financial and monetary system follow from these objectives and preferences of OPEC? To begin with, the current account and financial surpluses of the "high absorbers" are, because of their short-term nature, generally bearable within the existing system and pose no greater adjustment problems.[49]

The five nations that are "structural" surplus countries, accounting for an important share of world oil production, will own most of the $500 billion of foreign assets that OPEC is projected to have acquired by 1985.[50] It is with them that the dialogue on oil production, recycling, and adjustment must mainly take place. What are the limits of such a dialogue? The interest of consumer countries is first in a secure oil supply and the avoidance of sharp price fluctuations. In addition, most of the developing countries face the problem of how to finance their current accounts deficits, since that is a precondition for sustained economic growth. The interest of the OPEC surplus countries is in security, real value stability, and in an adequate return on their assets. Both groups of countries share the interest in a stable international economic system.[51]

From the point of view of financial stability, investments in national financial markets are preferable to investments in international (Euro-) markets. It is therefore important that the industrial economies grant continued, uninterrupted access to their financial markets (regardless, for instance, of short-term fluctuations of their exchange rates) and that they also try to refrain from all politically motivated interference with foreign assets. OPEC leaders, it appears, considered the temporary seizure of the official assets of Iran by the United States as a potential threat to their own financial assets, even those who had fully understood the reasons. Repeating such an action could well bring about a major crisis of confidence and might even cause some surplus countries to reconsider their oil production policy.

Continuing the efforts for an internationally coordinated, prudential control of the international banking system is desirable. In addition, negotiations on a jointly organized international safety net of commercial banks, to cope with short-term liquidity problems, appear sensible in view of the rapid expansion of international credit relations.[52] Also, whereas monetary authorities should avoid giving any formal backing to individual banks, there must be some assurance that the stability of the international financial system will be safeguarded. Therefore, the question of the "lender of last resort" in the case of a major international crisis in confidence needs an unambiguous answer. This issue should not just be waved away with the remark that the banking system has stood the test so far.

Along with a strengthening of the short-term end of the market, OPEC should be encouraged to invest a greater portion of its surplus funds in long-term assets, particularly in the productive sector of the oil-consuming countries' economies. This would certainly enhance the stability of the international financial system, but it would also be to the advantage of the oil-importing countries to have OPEC take a more direct and active interest in the performance of the world economy. For the same reason, greater diversification of the financial engagements of OPEC — to include, among other things, equity participations in and loans to industrial enterprises — is preferable to a heavy concentration on direct lending to governments where both economic risks and opportunities are virtually nonexistent. Outright restrictions against the acquisition of equity participations should therefore be lifted worldwide to encourage OPEC to invest directly in the economies of their customers. Moreover, tax and administrative regulations putting foreign investors at a disadvantage (compared to domestic investors) should be scrutinized and, wherever appropriate, changed. Finally, the oil-importing countries' overall economic policies should be designed in such a way as to help generate adequate returns on investment. This leads us to the final point of appropriate adjustment policies.

The Need for Adjustment and Policy Coordination

While the financial aspects of the oil problem have attracted widespread attention, they are perhaps amenable to solution. The readjustment of production, consumption, and savings patterns presents the greater challenge. Faced with high unemployment, overcapacity in traditional industries, and current account deficits, governments may try to shift at least part of the adjustment burden to other countries. Certainly there will be great domestic pressure to do so, as evidenced by the campaign of U.S. car manufacturers and the automobile workers union to restrict imports of Japanese cars. Subsidies to domestic production, specific policies of export promotion, and shielding, through various devices, the domestic market from foreign competition — these policies seem to offer quick relief and at the same time, merit the approval of business and trade unions alike.

It is obvious, however, that such an approach, if pursued by a greater number of countries, would not only be self-defeating but would, in the end, leave all countries in a worse position than if they had pursued forward-looking, positive adjustment strategies. Above all, a selfish "negative" approach would deprive the world economy of the gains in real income that free trade and the progressive international division of labor can be expected to generate, and have generated, during the postwar period. Moreover, insofar as industrial countries manage to shift part of their deficit to the OIDCs, this "success" will be paid for by an increased fragility of the international financial system.

A positive adjustment strategy will start from the tenet that, with petroleum becoming increasingly scarce and expensive, all countries — industrial as well as developing — will have to apply a deliberate conservation policy. The best overall setting for this goal is a functioning price system under which changes in the relative price of oil are allowed to work themselves through to the final users. A tight monetary policy that inhibits oil price increases from giving rise to major general wage and price increases helps this aim. At the same time, a general climate of price stability is likely to make for moderation in OPEC's price demands.

Assuming that consumers will indeed be fully exposed to high and rising oil prices, they should react by economizing in the use of petroleum and petroleum-based products (mainly chemicals) and by substituting other goods for them. At the same time, manufacturers will be induced to apply less energy-intensive production methods and to substitute the relatively cheaper production factor, human labor, for the increasingly expensive factor, energy. Energy productivity will thus be raised. To make the program fully effective, no sector of the economy can be shielded from the impact of rising oil prices. Our experience since the mid-seventies indicates that there is indeed a great savings potential, which, if supported by a more vigorous policy, could contribute significantly to solving future growth and employment problems.

Energy savings must be complemented by programs to substitute for oil-related energy production, methods based on other energy sources. This applies especially to the generation of electricity and heat, perhaps also to the production of gasoline. Robert Stobaugh's projections on the contribution of different sources of energy to total energy supply in the 1980s, while realistic, are by no means the only conceivable outcome (see Chapter 2). Deliberate policies to stimulate the development of specific energy resources could change the outlook for the better.

Adjustment policy under conditions of rising real oil prices would thus have to be largely supply oriented and consist of a number of elements: control of inflation; a willingness to let the price mechanism work for oil in domestic markets; deliberate measures to encourage change in the national

economy and, in particular, the introduction of new techniques that will
increase energy productivity; the application, where appropriate, of specific incentives (e.g., higher oil taxes, more stringent energy savings standards) to accelerate the move away from oil; and the development of alternative sources of energy.

Incentives for saving oil and substitution are only one side of the coin,
however. The other is the quick adjustment of national economies to the
changing pattern of domestic and international demand. We can expect,
for example, some reduction in demand for consumer goods and an increase in demand for capital goods in the OIDCs. While no major change
in the demand structure for OPEC countries is likely to occur, total demand from these countries is likely to rise quickly. Finally, the reallocation
of international production — the rise of the newly industrializing countries and the entry of other developing countries into the world economy, a
process that gained momentum in the seventies — will continue in the decades to come. These changes should be allowed to work themselves
through in a competitive framework. Any attempts to shield national
economies from the impact of change are likely to multiply the future adjustment burden.

The adjustment of national economies to the changes in real oil and energy prices and in international demand and supply patterns requires
major investment efforts. Economic policies conducive to growth, together
with tax and income policies suitable to generating adequate returns on
investment, may help the task and at the same time contribute to achieving
tolerable employment levels. But it may well be difficult to bring about a
national consensus on such policies. Labor may not be prepared to accept a
"deterioration" in its income, compared to that of business, at the same
time that it is already subject to a transfer of income to the oil producers
(through terms-of-trade changes). Yet, there is no workable alternative.
Capital imports, as a counterpart to current account deficits, may bring
some relief by allowing the oil-consuming countries to maintain a level of
spending that they could not sustain if OPEC insisted on any immediate
transfer of real resources. But in the end, the adjustment has to be made.

Change may also be facilitated by international policy coordination.
The major issues between oil-exporting and oil-importing countries are to
smooth the rise of oil prices, to make the adjustment processes in both
groups of countries as transparent and foreseeable as possible, and to
strengthen the basis for recycling both for the parties directly involved
(lenders, borrowers, and official and private intermediaries) and for a stable international economy at large.

The second aspect of adjustment involves the relations between the industrial and the developing countries. The former must be prepared to
honor the consequences of the latter's adjustment policies — namely, ris-

ing imports — even if certain industries or the current account is affected. This follows not only from the underlying economic philosophy of the industrial world but also from the consideration that a continuous growth of the current account deficits of developing countries would be a threat to the stability of the international economic system. Since the large group of low-income developing countries will generally not be able to increase their exports sufficiently, official development assistance needs to be maintained and even be raised. Adjustment by starvation is an unacceptable answer to the oil price problem!

The third aspect of adjustment concerns relations among the industrial countries themselves. It is imperative to avoid aggressive, mercantilist "beggar thy neighbor" methods. At the same time, successful efforts must not be penalized by protectionist measures. Changes in exchange rates should be allowed to reflect the relative success of national adjustment policies.

Will countries succeed in approaching the common problems with the necessary cooperation rather than confrontation? Much depends on how that change is negotiated. If oil prices continue to move in large jumps and if resorting to protectionism becomes more widespread than assumed, the consequences for the international economy would obviously be serious and the effects on most OIDCs disastrous. The major oil producers and the major industrial countries must live up to their joint responsibility for the world at large.

11

Cohesion and Disruption in the Western Alliance

by Robert J. Lieber

With the single exception of the East-West military balance in Europe, the availability of energy — both its supply and price — has become the most important security issue in the Atlantic area. At the same time, major questions of international energy policy have created considerable division within the European Community and among Europe, the United States, and Japan. The disruptive impact upon these relationships has come from many directions, such as competitive bidding for oil supplies, opposing reactions to the Arab-Israeli conflict, differences over the export of nuclear power technology and facilities, tensions over policy toward Iran and Afghanistan, and disagreement over imports of Soviet gas. These strains can run very deep. The potential for real animosity was succinctly conveyed in a front-page headline of *Le Figaro,* which described energy conflicts between the United States and Europe as *"La Guerre de petrole"* — the oil war.[1]

As the energy issue has become more important and more complex, the risks of major disruption have grown. Yet simultaneous pressures also exist toward cooperation among the major oil-consuming countries. For most countries, no mix of essentially national policies, whether in domestic energy strategies or in international energy diplomacy, will provide an effective means of coping with the conflicts and strains.

This study considers energy supply and price problems as likely to pose risks for the remainder of this century. A volatile oil supply pattern makes the international political system vulnerable to future "unexpected" disturbances or shocks.

Our premise stands in contrast to two other common, if misconceived, assumptions. First, in the five years between the end of the 1973–74 crisis and the onset of the Iranian revolution, many assumed that the energy crisis was a one-time event that would be banished by a series of specific and sometimes almost magical solutions.[2] This optimism was interrupted by the second oil shock in 1979–80. But the subsequent oil surplus of 1981–82 rekindled wishful thinking that the energy problem had become a thing of the past — now no more than a two-time event.

Another common assumption was that the October war was the direct cause of the energy crisis, and that a resolution of the Arab-Israeli conflict would solve the international energy problem. This notion, partly fostered by the relatively quiet interlude from the end of 1973 to the end of 1978, should have been dispelled, however, by the second world energy crisis, unleashed after the onset of the Iranian revolution. In retrospect, the momentous events of war and revolution, as in 1973 and 1978–79, are most accurately viewed not as causes but as catalysts of the energy crisis. The fundamental causes have been the vulnerability of the international oil supply system, the volatility of its supply and demand balance, and the risk that it can be disrupted by "unexpected" events. These include not only wars and revolutions but also technical accidents, bad weather, natural disasters, assassinations, sabotage, and any other random event capable of disrupting even temporarily the production, delivery, or refining of petroleum.

A resolution of the Arab-Israeli conflict or the peaceful consolidation of a postrevolutionary regime in Teheran would still leave the energy crisis essentially intact. Most OPEC oil producers would remain vulnerable to internal turmoil and regional conflict. Conflicts would continue between radicals and moderates, modernizers and traditionalists, Sunnis and Shi'ites, Egyptians and Libyans, Syrians and Iraqis, Persians and Arabs. Also untouched would be the complex political, economic, and even sociological elements that shape oil production, exports, and consumption behavior.

In short, by appreciating the crises of 1973–74 and 1978–80 (dubbed Crisis I and Crisis II for the remainder of this chapter) as catalysts rather than causes of the energy problem, our attention can be directed to the fundamental factors of a vulnerable oil and energy supply system and the resulting potential risks of disruption. With such an understanding, we will focus on the international political consequences of these disruptions and the implications for relations among the advanced industrial states of the developed world.

The Lessons of 1973–1974

The modernization of European economies from the early 1950s to the late 1960s was fueled by plentiful and inexpensive oil. Most of Europe went from an energy supply system based largely on coal, much of it domestically produced, to one resting on oil, virtually all of it imported. Meanwhile, periodic efforts at developing a European energy policy foundered on seemingly unbridgeable differences. These included free market versus government-mandated *dirigiste* strategies, state versus private oil companies, high versus low consumer prices, anxieties over export competition, differences of interest between countries with domestic energy resources and those that must import energy, and disagreements between advocates of cooperative approaches and those disposed to more individualistic strategies. Above all, no means seemed to exist for reconciling the potentially incompatible goals of cheap *and* secure energy, both of which received lip service in European Community policy statements.

There were other reasons for inattention by the Europeans — and by the Americans and Japanese. The major oil companies seemed to be strong enough to maintain a stable system of production and distribution. Supply and demand balances were not considered a significant problem. And who would have predicted the circumstances under which the producer governments could wrest away and effectively exercise price-setting powers? The major oil-consuming countries were also lulled by their ability to withstand temporary oil dislocations following the 1956 and 1967 Arab-Israeli wars, in circumstances where physical production capacity around the world, including in the United States, exceeded demand.

With the outbreak of the 1973 Arab-Israeli war, however, the Europeans suddenly found themselves in disarray. The Organization of Arab Petroleum Exporting Countries (OAPEC) adopted an embargo policy on October 17, 1973, effectively splitting the West into three groups. In one category were those countries to be totally embargoed — the Netherlands, along with the United States — for their real or imagined support of Israel. In a second category were two privileged countries, France and Britain, to be essentially exempt from import cuts; and in a third category were the other EEC members, including Germany and Italy, which were to experience progressive reductions of 5 percent a month.

The Europeans reacted by pursuing three different, though not always mutually exclusive, strategies. First, governments such as those of Germany and the Netherlands hoped to encourage a coordinated EEC policy. Initially, these efforts were simply overwhelmed. Next there were those, such as France and Britain, that sought to respond on a mainly national basis. That meant signing bilateral deals in which they would obtain assured supplies of oil in exchange for exports of industrial products and technology, weapons, and investments. However, many of these deals failed because of the producer governments' reluctance to lock themselves into exclusive bilateral agreements, and virtually none of the oil supply and price objectives was achieved. Finally there was support, particularly in the United States and in some of those countries receptive to a European strategy, for wider cooperation among the major oil-consuming states.

The lack of initial consumer cooperation was epitomized by the various reactions to OAPEC's oil boycott of the Netherlands. In contravention of EEC regulations, France and Britain initially let it be known that they were willing to cooperate with the embargo, even though a determined and unified European response could almost certainly have prevailed without additional deprivation of oil.[3] After several weeks of intra-European re-crimination, the nine EEC countries finally moved to support the Dutch, but only following strong pressures from the United States and Dutch threats to curtail natural gas shipments to France, Belgium, and Germany — as well as a growing realization that a world oil shortage would be manageable.

By the beginning of 1974, anxieties over physical supply of oil were giving way to concerns over explosive increases in price. The major oil companies were able to manage a global shortfall of roughly 7 percent by allocating supplies on a pro rata basis. Even state oil companies followed this pattern, albeit over the objection of national governments. Cooperative behavior proved easier on the more familiar grounds of economic and financial policy than on the question of supply. At the Energy Conference in Washington, D.C., in February 1974, U.S. policy-makers succeeded in moving the EEC away from positions around which the French had previously secured a loose understanding. This shift reflected a consensus that managing the financial effects of the energy crisis required broad, multilateral mechanisms, including the use of the Organization for Economic Cooperation and Development (OECD), the International Monetary Fund (IMF), and petrodollar recycling arrangements. In these efforts, a major role for the United States was inescapable.

The first crisis, together with the quadrupling of oil prices and the dangerous underlying supply insecurity that it revealed, led to a reassertion of American leadership among the OECD countries. There had been a growing belief, until Crisis I, that Europe and Japan were economic superpowers on an equal footing with the United States in a world in which traditional military and resource strength mattered far less. This view was premature. The reassertion of U.S. leadership was facilitated by the underlying strengths of the United States and continued inequality in European-American interdependence. It also resulted from the absence of political and institutional unity and the lack of essential agreement among the EEC nations.[4]

In the end, the following conclusions could be derived from Crisis I:

• Relatively minor reductions in oil supplies could create the conditions for a wild competitive bidding scramble and enormous price increases.

• Individual national efforts had limited long-term value in securing oil price and supply guarantees, though they would be used on a short-term basis in response to supply anxieties.

• Consumer cooperation was easier to achieve in dealing with problems of price than of supply.

• The European Community itself was too limited a group in which to seek a consumer country response to the energy crisis, since it excluded such major participants as the United States and Japan.

• Multilateral cooperation offered advantages in dealing with the energy crisis, particularly in its price effects, and the United States was favorably situated to organize a common response.

The Lessons of 1978-1980

The second crisis occurred after a lull of nearly five years, during which the urgency of the energy problem seemed to recede. At a significant cost in inflation and reduced economic growth, the oil-consuming nations had coped with the price and balance of payments effects of the oil price rise. Indeed, despite continued official OPEC increases, the *real* oil import price (calculated by deflating by the prices of OECD manufactured exports) actually declined by 13 percent between 1976 and 1978.[5] Meanwhile, the overall OPEC balance of payments surplus, which had been at a seemingly unmanageable level of $60 billion in 1974, had almost disappeared.

National responses to these developments diverged between the United States, on the one side, and Western Europe and Japan on the other. The Europeans and Japanese continued to be acutely aware of their dangerous vulnerability; as a result of conservation, pricing policies, slower economic growth, and a move to other sources of energy, the total oil requirements of the OECD countries other than the United States actually declined by 2.3 percent between 1973 and 1978.[6] By contrast, in this same period, total U.S. oil requirements actually increased by 12 percent. Even more damaging, however, were the oil import figures. While the other OECD countries (excluding Britain and Norway, the new oil producers) experienced a 2 percent decrease in net oil imports between 1973 and 1978, U.S. imports soared by 28.5 percent as a result of both greater consumption and declining domestic production. Coming at the same time as a sharp decrease in U.S. oil imports from Canada, this trend resulted in a significant increase in U.S. dependence on Middle Eastern oil.

Since the United States accounted for almost a third of total OECD oil imports, the increase in U.S. demand stimulated a growing resentment that America's failure to conserve and to exploit its indigenous energy supplies was jeopardizing Europe's economic security. The American performance tightened the supply and demand balance for OPEC oil, contributed to the eventual upward price pressure, and left the international oil supply system more vulnerable to unforeseen shocks. Huge U.S. balance of payments deficits, due to high oil import levels, were also inflationary and potentially destabilizing.

It was against this background that the Western countries faced the Iranian crisis in late 1978. The upheaval in Iran first reduced and then, during January and February 1979, virtually halted oil production. Production resumed in the early spring of 1979 at a level roughly 2 million barrels of oil per day (mbd) below the precrisis average of 5.3 to 6.0 mbd.

The lessons of 1973–74 pointed to the value of greater cooperation among the Western oil-consuming countries. But significant obstacles stood in the way. The most important one was the eruption of frenzied bidding for oil on spot markets, particularly in Rotterdam. The spot market trading had previously amounted to no more than a few percent of international oil sales. It had performed a technical and market-balancing function, based mostly on oil not already moving under long-term contracts and applying largely to refined products. But the rapid rise of crude prices above $30 per barrel quickly outdistanced the official OPEC figure, which had been $12.70 in December 1978. Spot bidding to offset the Iranian production loss and feared shortages touched off a chain reaction. The rising prices tempted some OPEC producers to divert oil previously under contract to the spot market. In turn, this forced the major oil companies either to turn to the spot market themselves or — more often — to reduce deliveries to their own traditional customers. These customers, in turn, whether private or government-related companies, then entered the spot market in search of additional supplies, thus further bidding up prices.

Although the United States sought to discourage spot market purchases and France advocated restraints, IEA and EEC countries were unable to agree on such measures. The Germans and Japanese were particularly active on the spot market, in part because of their buoyant currencies, strong balance of payments positions, and competitive economic strength. In the case of Germany, this also reflected a market-oriented economic policy and skepticism toward regulation. For the Japanese, it was due to fears of extreme vulnerability as well as the disruption of previous supply patterns.

IEA mechanisms proved irrelevant because shortfalls were below the 7 percent threshold for invoking emergency oil-sharing. While the IEA secretariat had obtained assurances from its members that they would honor their commitments should this become necessary, that stage was never reached. Overall, the oil supply gap was modest. According to IEA calculations, the shortfall between OPEC supply and demand during the second quarter of 1979 was only about 4 percent, and virtually all of this was accounted for by stockbuilding. Indeed, for the year, world stocks actually grew by 1.2 mbd.[7] Other producers, both within OPEC and outside, had made up the Iranian shortfall on a temporary basis, with OPEC production for 1979 actually up by 1.1 mbd — from 30.5 to 31.6 mbd — a 3.6 percent increase over 1978. Nonetheless, the sporadic distributional disruptions and the deeper political and economic anxieties caused governments, international oil companies, businesses, and even individuals to seek to build their own oil reserves.

Japan and Germany were by no means the only countries hesitant to enact practical measures of cooperation. In one vivid instance, in May 1979, the United States suddenly appeared to have given its oil importers a $5 per barrel import subsidy. In fact, the "subsidy" was quite limited, applying only as a temporary measure to heating and diesel oil imports from Caribbean refineries, covering less than 1 percent of net U.S. petroleum imports and involving no actual government subsidy but an adjustment through the complex Entitlements Program. Nonetheless, before it was accurately explained, it provoked an explosive reaction in Europe.[8] While the uproar was out of proportion to the cause, it reflected the extreme sensitivity and vulnerability of the Europeans to energy issues. In addition, American policy-makers were maladroit in failing to consult with their IEA partners.

Cooperative efforts did occur. The most important ones took place at two summit meetings in June 1979. The first of these, at Strasbourg, brought together the nine heads of state and government of the European Community. They sought to hold overall EEC petroleum imports from 1980 to 1985 at a level not to exceed that of 1978. At the subsequent Tokyo summit, the seven leading industrial powers — the United States, Japan, Germany, France, Britain, Italy, and Canada — adopted oil import goals for 1985. However, the Americans and the Europeans disagreed over whether the EEC would be committed to an overall figure, thus allowing it to benefit from increases in Britain's North Sea oil, or to individual targets for each country. Resentment also arose among the five smaller EEC nations over decisions being made in their absence.

Not until early December 1979 did the EEC agree on import goals within its overall Tokyo target. Five of these national figures (for Belgium, Ireland, Italy, Luxembourg, and the Netherlands) actually involved increases between 1980 and 1985 that were balanced by net reductions elsewhere, particularly through increased British oil production. This difficulty in reaching an internal consensus has characterized EEC decision-making on a wide variety of issues for two decades. In this case, it made its members reluctant to consider further changes at the subsequent IEA ministerial meeting. On December 10, 1979, the members of the IEA — including all of the EEC, except France — agreed on their 1980 import "ceilings" and 1985 "goals." They also initiated additional measures to gain better information about the spot market, to discourage unnecessary spot market purchases by government-related and private companies, to develop a "code of conduct" for market participants, and to seek greater governmental influence over stock levels. Their most important commitment was to develop plans for adjusting oil import ceilings and goals in response to a tightening of market patterns in the future.[9]

Most of these measures remained limited. Other U.S. efforts for an additional reduction in 1980 import ceilings and for sanctions to enforce them failed. There were a number of reasons for the failure — domestic political conflicts within the various countries, the legacy of energy-related conflicts and differences of interest between the United States and Europe over the previous six years, a perception that America's past energy performance had made it disproportionally responsible for common energy difficulties, and problems in partial overlap and incomplete membership among the various available forums. Above all, governments reacted too slowly to the energy situation. As summed up privately by a leading IEA official, action continued to be taken on a *sauve qui peut* — save one's self — basis, although this was veiled by nice-sounding communiqués issued at high-level meetings. There was no real effort to face the situation as an international community.[10]

Disruptions in the Western Alliance

Differences among the Europeans, North Americans, and the Japanese have been pronounced over the oil supply and price and in energy diplomacy. Both Crisis I and II included a supply phase with disorderly competitive bidding for crude oil. In 1973–74, there were also indiscriminate efforts to achieve bilateral agreements with oil-producing countries without regard to previous supply patterns or concern for other consumers. In 1978–79, competitive bidding for oil supplies on the Rotterdam spot market demonstrated the inability of the major consumers to cooperate in protecting themselves against disruptive price increases.

Supply Anxieties

The reasons for disarray stem from anxieties over supply. At their most extreme, as in the early weeks of Crisis I, they bring back memories of wartime deprivations as well as stimulate fears that part of national economic and social life could shut down because of supply shortfalls or an inability to pay for available supplies. Add to this the responsibilities that most postwar governments have assumed for the successful management of the economy and welfare state — upon which their political survival depends — and there is ample reason for lack of cooperation.

Export Competition

As a result of increases in the price of imported oil, the OECD countries have faced major balance of payments problems. This has caused them to intensify their efforts to increase exports in order to pay for the oil they must have. The sudden and uneven nature of oil price increases — a fourfold rise in the year after the 1973–74 crisis, followed by a slight decline in real terms and then an additional 170 percent increase in 1979–80 — have made these pressures harder to accommodate in an orderly way. This necessity to increase exports came at a time of increased competition from the more successful of the newly industrializing countries. It also took place in the midst of problems of declining older domestic industries, industrial restructuring, and domestic stagflation. These factors have fed protectionist impulses. This has been most clear recently in the case of Japanese exports to Western Europe and the United States, but it is hardly the only example.

France, for example, has feared being caught in a difficult competitive situation — between the most efficient or powerful of the modern industrial economies and the most aggressive of the newly industrializing countries. This helps to explain the French commitment to the export of nuclear power plants and technology, including the 1975 decision to sell an Osirak research reactor to Iraq, fueled with 65 kilograms of 93 percent enriched uranium — which would have been enough to construct nine nuclear devices.[11] Differences between France, Germany, and the United States over nuclear nonproliferation policy in 1977–79 owed at least as much to underlying export market concerns as to differences over the risks of proliferation. Export competition is also reflected in French weapons sales: During the decade from 1969 to 1978, their value increased at more than twice the rate of France's industrial exports.[12]

Differences between OECD Producers and Consumers

The advanced industrial countries have far from identical interests on energy issues. Norway and the United Kingdom, as significant oil producers, see questions of price and resource depletion rates differently from their oil-poor neighbors. The continental energy resources of the United States and a relatively lower degree of overall energy import dependence provide another perspective, as does Canada's rich resource base, the conflicts between the provincial and federal governments, and its complex attitudes toward the United States.

The differences have led to strains within the European Community. For France, Germany, and Italy, the maximum production of British and Norwegian North Sea oil and gas at minimal prices can be an important contribution to the stability and security of Europe.[13] But the British look forward fearfully to progressive oil depletion and worry about the "reentry" problem — being forced to come back into the world oil market as a net buyer in the early 1990s if production were to proceed too quickly. North Sea oil revenues have also tended to push the pound higher, further eroding the competitiveness of the U.K.'s export sector and contributing to "deindustrialization." As a result, the British government has sought to delay the onset of "reentry" into the mid-1990s or beyond by following a more restrained production policy. And on questions of price, the British National Oil Corporation (BNOC) tended in 1979 and 1980 to follow closely the leadership of Libya and Nigeria in raising their rates for desirable light crude oil.

The Norwegian example is even more striking. A far smaller country, and thus more affected by the financial impact of investments in and revenues from North Sea oil and gas, Norway faces some problems parallel to those of the less populous and less developed OPEC states — the so-called low absorbers. To control inflation, avert the deformation of its own economy, and to prevent the destruction of other export sectors, including fishing, Norway has strong incentives to restrain the expansion of its gas and oil exports.[14] Norway has thus opted for association rather than full membership in the IEA. Its purpose is to retain control over how much domestic production to make available in an emergency.

The different perspective of producers and consumers has also played an important role in the clash between the United States and Western Europe, particularly West Germany and France, over plans to import additional natural gas from the Soviet Union. The disagreement put a good deal of stress on the NATO alliance in 1981 and 1982. To the Europeans, this gas is valuable both in terms of security — an important step in the diversification of energy supplies — and for the economic stimulus it provides to their industries. Some in the U.S. government oppose it because they see it creating a new vulnerability; others regard it as an undesirable economic and technical boon to the Soviets and a source of valuable foreign currency to an otherwise hard-pressed USSR.

Not even the principal importing countries shared full agreement in Crises I and II. Japan, the major country most dependent on oil imports, was among those most reluctant to cooperate. Japan's vulnerability, along with the impact of disruptions in its oil supplies, affected its position on oil import ceilings, competitive bidding, and a host of more subtle matters concerning political relationships with oil producers. However, with the winding down of Crisis II, the Japanese began to show a greater willingness to cooperate.

The unity of the Western alliance was also adversely affected during the 1970s by the large increase in the U.S. oil imports, which reduced the American capacity for alliance leadership. And U.S. production decisions and energy price controls have been criticized by other OECD consumers. Legislation preventing the export of Alaskan crude oil to Japan was seen as uneconomic and excessively nationalistic.

Beliefs

A number of commonly held beliefs have worked against cooperation. In the period between the two crises, many observers discounted the existence of an energy problem. They pointed to the availability of sufficient oil, a modest decline in the real price, the successful management of balance of payments problems, and the near disappearance of the OPEC balance of payments surplus. Some argued that particular solutions could solve any long-term problem: nuclear power, the breeder reactor, technical or political breakthroughs, coal, solar power, synthetic fuels, oil from China, Mexico, or Alaska, an unfettered reliance on the price mechanism to call forth vast supplies of oil, or the disintegration of OPEC. As late as March 1979, 68 percent of the American public described the oil crisis as a hoax perpetrated by the oil companies.[15] This made it difficult to achieve coherent energy policies. The disbelief in the problem returned with rapidity in 1981, as oil demand declined in the aftermath of a price jump almost three times as large as in the first oil shock.

Although these sentiments were more common in the United States, European and Japanese perspectives also impeded cooperation. In the case of Germany, this meant a view of the market mechanism as an efficient device, not merely necessary but perhaps nearly sufficient for coping with the energy problem. Although this position was more characteristic of the Christian Democrats (CDU) and Liberals (FDP) than of the Social Democrats (SPD), the nature of the SPD-FDP governing coalition ensured a strong orientation in that direction. Hence, when the Iranian crisis broke, the Germans not only were willing to pay whatever price was necessary to obtain oil on the spot market, but they actually added significantly to their oil stockpiles. Indeed, German crude oil stocks at the end of 1979 were approximately 12 percent higher than they had been the year before.[16] This had two damaging effects: First, it helped to bid up prices for all consumers; second, it contributed to forcing the burden of scarcity onto the shoulders of others rather than sharing it fully and equally, as would have been the case had the shortfall reached the 7 percent threshold for the emergency sharing system of the IEA.

On the domestic side, these beliefs did not hinder the development of a useful German coal policy. In addition, conservation incentives were pursued and strengthened. Nonetheless, a hostility to autobahn speed limits was held with an emotional intensity similar to that toward handgun control in the United States. A failure to impose these limits also symbolized an unwillingness to treat the energy problem as anything but business as usual.

In short, a variety of feelings affected the ability of the consumer countries to achieve consensus either on demand restraint or on more ambitious international measures.

The U.S. Role

After 1973, American policies and practices, many of them unintentional, often made cooperation more difficult. Certainly its central role in the creation of the IEA was a major contribution. Yet successive U.S. administrations were unable to implement appropriate energy policies. This stemmed from a series of factors, beginning with the misconceived Project Independence of the Nixon administration, which foresaw the United States free of energy import dependence by 1980.

Questions also existed about U.S. policies at the time of the OPEC price increases during the first energy shock, and these left an enduring legacy of allied mistrust.[17]

Other U.S. energy policy choices also contributed to difficulties. For example, the entitlements program — apportioning the overall costs of domestic and imported oil among American refiners — had the effect of encouraging oil imports. In the late 1970s, policies placed too much emphasis on synthetic fuels and did not devote enough resources to energy efficiency and renewable energy strategies, which would have offered more effective short- to medium-term results.

During the years from 1974 to 1978, the significant growth of U.S. oil imports was particularly damaging for cooperative relationships. This gave rise to European objections that Western energy policy depended less on their own actions than on the United States doing something about its gas-guzzling automobiles and its overheated and overcooled buildings. The European Community energy commissioner, Guido Brunner, described the high level of U.S. oil consumption as a "danger" for the world economy and the international monetary system and warned that in the medium and long term, neither could bear the weight of U.S. oil imports at their existing (1979) levels.[18] The overall effect of growing U.S. oil imports impeded cooperation and diverted attention from measures that America had taken, notably in automobile mileage standards. It also gave rise to a view that multilateral conservation measures and ceilings or reductions in oil consumption and imports should take into account the leaner European energy patterns.

The reduction in American oil consumption and imports since 1980, as well as the easing of the international oil supply pattern, dispelled much of the earlier European criticism. But the advent of the Reagan administration brought new tensions to the fore. Other countries are now concerned by what they view as the complacency and "energy isolationism" of the administration, in particular its indifference to international cooperation and communication on energy. Moreover, the Europeans have been disturbed by the inability of the American government to see the connection between the energy issue and the serious economic problems of the 1980s. A good deal of worry has arisen over an apparent American desire to depend on market mechanisms even during oil disruptions. One prominent Western European warned privately that the Americans, despite the clear lessons of 1979 and 1980, wanted to rely on the market during a crisis, at the very time when it would be operating at its "crudest" and normal market forces would most likely be overwhelmed by panic bidding on the spot market. Other oil importers fear that the obvious weakness of the American commitment to cooperation in an emergency could in itself give a powerful impetus to panic buying and another oil scramble.

Institutional Obstacles

Existing organizations offer useful forums, and the IEA emergency oil-sharing system provides a potentially crucial basis for limiting the damage from a major supply interruption. But until the onset of the Iran-Iraq war, there had been no effective means to limit competitive bidding and disruptive price increases. In Crisis II, the 7 percent threshold for triggering IEA oil-sharing arrangements proved irrelevant in the face of a temporary 4 percent shortfall between supply and demand in the international oil market. Consequently, a 170 percent price increase developed in little more than one year.

For the consuming countries, the absence of a means to avert competitive bidding produced a classic prisoner's dilemma. Each oil company, whether state or private, and each government found itself in a threatening situation; to exercise restraint and forgo bidding for scarce spot market supplies risked leaving it in a worse position if other possible bidders did not cooperate. They thus were pulled into the bidding. However, in the absence of additional supplies, all the participants paid substantially higher prices for the same amount of oil than if they had managed to restrain their competitive bidding in the first place.

Institutional obstacles to cooperation have also included the problems of agreeing on oil import targets (as in the June 1979 Strasbourg and Tokyo summits), limitations on energy efficiency policies, and difficulties in achieving oil import reductions. Basic notions of national sovereignty are particularly at issue here. The pull of domestic considerations also works against cooperation and institution building. Thus, important political and economic burdens are involved in deciding whether to pass higher energy prices directly to users, as recommended by the OECD.[19] Indexation or a predictable yearly increase in energy prices would have contributed to a steadier reduction in energy demand without the dislocations that sudden price shocks have brought. Yet, even if there were mechanisms to cushion the effects on lower-income groups, the domestic implementation of such policies would be difficult. For instance, French electricity prices were kept at virtually the same level in real terms at the end of 1979 as they were in 1973 in order to restrain inflation and maintain industrial competitiveness. Politically, price increases can also be anathema, as illustrated by the 1979 defeat of the Canadian prime minister, Joe Clark, after his government proposed a 25-cent-per-gallon gasoline tax.

Yet another obstacle to institutional cooperation is the "free-rider" problem. The benefits of less constraint in the international energy supply situation would be shared by all oil-consuming countries in both eased access to supplies and relief from price pressures, whether or not individual states chose to cooperate. It is thus potentially harder to gather the necessary range of participation.

Finally, lack of leadership constitutes an obstacle to institution building. Not just in energy, but in the whole range of international economic relationships, there has been a relative decline in U.S. power and the erosion of mutual confidence. The postwar establishment of the Marshall Plan, GATT, OECD, IMF, and NATO was largely due to the American economic and military hegemony at the time of their inception. In political and economic terms, these relationships were a "public good" for which America was able to afford a certain price and from which it surely benefited. But with the United States less able to exercise leadership, much of the impetus for agreement in economic issues has been reduced. This is due not only to a decline in relative power but also to increased questions about the judgment of American leadership. Nor is there an obvious principle of economic management to replace hegemony in grappling with a series of issues including not only energy, but monetary problems, trade, industrial structure, LDC relations, and other matters.[20]

Pressures toward Enhanced Cooperation

The scope of the international energy problem, particularly that of oil, remains. Long-term planning for oil emergencies requires cooperation among all the major consumer states. Whatever the issue — limiting OECD oil imports, emergency oil sharing, plans for energy conservation, agreements on mandatory consumption cuts in an emergency, or even measures to act directly in a crisis in the Arabian/Persian Gulf — it is difficult to foresee viable arrangements in any narrower context. The same considerations apply in the economic realm — balance of payments problems, petrodollar recycling, measures to restrain protectionism, maintenance of international financial stability, and responses to the energy-related economic problems of individual countries such as Turkey. All these are more readily approached on a multilateral basis. Diplomatic and military elements play a part, too. There have been fragmentary signs of an emerging division of labor among the United States, Britain, France, and Germany. These involve allied responses to military-political instabilities in the Arabian/Persian Gulf, the Mediterranean area and in Africa, partly in response to events in Iran and Afghanistan as well as to concerns about the security of Saudi Arabia and other oil producers. Even an eventual producer-consumer dialogue or negotiation would prove more fruitful if it were not confined to individual consuming nations or regions.

The consumer countries' response to the oil shortfall following the outbreak of the Iran-Iraq war in September 1980 showed a far greater awareness of the need for cooperation than was the case in Crisis I or II, though the availability of unusually high oil stock levels was also quite important. In its ministerial meeting on December 9, 1980, the IEA established a series of measures aimed at avoiding the disastrous competitive bidding and market disruption that followed the Iranian revolution. Members agreed to draw down their stocks at a level 2.2 mbd greater than normal in an effort to ride out what they hoped would be a temporary shortfall. They also agreed to aid those countries with less favorable stock balances and to reduce their combined demand for oil on the world market by 10 percent in the first quarter of 1981.[21] The IEA secretariat was given the task of monitoring these actions and of consulting closely with the oil companies.

This elaborate exercise in consensus and persuasion, particularly vis-à-vis coordinated stock management and oil company cooperation, was a significant step forward by the consuming countries. It reflected their recognition of the severe macroeconomic effects that another oil price explosion would bring. It proved the value of the IEA as a forum for coordination in a crisis — or, more correctly, in averting a crisis. It also obtained the active cooperation of France. The success of this response, however, rested in important part on the existence of high oil stocks, spare production capacity, and a downward trend in oil consumption in the consuming countries.

American Responsibilities and Problems

The balance between cohesion and disruption will be heavily influenced by two key factors: the nature of the U.S. role and the type of energy crisis that unfolds. Unless there is a practical demonstration of U.S. leadership, a coherent and united consumer nation response remains remote. Yet continuing cohesion among the oil-consuming countries, let alone the development of an effective response, had been jeopardized periodically by problems of U.S. performance and policy.

During the 1970s, several features of the U.S. energy policy weighed directly upon the nation's ability to cooperate internationally. The first was a disproportionate growth in American oil imports from the Arabian/Persian Gulf, due not merely to the decline in U.S. production and growth in consumption but also to the ending of oil imports from Canada and the stagnation of Venezuelan oil production. This not only made America more dependent on supplies from a politically unstable region, but it also put the Americans into greater potential rivalry with the Europeans and Japanese for the same oil.[22]

The U.S. had a much higher level of energy consumption per unit of gross domestic product (GDP). Although the United States succeeded in reducing this ratio by 12 percent between 1973 and 1979, the coefficient of tons of oil equivalent per $1000 of real 1979 GDP was 1.01 compared to a figure for Germany of .58; France, .51; Japan, .60; and Britain, .89.[23] These figures help to explain why initial improvement in U.S. energy performance in 1979, in terms of reduced oil consumption and imports, was not greeted in Europe with the same unbridled enthusiasm that it received in Washington. However, the European response to U.S. efforts became more favorable in 1980. This was due to the mutually damaging experiences of Crisis II as well as to the continuing decline in American oil imports.

The nature of U.S. antitrust laws, particularly the Sherman Antitrust Act and the Clayton Act, created obstacles to the cooperation of American oil companies with IEA efforts to exchange information and restrain competitive bidding. The antitrust laws are often more of a procedural hindrance than a substantive barrier, since relevant information is exchanged both in formal meetings, under legal authorization, as well as in informal ways. However, in September 1980, the Justice Department sought to end an antitrust waiver allowing U.S. oil companies to report data to the IEA. This would have hindered the IEA's ability to function in a crisis, and the effort was only dropped after the outbreak of the Iran-Iraq war.

Finally, there remains the matter of the U.S. Strategic Petroleum Reserve (SPR). Although envisioned as a 750-million-barrel reserve supply, equivalent to over 90 days of imported oil, by the spring of 1980 this trouble-plagued program had accumulated only 91 million barrels (i.e., just 11 days of total imports) in underground salt domes in Louisiana and Texas. In early 1980, under reported pressure from Saudi Arabia and to cut $850 million in its domestic budget-balancing and anti-inflation program, the Carter administration postponed any further additions until June 1981. At the time, the small size of the SPR set limits on America's maneuverability and on its leadership role. In the event of even a short-term oil shock, an inadequate SPR would have left the United States less prepared to cope with a crisis, more vulnerable to external political pressure, and less able to act as the focus for allied consumer actions. However, this position on the SPR was subsequently reversed through congressional action, and additions to the SPR were resumed in September 1980. By the autumn of 1981, the strategic stockpile had grown to 200 million barrels and was projected to reach 250 million by September 1982. The goal of 750 million barrels thus remained distant, but the additional stocks were slowly improving American ability to weather a sudden oil emergency.

The United States has adopted a series of energy policy measures that have had an impact. Oil and gas decontrol are also bringing about reduced consumption. However, there remain serious risks in both the domestic and international arenas. In the absence of more ambitious measures to assure long-term reductions in oil consumption and imports as well as to deal with specific problems and emergencies, the basis for cooperation with Europe and Japan will remain fragile.

Lessons from Europe — The Limits of Bilateral Diplomacy

Much of the preceding analysis has dealt with the overall international energy pattern, the nature of European-American relations, and the American role. However, considering the French experience as well as that of the Europeans in the Euro-Arab dialogue points to limitations on specifically European initiatives at both the national and regional level. This also suggests why the energy problem creates a need for the consumer countries to seek multilateral solutions, despite the factors working against such cohesion.

The Case of France

Of all the major Western oil-consuming nations, France has traditionally pursued the most willfully independent course of action.[24] Equipped with a strong, centralized and *dirigiste* state apparatus and driven by the nationalist legacy of Gaullism, successive French governments have sought in both domestic and foreign policies a maximum degree of national autonomy.

Internally, particularly since the 1974 inception of its Messmer Plan, France has pursued the most ambitious nuclear power program of any Western country. Overall, it may well have become the most advanced at all stages of the nuclear cycle. Owing to the nature of the French political system, governments have also managed to override environmentalist opposition to nuclear power, despite a balance of public opinion only slightly more favorable than in Germany. For instance, a 1979 poll showed the French public divided 47 percent in favor and 43 percent opposed, while the German figures were 37 percent and 41 percent.[25] If the objectives of the Ministry of Industry for 1990 are adopted and attained — and they may or may not be — 73 percent of France's electricity and 26 to 28 percent of its primary energy supply would be provided by nuclear power.[26]

The French have also undertaken an ambitious energy conservation effort. As part of this policy, gasoline prices have risen substantially — to $2.60 per gallon of regular in March 1982, half of which was accounted for by taxes.[27] Specific government conservation programs were expanded by the socialist government of President François Mitterrand, but remain far below the 30-billion-franc annual investment in nuclear construction and electric transmission facilities. While the French have achieved results with their energy efficiency program, cost-effective opportunities have by no means been exhausted.

As a result of French policies, oil imports, which made up 67 percent of primary energy consumption in 1973, declined to 49 percent in 1981 and are forecast to fall still further, to 45 percent in 1985 and 30 percent in 1990. Yet, despite these and other ambitious measures, by 1990 the country will continue to remain dependent on external energy supplies — including imported Middle East oil, though increasingly also natural gas, coal, and uranium — altogether about 55 percent of its primary energy requirements.[28]

Internationally, a vigorous energy diplomacy has not brought fundamental respite from the energy problem. France has certainly tried — the forceful promotion of the European-Arab dialogue within the European Community, the pursuit of bilateral deals with individual oil producers, an essentially pro-Arab stance toward the Israeli-Arab conflict (until the May 1981 election of Mitterrand), the blocking of a favorable EEC response to the Israeli-Egyptian peace agreement, a special relationship with Iraq, extensive weapons sales programs, the export of sensitive nuclear facilities, a cautious position on terrorism, and the provision of generous terms of temporary residence for the Ayatollah Khomeini. Yet the payoffs have been meager. There have been no advantages in oil price. Nor has France gained major additional supplies, although increases from Iraq had been achieved before the Iran-Iraq war.[29] Even commercial links have not benefited substantially. Thus, France's share of OECD exports to the OPEC countries, which stood at 10.7 percent in 1972–73, actually declined slightly, to 9.8 percent in 1980. By contrast, the share held by the Netherlands, often considered to be more sympathetic to Israel, actually increased from 3.4 percent to 4.0 percent.[30] Only in weapons sales did France make major gains. Even providing hospitality for the exiled ayatollah did not prevent the cancellation of major industrial and nuclear contracts with Iran after his return.

In April 1980, France found itself, along with the United States, on the receiving end of a determined Algerian policy to raise liquefied natural gas prices to a level corresponding to that of oil. When the French resisted this doubling of gas prices, Algeria suspended deliveries.

In early 1982, France finally concluded massive gas contracts with both the USSR and Algeria. Each of these is to provide more than a quarter of France's total gas supply. The latter deal also included provisions for the Algerian purchase of $2 billion worth of goods from French industry. The price for Algerian gas remains well above world prices, but includes, in effect, an indirect subsidy from the French treasury, via Algeria, to French industry.[31]

Despite determined efforts, France has experienced the limits of what it can achieve on a national or independent basis. As a result, the French government took a more cooperative posture in addressing the second energy crisis and the Iran-Iraq shortfall. Within the EEC and at economic summits, the French proved more interested in measures aimed at conservation, spot market controls, and other collective agreements.

The Limits of the European-Arab Dialogue

If Europe were ever to have gone it alone on a regional basis, the Euro-Arab dialogue would have seemed the logical route. Initially, in the midst of Crisis I, the European Community's Copenhagen summit meeting in December 1973 produced proposals by the British and French for a dialogue with the Arab League. The objective was to restore oil supplies, intensify trading relationships, and to establish a triangular arrangement in which Arab petrodollars would subsidize the Egyptian purchases of European exports (thus recycling the funds so that Europe could buy additional high-priced oil). After its March 1974 adoption by the EEC Council of

Ministers, this dialogue quickly ran into difficulty. First, there was fierce American opposition, which placed eight of the EEC members in the potentially untenable position of having to choose between the United States and France. Next, the Arab League proved to be an unwieldy partner in contrast to the smaller and more specialized OAPEC. In addition, oil producers among the Arab League countries were flatly unwilling to include oil production and price among the topics for discussion. Finally, the dialogue itself was delayed for more than a year over the issue of Palestinian representation.

Individual European countries continued to try to outbid one another in courting Arab opinion. At times, this reached humiliating proportions, beginning with that initial unwillingness of the European Community to support the Netherlands during the first weeks of the oil embargo, and in a number of incidents, embarrassing to moderate Arab regimes, in which European governments released terrorists rather than enforce their own national laws.

Even after several years, there was no fundamental change in the Euro-Arab dialogue. On the European side, individual countries often pursued their own contacts. The EEC was also hampered by the need to reach unanimity through its political consultative process. Even though their positions on the Arab-Israeli conflict tended to converge over time — in the general direction of the French stance — the EEC was not in a strong position to deliver anything tangible. And while the Palestinian issue remained an irritant, the energy crisis was not coterminous with the Arab-Israeli dispute — which was underlined by the nature of Crisis II and the aftermath of the Iranian revolution as well as by French experiences in dealing with Algeria. Nor did other European proposals prove more fruitful. President Giscard d'Estaing of France did propose a "trilogue," involving the principal European, Arab, and African countries in a discussion of economic, cultural, and political cooperation. But this proposal never received the necessary impetus. A dialogue between the EEC and the Arabian/Persian Gulf countries also stalled over Iraq's reluctance to participate.

From the Arab side, there were problems of comparable magnitude. The Arab League was not synonymous with OAPEC because it grouped both producers and consumers of oil. Since neither body was equivalent to OPEC, no general understanding about oil policy was foreseeable. In addition, not only did the EEC itself prove a cumbersome negotiating partner, but the absence of the United States and Japan made it an incomplete one as well. The Egyptian role also was a major source of intra-Arab disagreement following the March 1979 Israeli-Egyptian peace treaty.

In sum, though episodic contacts continued in a variety of forms, the

Euro-Arab dialogue offered no major route for a separate European approach to the international energy crisis.

Crisis Scenarios: "The Crunch" versus "the Squeeze"

Two different categories of crisis can be identified, each one likely to carry different consequences. One, implicitly assumed in most discussions of an energy crisis, can be called "the crunch." Here, an abrupt reduction in the international oil supply is unleashed by some sudden event — a revolution in Saudi Arabia, a war in the Arabian/Persian Gulf, sabotage or other destruction of key oil production or transportation facilities, renewed Arab-Israeli conflict, or another Arab oil embargo. In this sense, both the October 1973 war and the Iranian revolution were the catalysts of modest oil "crunches."

By contrast, however, an energy crisis might also be precipitated by a gradual "squeeze," without any one dramatic event as catalyst. Here at some future time, especially in the case of Robert Stobaugh's Lower Bound, or "OPEC conservation" scenarios, a host of factors could come into play —

- structural problems of international energy supply and demand;
- the short-term relative inelasticity of oil consumption patterns;
- relationships between economic growth and increased energy consumption;
- the possibility that Soviet oil available for export will decline or disappear;
- increased domestic oil consumption within OPEC;
- the oil needs of the low-income developing countries;
- short-term limits on increased energy production in the United States.

Eventually, higher oil prices and stagflation would close the gap between oil supply and demand. However, this would hamper economic growth in both the advanced industrial and the less developed countries. There also would be reason to anticipate increased unemployment, growing balance of payments problems, potentially destabilizing international financial crises, social unrest, and the risk of political upheavals at both the domestic and international levels.

Paradoxically, "crunch" problems seem to allow for greater possibilities of cohesion than would "squeeze" scenarios. A sudden shock, such as a replay of 1973–74 but with an oil supply shortfall of 10 percent, would present the kind of disturbance with which the IEA emergency sharing system was designed to cope. The IEA mechanisms provide for a nearly automatic organizational response to come into play as soon as the secretariat formally identifies a shortfall to even one member of at least 7 percent. The entire presumption of the system is toward implementation of the oil-sharing scheme. Thus, positive government actions are not required to initiate such measures, but are required to halt their momentum and only then on a weighted majority basis.

Not only would emergency oil sharing embody the IEA's raison d'être, but the domestic political interests of the twenty-one members might well tend in that direction. Administrative and political elites, as well as the general public, have begun to recognize the gravity of the oil problem. Although agreement on specific programs may be inadequate, emergency oil sharing, based on formal international obligations dating from 1974, should be acceptable in domestic political terms.

A lack of cooperation among the oil-consuming states may be more likely in a squeeze than in a crunch. With shortfalls at less than the 7 percent IEA threshold, no dramatic event to rally public opinion, and mounting pressure for competitive bidding, hoarding, and bilateral deals, the environment for international joint action would be unfavorable. Thus, a squeeze of modest proportions may touch off damaging oil price increases while failing to stimulate an effective cooperative response among IEA members.

There are additional complexities. If, for example, a future crunch were very severe (e.g., several times the magnitude of the 1973–74 shock), it could conceivably overwhelm the IEA machinery for cooperation. Or, if the crisis were prolonged excessively or perceived as targeted at one country, the pressures toward disruption could intensify.

Military and Economic Security

How might the energy problem affect Western security? First, a more intractable energy and resource picture, together with economic difficulties, implies an overall situation conceivably as troubled as that of the 1930s. It remains to be seen whether the major oil-consuming countries react with the disarray and conflict reminiscent of that period or with the cohesion and purpose of the immediate postwar period — or in some other pattern entirely. In military terms, however, much will depend on the nature of perceived threats, particularly those from the Soviet Union.

A pattern of energy squeeze means greater difficulties for peacetime cooperation in security affairs. In particular, the capabilities of NATO countries would be reduced by the resources needed to meet national demands. Increased energy costs, massive capital investments in energy production and conservation, reduced economic growth, and the pressures engendered by stubbornly high levels of inflation and unemployment all limit the growth of the national economic pie and increase conflicts over its distribution. For now, at least, increases in defense spending — or even the maintenance of existing levels — can be seriously challenged by those who want to see government money put to more immediately productive use. There may also be tradeoffs between greater defense spending and national investment strategies aimed at reducing energy import dependence and thus reducing security vulnerability.

Western security cooperation is also heavily influenced by the East-West climate. Thus, the perception of increased Soviet threats might create a climate for greater allied cooperation in energy. In contrast, a substantial period of calm in East-West relations might reinforce a tendency toward less willingness to cooperate militarily and toward increasingly disruptive competition among oil consumers.

European and Japanese perceptions of the United States are at least as important as their view of the Soviet Union. The American ability to guarantee the security of Europe and Japan has provided the cement for the Western alliance system. It has put limits on disruption in these ties, despite perennial tensions and disagreements. However, the exacerbation of U.S.-European tensions — for example, over the import of Soviet gas, East-West trade, and responses to the repression in Poland — can indirectly weaken the credibility of security guarantees.

There may also be less enthusiasm among America's partners for its stress on military preparedness in the name of securing oil supplies. The January 1980 Carter Doctrine did not elicit substantial European support, despite greater European dependence on the Arabian/Persian Gulf.[32] This may have had as much to do with lack of confidence in U.S. judgment as in differences about the efficacy of force (though the unpublicized French role in providing advisers to help Saudi Arabian forces drive terrorists out of the Grand Mosque in Mecca in 1979 is one indication that the European role will not be entirely passive). If the United States were to become substantially less dependent on Middle Eastern oil supplies, American bargaining power and credibility would be increased vis-à-vis its allies, adversaries, and OPEC, while giving weight to a U.S. demand that the Europeans and Japanese play a greater role in sharing military burdens. Nonetheless, allied perspectives will never be identical.

A division of labor among the Western allies in Middle East and Mediterranean security matters may well evolve in a de facto manner. Multilateral or European guarantees to Arabian/Persian Gulf states are likely to be more politically palatable than unilateral American commitments. The British role in some of the Gulf states, the French presence in Iraq and

North Africa, and the German role in Turkey suggest ways to mitigate instability and to provide alternatives to a Soviet presence. Over the long run, however, this will be difficult. From a European and Japanese perspective, the Soviets are by no means the primary threat to oil supplies and regional stability. The Europeans and Japanese tend to emphasize more their concern over regional conflicts within the Middle East, including Arab-Israeli tensions as well as threats to the internal stability of the major oil producers.

To have the economic security of the major oil-consuming states dependent on the domestic stability of a half-dozen states in a gravely troubled region presents a dangerous vulnerability. While security commitments, sharing burdens, a division of influence, and other measures can be helpful, it is difficult to be sanguine about the long-term efficacy of these measures.

Finally, there remain economic security questions. Beyond the budgetary implications of constrained economic growth and increased competition for government spending, what are the possible consequences of international economic and financial instability stemming from future energy shocks? The fluctuating OPEC petrodollar surplus and the awesome level of debt of the East European countries and the low-income developing countries suggest that in future years, the system may not be immune to problems of dollar instability, loan defaults, and some form of financial panic. The reverberations from events such as these, together with pressures toward economic autarky and political upheaval that they would engender, make the comparison with the scale of problems in the 1930s a reasonable exercise.

Policy Choices

Despite the respite following the second oil glut, there remain real energy security risks. In responding to these, the major oil-consuming countries have a number of policy options available. These can allow them both to cope with an energy crisis and to lessen the damage to their relationships with one another.

Maximum Efforts at Consumer Solidarity

The United States, Europe, and Japan will never fully agree among themselves. This is not only a matter of interest and geography but of policy and political inheritance. Nonetheless, maximum efforts at agreement among the consuming countries remain imperative. These require a particular

American effort. Despite a relative decline in U.S. potential, the American role continues to be central. The European Community remains too divided and amorphous concerning energy policy, and no other group will spontaneously generate the necessary response. The importance of the United States as the world's largest importer and consumer of oil, as well as American domestic energy resources and technology, also dictates a profound role in consumer coordination.

Despite the great need for cooperation, domestic obstacles could mean that a business-as-usual orientation will not be surmounted — short of further jarring crises in the 1980s. Extensive cooperation among the energy-consuming countries is a necessary condition — though not necessarily a sufficient one — for coping with the energy problem.

While it may be that no effective consumer response will succeed in fully guaranteeing the flow of oil, less inclusive efforts are even more likely to fail. The stakes involved, the supply and price problems, and the potential nature of producer-consumer bargaining all create a need to seek solutions at a multilateral level in which Europe, Japan, and America will find themselves necessarily taking part.

Priorities for Decreased Dependence on OPEC Oil

Although the term "energy crisis" is commonly employed, the cutting edge of the problem has been and will be oil. Energy priorities must be determined by the criterion of reducing dependence on imported OPEC oil. However, the problem is by no means limited to these imports; overall reductions in oil consumption are inherently a major requirement, since any domestic consumption ultimately affects the rate of imports and hence of dependence on potentially unstable sources of oil. This suggests the need for greater efforts at measures already under way on a smaller scale, particularly encouragement to oil exploration among the LDCs outside OPEC as well as increased exploitation of coal and a far more extensive and coordinated commitment to energy efficiency and conservation.

International Energy Crisis Preparedness

Even in the fortunate absence of another crunch, an intensifying squeeze remains a long-term possibility. However, advance preparations can help to mitigate the worst divisive pressures from either type of crisis.

First, the readiness of the IEA's emergency oil-sharing system should be strengthened. Its credibility will be enhanced the more it appears that the system is workable. Simulated crisis testing (of the sort employed in October–November 1980) is thus desirable. So are efforts to iron out possible problems over the pricing of the shared oil as well as measures to discourage cheating in an emergency. While OPEC or OAPEC members are unlikely to react with enthusiasm to such a display, it can hardly prove more damaging than the lack of cohesion that major consumer states have often exhibited. Public presentation of the system as essentially a defensive measure to avert economic and social disruption, rather than as an offensive threat, is also appropriate.

Second, commitments on the development or identification of emergency standby oil production ("surge") capacity should be sought from all quarters; that is, from OPEC and from non-OPEC producers such as Norway, Britain, and Mexico. Some producers will be reluctant to increase production on a regular basis, because of such considerations as resource depletion, low absorptive capacity, and maintenance of asset value. But others may be persuaded that responding to a genuine emergency is in their interest in order to avert dangerous crises that will reverberate far beyond the consumer countries affected. Under certain circumstances, the investment costs of additional but "mothballed" emergency production capacity might be jointly financed.

Third, specific preparations for crisis planning should be undertaken, possibly through an enhanced summit machinery among the seven leading industrial countries. These would include arrangements to deal with a wide array of political, economic, financial, energy, and — conceivably — military matters.

Finally, the possibility of military force cannot be ruled out. Direct interventions should only be made with care, either on the invitation of a threatened oil-producing state or in some other major emergency — for example, to keep open the Strait of Hormuz. At times there has been an excess of conjecture about the use of military force. For one of the most likely problems — that of domestic political instability — U.S. or allied forces may be of limited value. This is not to say that force is always irrelevant or unusable. Noncrisis deployments of U.S. or other Western forces may also serve a purpose, for both practical and symbolic reasons. But here, too, great diplomatic discretion is required.

Domestic Energy Crisis Preparedness

Each of the major consuming countries will need to improve its preparations for dealing with oil shortfalls. Above all, more emphasis must be placed on the demand side of the energy equation. This is particularly true for the United States, although the 1980 Saint-Geours report for the European Community suggests that substantial long-term energy efficiencies, on the order of 20 to 30 percent, are available in Europe. Nonetheless, in one important respect, Europe and to a lesser extent the United States are less able to cope with a crisis than they were in 1973. That is because some of the easier and less painful measures have already been taken and also because their economies are in less robust condition than at the time of the first shock.

Even without a crunch, the oil problem inflicts substantial economic burdens. IEA economists estimate, for example, that the real cost of oil "saved" through lower economic growth comes to $300 per barrel. The consequences follow in unemployment, stagnation, inflation, and social strain. There are much cheaper ways to save oil.

In any case, preparations for coping with an oil shock must include viable rationing or tax arrangements, emergency demand reduction and fuel switching measures, a full readiness to comply with IEA emergency sharing, and increases in oil stockpiling.[33] In this light, U.S. increases in filling the Strategic Petroleum Reserve have been a positive step.

A Readiness for Consumer-Producer Negotiation

There is little evidence that a meaningful dialogue will soon occur. However, it is in the interest of the oil-consuming states to try to keep such a window open. In may provide a means for deemphasizing the confrontational aspects of international energy relations. In the long term, negotiation could offer a basis of mutual interest for the cooperative management of an extremely difficult problem.

In the abstract, an agreement might be based on guaranteed oil production and price levels, assurances to maintain the real value of OPEC investments, technology transfer, petrodollar recycling, systematic aid to the most needy LDCs, and the use of excess Western industrial capacity to meet LDC needs while stimulating economic growth. Some of these measures might parallel recommendations in the Brandt Commission report.[34] However, in a world of conflicts — of ayatollahs, upheavals, strong passions in the developing world about real and imagined injustices in North-South relations, divisions within OPEC and among OECD countries, and of interest group demands — the prospects are not favorable for serious consideration until after one or more additional oil shocks.

A Final Observation: Energy and the Nation-State

The gravity and the intractability of the energy problem underlie the fact that the subject of international oil, like that of nuclear weapons, has moved into a class of its own. And, as with nuclear weapons, it raises the question of the serious lack of fit between a world based on nation-states and those vital issues that penetrate national borders and that escape national control partly or entirely.

We live in a world in which a few individual states hold exclusive title to the world's oil. Thus a Saudi Arabia, a Kuwait, a Libya — or a Norway, for that matter — will be bound to make decisions that reflect the real (or imagined) interests of its leaders and populations of a few million people but that fundamentally affect the lives of more than four billion people elsewhere. In short, decisions about oil development, production, and price are of concern to more than the populations of the oil-producing states. This does not mean that we can turn a blind eye to the finite nature of oil as an energy resource or ignore the need for a gradual transition away from oil. Nor does it absolve major industrial states from the need to undertake essential and thoroughgoing measures to improve their own energy efficiency.

Oil is a resource that has become nearly as essential as food or water. Decisions to push explosive price increases — or to constrain production excessively — would carry far-reaching consequences that cannot be avoided merely because they reflect exclusively national prerogatives. Some analyses of international energy problems have been exceedingly careless. At one extreme, there are those who have foreseen or advocated simplistic solutions, based exclusively on technological breakthroughs, the discovery of new Saudi Arabias, decontrol alone, or other chimeras. They share the faulty assumption that there is no essential oil constraint. At the other extreme, however, are those who focus on the real and often imagined sins of the major oil-importing countries — their past status as colonizers, the international operations of their economies and multinational enterprises, and their appetite for oil. From this viewpoint, the actions of OPEC and the oil-producing states are portrayed as not only justified but even desirable in an effort to redress a North-South imbalance. In fact, the latter view is at least as myopic as the former. In contrast to the "my OPEC right or wrong" school of thought, the consequences of recent price increases have been especially harmful to the developing countries; their principal victims within the Western world have been working-class populations and those already least privileged in society.

Over the longer term, therefore, any management of the global consequences of the oil problem will need to recognize these considerations. In the future, a failure of the major importing states to maintain cohesion, and to achieve acceptable understandings with oil producers, could have dangerous consequences. If the second oil surplus were to be as temporary as the 1975–78 interlude, eventual resumption of energy price trends established in 1973–74 and renewed in 1978–80 could bring major economic disruptions. Ultimately, no group of states is likely to allow itself to be pushed up to or beyond the limits of economic survival without responding in some profound, unpredictable, and conceivably violent manner.

In short, the political implications of the energy crisis are not only global, they are fraught with potential peril for us all. How far we are from a serious crisis, and how close we get, should become clear in the course of this decade.

12

Energy and the Power of Nations

by Ian Smart

The sense of international change and turmoil, with energy at the center of the storm, is one of our legacies from the 1970s. Within the larger context of world energy affairs, problems of oil politics and oil supply have played a special role in provoking that unease. At the beginning of the last decade, as OPEC reached maturity, as United States oil production began to decline, and as Britain finally withdrew from the Arabian/Persian Gulf, control over many of the world's most important reservoirs — in Africa, Latin America, and in the Gulf area itself — began to pass decisively from private companies to local governments. So, as a result, did control over the price of oil. The new system of control then proved its strength, and reinforced it, in a series of confrontations culminating in the dramatic price rises of 1973–74 and in the simultaneous action by Arab producers to restrict production and trade. Meanwhile, a separate, if associated, sequence of events demonstrated the extreme vulnerability of the world oil market to international conflicts: to an Arab-Israeli war in 1973, to an Iranian revolution in 1978–79, to a war between Iraq and Iran at the end of the decade. With the total oil imports of the OECD countries growing in the same period by 30 percent (and those of the United States alone by more than 100 percent), it was hardly surprising that so many people in the developed world entered the 1980s with the conviction that the world oil situation was out of control, that their energy security was gravely threatened, and that this was symptomatic of a wider international malaise.

Most contemporary judgments about energy and politics are based on a simple assumption that those who control the supply of fuel always have a decisive advantage over those who do not, that it is the relative ability to produce energy resources that, by itself, yields a ranking of individual nations. Yet, only a few years ago, it was common to emphasize instead the relative consumption of energy as an index of development, affluence, or status. Primitive Leninist slogans in support of electrification — "Soviets plus electrification equals communism" — represent an extreme case. Much more commonly, however, in the West as well as the East, the per capita energy or electricity consumption of different countries or regions was offered as a guide to their respective levels of economic and social development. The higher the consumption, the higher the development. It has taken little more than ten years, therefore, for us to shift our assumptions to convert the maximum use of energy from a source of pride and an indication of prowess to a reason for distress — and to substitute, as a criterion of achievement, the maximum control of energy supply.

Our task here is to analyze the relationships between energy and the power of states. Energy is an idea to which myths are persistently attached. Above all, it is encumbered by myths of homogeneity. Energy is not the same as oil alone. The importance of a single fuel that satisfies almost half of the world's total energy demand is not in doubt. Nor is the importance of the strategic issues with which the international flow of oil is associated — for instance, in the Arabian/Persian Gulf. However, we must not lose sight of the additional roles that other fuels, together with technological components of the energy supply equation, will play in international politics.

Power is also an elusive idea.[1] Early in this century, a visionary British chemist asserted that the laws of energy conversion "control, in the last resort, the rise or fall of political systems, the freedom or bondage of nations, the movements of commerce and industry, the origin of wealth and poverty and the general physical welfare of the race."[2] So, in the last resort, they may. In the meantime, the connection between energy and political, military, and economic power is far from self-evident. We begin, therefore, with the question of what kind of contributions energy may make to a country's power.

Energy Assets

The first way in which energy contributes to a country's power is the most obvious today: assured access to energy natural resources. If foreign supplies of fuel are liable to manipulation or interruption, the possession of an indigenous alternative clearly contributes to a country's strength. However, one country may enjoy a potential advantage over another with comparable domestic reserves if it also has ready access to foreign supplies, the use of which will enable it to save more of its own resources for the future.[3]

Further, the asset of indigenous fuel cannot be valued without comparing the cost of exploiting it to the cost of alternative fuels from abroad. A country paying $30 a barrel for imported oil, and experiencing no difficulty in getting it, may have a real advantage over another that is forced to produce indigenous oil at $40 a barrel. The significance of that advantage depends, however, on the difficult problem of attaching a value to security of supply.

One reason for the difficulty is that the value of a secure supply is inevitably circumstantial, varying from place to place and time to time.[4] In 1973–74, for example, the British economy was more seriously weakened by a national coal strike than by reductions in the flow of Arab oil. In the case of the Netherlands, the development of domestic gas reserves, as both a substitute for foreign oil and an export commodity, has been called a cause of the "Dutch disease": the inflation due to gas revenues being channeled into additional consumption. More recently, as indigenous oil supplants imports, attention in Britain has begun to focus on the so-called Forsyth Kay effect, the essential argument of which is that the chief economic benefit of oil income can come only from changes in the sterling exchange rate, which make imports cheaper, thus causing unemployment in domestic industries.[5] When unemployment benefits alone threaten to exceed the total government revenue from oil, that is bound to prompt doubts in Britain about the real value of the energy security ostensibly provided by indigenous oil.

The second category of national energy assets consists not of fuels but of technology: of the information, skills, and equipment needed to produce, transport, and convert energy. Almost all the energy used in a modern society passes through several industrially based conversion processes — in refineries, furnaces, boilers, engines, or appliances for heating or cooling — and the ability to make and use processing equipment represents a crucial energy asset in its own right. So does the ability to use technology to find and develop fuel resources — to explore for oil, for example, and to extract it once it has been found. Developing and other countries have come to attach a high importance to the availability of energy technology and the terms for its transfer. Indeed, it is now commonly argued that access to technology is no less important than — and may be part of the price for — access to the fuels themselves. However, as is clear with oil today, control of energy technology, together with the relative power it confers, is inherently liable to be dissipated more quickly than power based on control of fuel supplies, whether by deliberate transfer or by some unplanned process of international diffusion.

The capacity to develop and exploit energy technology, like the capacity to exploit natural fuel resources, depends not only on knowledge and skill but also on a third category of energy-related asset: access to adequate financial resources. Wealth cannot, of course, be an energy asset in any exclusive sense; money is quintessentially fungible. Clearly, however, countries differ in their energy circumstances, not only in their dissimilar access to fuels and technology, but also in their differing abilities to pay for the required goods and services.

The capacity to invest in adapting energy supply or use is especially important, because it is the only way a country can acquire what should be treated as a fourth type of energy asset: energy congruence. This means an economically and socially efficient relation of energy supply to demand — not only a physical "balance," but also a pattern of procurement, conversion, and consumption that assists in the attainment of national goals. Such an asset presumes, but is more extensive than, all the others already mentioned: assured supply, technology, and money. The relative congruence of a country's energy system depends also upon its geography and history. Geology and economic history determine the extent to which a country is or has been a major producer of fuels and thus help shape its energy system and habits of energy use. The historical processes that have settled the character and distribution of a country's industry and population will also have raised greater or lesser obstacles to achieving energy congruence. Above all, however, congruence is the product of recent and contemporary policies.

Energy assets cannot be analyzed outside a larger context of national strengths and weaknesses. Governmental and institutional competence is one part of the larger setting. Some governments, parliaments, and policy-making bodies are wiser or more competent than others. Another part, as indicated by the reference to energy financing, is the general economic strength of different states. Yet another, it might be argued, is their relative military strength — although its relevance to contemporary energy affairs has still to be considered. But the most elusive component of all is the apparently diverse ability of different national communities to adapt to changing energy circumstances.

The fact is some consumers (or producers), such as Japan, appear to have a higher propensity than others to respond effectively to change in price or in the pattern of supply or demand, to react to market signals or official guidance, or to mobilize in support of collectively beneficial policies and to act cohesively for a common good. The relationship between energy and the potential power of nations is a function of qualities as well as quantities.

Any attempt to consider the impact of energy factors on the international distribution of power in the rest of this century must start from a view of how energy supply and demand will develop in that period. This we have done with Robert Stobaugh's Upper and Lower Bound scenarios. Their implications for the future point both to higher prices for energy and to a reduction in the welfare (or welfare expectations) of consumers — in standards of living or, at the limit, in population itself. That is an alarming prospect. But it is also a prospect filled with uncertainties: the uncertainty about how the energy situation will actually evolve, and the uncertainty, considered in a number of chapters, about how the various national societies, as well as the international society, will respond.

Trends in energy production and use have always been difficult to forecast. The 1980s and the 1990s are fated to be more than usually unpredictable as a divided world confronts the problem of a stagnating oil supply and begins, perforce, to turn to other sources of energy. In such a period, a particular advantage must lie with the countries that display the greatest ability to adapt and to manage uncertainty itself. When, as is the case today, it is impossible to be confident that any one line of action will remain appropriate throughout the coming decade, the highest value must be attached to a capacity for flexible adjustment. The energy assets discussed earlier can all contribute to that capacity.

Resources, Technology, and Military Force

Direct control of or access to natural energy resources such as oil is obviously one component of potential power. Direct or indirect resource control may provide either international influence or immunity to the influence of others. On the one hand, especially when the margin between world energy supply and demand is a narrow one, increasing importance seems due to the control of surplus resources that can be exported or withheld at will. On the other hand, rising significance is almost everywhere attached to energy independence: to the ability to satisfy national needs from nationally controlled resources. A world of autarkic national energy economies, even if autarky confers only a negative power to resist external pressure, is one in which potential energy exporters have little if any influence.

A country such as Norway or, perhaps, Canada may represent the model of an "autarkic exporter," with energy surpluses of a particular sort to export while needing to import no other significant energy goods or services. To the extent that power flows both from the command of exportable surpluses and from domestic self-sufficiency, the power of those countries within the limited field of energy would seem to be unalloyed. Other countries, like Japan, while almost totally dependent on imported fuel, will nevertheless be in a position to export energy technology, equipment, and services as well as to exploit their larger economic strengths. The United States will continue to import oil but will also have technology and services and some fuels, such as coal, to export. Other countries, like Saudi Arabia or Mexico, will be "autarkic exporters" in regard to fuel but will continue for some time to rely on foreign technology or services. Yet others, of which Argentina or India may be an example, will have no surplus energy goods or services of significance to sell abroad but will produce enough energy to become increasingly independent of foreign suppliers. In all those cases, energy assets can be expected to confer some measure of potential power, but always of a restricted or qualified — and sometimes of a merely negative — kind. That still leaves the pure "dependent importer" model: countries such as Sri Lanka or Tanzania in the Third World or developed countries such as Denmark or Portugal, which have little to offer internationally in the energy field but which need a great deal, and which lack, therefore, any element of energy-based power.

Positive power — the ability to exert influence — flows in part from the ownership of exportable energy goods or services. Negative power — the ability to withstand such pressure — will derive in part from energy self-sufficiency. Any shifts in the balance between states during the 1980s and 1990s will partly depend, therefore, on the extent to which additional countries achieve effective energy independence. A great deal will also depend on the extent to which other countries succeed in converting their exportable energy assets into positive power by extorting some extraordinary concession from those who wish to import from them. The crux will be whether relevant governments engage successfully in international energy power plays, where a power play represents a policy demand based on threatening or implementing an enforced change in the supply or price of energy at the expense of, or against the will of, customers and beyond the limits that would be set by a notionally free market.

It is common that those who control the international supply of oil can, and sometimes do, engage in such power plays. International oil companies were long accused of doing so in order to maximize their profits. Today, the governments of OPEC or other oil-exporting countries are charged with such action in pursuit of financial gain or, as with Arab export restrictions in 1973–74, in an effort to extract political concessions. Certainly, the Arabs' use of the "oil weapon" to manipulate the attitude of Western governments toward Israel has been a particularly obvious case. However, oil is not the only fuel to have been so used. Exporters have also sought, on occasion, to exploit their control of natural gas or uranium, providing an example that coal exporters may imitate in the future. One example is provided by the attempts by some exporters to restrict or manipulate uranium exports, technologies, and services as a means of inducing customers to accept nonproliferation policies.

The comparable political use of control over other energy technologies has been less widespread, but the long argument over Western exports of oil and gas technology to the Soviet Union offers one striking case. Conversely, it is trade in energy technology, rather than in fuels, that provides the best examples of a reverse power play; that is, of policy leverage exerted by importing countries as a condition of access to their markets. However, the policy advantages sought in that case have more often been economic than political and thus less obviously outside the usual context of a market.

The kind of power play so far discussed is when a nation uses an exportable surplus of some energy resource to induce its customers to conform to its own policy preferences. A different kind of power play is that employed by governments that use instruments of national power unconnected with energy to secure access to, or to influence the price of, energy resources controlled by others. All sorts and levels of diplomatic support and persuasion are, of course, directed to that end, as are economic sanctions and restrictions, such as those that enforced the effective abandonment of Iran's oil nationalization policy in 1953. Beyond such cases, however, lies the possibility of the use or threat of military force.

Nothing generates more passion than the international relevance of military force to energy, and in particular to oil supplies. Obviously, physical control as well as legal ownership is a prerequisite of exploiting oil as an asset, whether as a fuel or as an instrument of international influence. Its owners, therefore, will contemplate using military force to protect it or to deter attacks on it, to the extent that they fear forcible attempts to deprive them of the opportunity to exploit it. Saudi Arabian units guard oil wells ·and loading terminals, just as British ships guard oil platforms in the North Sea. But the owners of oil may, in some cases, find it beyond their own capacity to oppose or deter the military threats they fear, and they may look to others for military protection — as Oman, for example, has looked to Western governments or, in the past, to the shah of Iran. Conversely, those who depend on foreign oil may seek to provide that sort of military protection against external threats but also, perhaps, against internal disorder. Thus, military force is seen by both owners and consumers of oil, on occasion, to be a necessary means of preserving control or access, just as others may see it as a means of seizing oil resources, destroying oil facilities, or compelling the owners of oil to provide or withhold it. The argument is immediately complicated, therefore, because military force is a possible means to a multiplicity of ends: protection, assurance, deterrence, sanction, coercion, and conquest.

One reason why so much argument has been lavished on these possible roles of military force is that so much exportable oil comes from countries that are militarily so weak. The area that arouses the most concern is the Middle East, including North Africa but with special reference to the Arabian/Persian Gulf, which yields a third of all the world's oil. There, several factors come together. First, oil is owned and its flow controlled by states that are militarily weak in relation to their major customers in North America, Western Europe, and Eastern Asia. Second, those same states are militarily trivial in relation to the Soviet Union, the superpower that stands on the edge of their region and is seen to be hostile to their major customers. Third, the Gulf area is demonstrably subject to regional conflicts affecting the supply of oil, as the war between Iraq and Iran has recently demonstrated. And fourth, a number of the oil-exporting states in the Gulf are widely believed to lack internal coherence and thus to be vulnerable to disruptive domestic conflict, as again in the case of Iran. Understandably, therefore, the relevance of military force to Gulf oil has attracted particular attention.

Certainly, the Gulf is politically and sometimes militarily volatile, although hardly to a degree unusual in the Third World. Moreover, the military policies of non-Gulf states, in presenting threats, offering protection, or in supplying arms to particular Gulf countries, sometimes help to exacerbate that volatility. But what makes it special is oil itself, and what has really changed during the last decade is not the military security and stability of the area, but its relative significance as an oil source. Some 55 percent of Western Europe's oil and about 70 percent of Japan's now comes from that area alone. In addition, the United States, which imported hardly any Gulf oil at the beginning of the 1970s, now obtains about 10 percent of its supply there. Moreover, these current levels of dependence must be seen against the background of greater apprehension about the oil supply as a whole, not least because so much of the control over it has passed from international companies to producer governments. On the one hand, there is the possibility that a powerful Western customer might resort to military force to ensure access to oil resources if subjected by local action, regional conflict, or revolution to what Henry Kissinger once called "strangulation." On the other hand, there is the possibility that a powerful neighbor such as the Soviet Union might use military force to manipulate or control the flow of Middle Eastern oil to its own domestic or international benefit.

There is nothing inherently unstable in an international situation where militarily powerful countries depend on those that are militarily weak for vital resources. Despite the memory of colonial rights that a few of them enjoyed in the past, West European countries should therefore see nothing eccentric — as distinct, perhaps, from uncomfortable — in depending on militarily inferior states for oil, although they will certainly feel less apprehensive to the extent that their suppliers pursue predictable policies and recognize a degree of reciprocal interest in sustaining the relationship. To the United States, the situation of dependence is less familiar, and more time may be needed to adjust to it. The reality, however, remains immutable. Military strength is a specific component of international power. So is control of a vital resource such as oil. When one is balanced against the other, the relationship may well be stable and mutually beneficial, provided that neither party exploits one form of power to an extent that provokes unrestrained exploitation of the other.

A powerful importing country, facing apparently crippling costs as a result of unfriendly activities by a Gulf oil exporter, might be tempted by anger or desperation to resort to military action. Such an action might eventually be more costly than its original predicament. In all probability, for example, the use of Western military force against one exporting country would be condemned and, to the extent possible, penalized not only by other Middle Eastern states and oil exporters but also by the Soviet Union. It might also be opposed by other allied or friendly states outside the Middle East, relations with some of whom might be seriously strained. Meanwhile, the target country of the action, with whatever international assistance it could gather, could respond by exploiting its own potential power to destroy the means of access to its oil and by seeking to prevent its production or export.[6]

A situation in which an external power intervenes to avert or quell a regional conflict in the Gulf, or to prevent or correct an interruption of oil supply as a result of insurrection in a Gulf state, may be acceptable if the governments concerned invite such intervention and if the intervention itself is conclusively effective. In other cases, however, the risks and costs of intervening militarily in such circumstances may hardly be distinguishable from those associated with military action against a single oil exporter. The recent history of comparable military intervention by external powers in other regions such as Africa, Latin America, or Southeast Asia suggests that local resentment and retaliation or wider international opposition is liable to deny success to the enterprise.

Still, the possibility of a military reaction to threatened "strangulation" cannot be ignored, and an awareness of the possibility, with its unpredictably disruptive consequences, may well inhibit hostile or adventurist action by oil exporters or their neighbors. Desperation knows no reason, and it is not surprising that many would be nervous about provoking it.

The case in which a Middle Eastern oil exporter or its customers may fear that a third party — the Soviet Union or another regional power — will make a military effort to gain control of its oil is not entirely dissimilar. Analogous questions can certainly be asked about the probable cost to an attacker. In at least two respects, however, the case of Soviet military action deserves separate attention. The Soviet Union, whatever its prospective energy problems, is not now, and is not likely to be, dependent on the flow of oil from the Middle East. Also, the Soviet Union, under the current circumstances, does not automatically have to weigh the immediate military reaction of another major power, since no such power is physically as close to the Gulf as it is. Within limits, Soviet forces might, therefore, think it possible to present the rest of the world with a military fait accompli, affecting an area, such as northern Iran, immediately adjacent to the Soviet frontier. Some would add that the Soviet Union, on the evidence of Afghanistan, is unlikely to see widespread international condemnation after the event, even by otherwise friendly states, as a significant cost. Others would say that the sort of contingent gain Soviet leaders might expect to obtain in the Gulf is a much weaker motive for military action than the prospect of crippling loss implied by "strangulation."

Whatever view one takes, it does seem that the Soviet Union may have to fear some additional penalty if it is to see the probable cost of military action against the Gulf oil states as clearly exorbitant. In other words, Soviet military strength may have to be specifically deterred. Even without that, the prospective cost of a Soviet military move against Middle Eastern oil, in terms of political relations with regional states and with Western suppliers of foodstuffs and technology, might already seem sufficient to deter a military adventure. But just because the disruption of oil flows represents a trivial cost — or even a benefit — to the Soviet Union, others may reasonably seek to raise the expected price of Soviet military intervention by other means, especially by demonstrating that it would invite an effective military response.

It is equally reasonable that an oil-exporting country that fears Soviet aggression should want the assurance of a deterrent military response. Stronger powers must therefore not only convey suitable commitments and warnings but also give them credibility by providing suitable military forces. The latter is bound to be a delicate exercise. Military forces that are seen as unnecessarily potent or indiscreetly deployed, even if advertised as a deterrent, can too easily be interpreted as potential instruments of coercion or conquest. It is nevertheless understandable that the United States, in the aftermath of Afghanistan, should have warned the Soviet Union of its opposition to any military interference with Gulf oil, and that appropriate military forces should be made available to back up that warning. The natural sensibilities of regional countries such as Saudi Arabia, as well as the need to avoid ambiguity or suspicion concerning American and Western intentions, argue for the greatest prudence in selecting and locating the forces concerned. In principle, however, this is a clear opportunity to employ military force not as an instrument of extortion in connection with energy supply but as a means of precluding such extortion.

On rare and unusually favorable occasions, it may be possible to use military capacity to assist the willing governments of energy-exporting countries to avert or contain local conflicts. More generally, its existence and plausible availability as a deterrent may well set negative limits to the rashness of either oil exporters or their possible aggressors. The crucial distinction turns out to be, therefore, between the roles that military force might play in regard to the control of energy resources. As a means of protection, deterrence, or assurance, it can play a useful — sometimes vital — part in "holding the ring" by convincingly ruling out forcible attempts to interfere with energy assets. Conversely, the costs and risks likely to be attached to using military force to change the pattern of oil supply (or price) are so high as to rule it out in any but the most desperate circumstances.

The power conferred by control over exportable energy resources often seems, in contrast, to be universally exploitable. Yet such power, far from being absolute, always depends on the attitude of other suppliers and is always subject to change with use and time.[7] An international supplier of fuels or energy services may have initial success in exerting policy influence over its customers, expecially in a "tight" seller's market, only to find its power declining as the customers evolve ways of escaping from that unpalatable subservience — by finding less exigent exporters, by reducing demand, or by developing domestic sources of the same supply. In other cases, an exporter that tries to extort political conformity from customers may be disappointed in the short term, because they prefer to draw on emergency stocks or other finite alternatives, but may find its political power growing as those alternatives are exhausted or become prohibitively expensive.

A clear case of energy-based influence that declines as a result of its exploitation already exists: The experience of Arab supply restrictions in 1973–74 has combined with higher prices to accelerate both the development of alternative sources, such as Alaska and the North Sea, and the reduction of relative dependence on oil as a whole. Such a change was particularly evident in the sharp decline in demand for OPEC oil between 1979 and 1981. A parallel case can be found in the field of nuclear energy: American exploitation of a uranium enrichment monopoly in the 1960s and early 1970s, in an effort to impose not only nonproliferation policies but also a series of unilateral changes in commercial terms, was instrumental in encouraging other countries — the Soviet Union, France, Britain, the Netherlands, West Germany, and Japan — to enter the world enrichment market and, by detaching part of that market from the United States, to diminish the power it could exert. And something similar may happen in the future to the power of certain natural gas exporters, such as Algeria, which have been demanding prices high enough not only to prompt customers to develop indigenous reserves but also to tempt alternative suppliers to enter the international market.

Oil exporters already recognize that many of their customers, especially in the industrial West, have built up substantial stocks of oil and products that would cushion the immediate effect of export restrictions or embargoes.[8] If those stocks were heavily depleted, however, as they might be if trade were interrupted for many months, the influence of exporters would clearly increase. An analogous situation may come to exist in the later 1980s as international trade in coal expands. In theory, indeed, it already exists in the case of uranium, where importers commonly hold from one to five years' worth of stocks, but would be in serious difficulty if those stocks approached exhaustion.

The essential distinction between these two cases is a difference of time. It may take only a few months, or even less, to deplete customers' stocks of oil or some other fuel, and thus to reinforce the power of those who control its supply. It takes much longer for importers to reduce permanently the potential power of an external supplier by developing indigenous sources, diversifying trade, substituting other fuels, and reducing consumption.

The Ability to Pay

A country's ability to withstand international pressure based on the control of energy resources that it needs to import is also a function of its capacity to invest in progress toward eventual energy autarky. Meanwhile, the immediate impact of energy factors on its general economic welfare, and thus its relative international power, depends partly on the capacity of its consumers to pay for their current energy supplies, whether imported or indigenous. As with power based on resource control, however, the relative ability of each country to pay for energy, from capital or income, is liable to be affected over time by the impact of energy costs and benefits on its own economy.

For many countries, the immediate problem is paying for energy imports. The budgetary cost of buying energy eventually weighs on the whole national economy, and those countries best placed to carry it are those whose economies are strongest. Import costs bear first, however, upon a nation's external balances, and the particular strength of export sectors whose earnings can be used to meet fuel import bills may then be as important, at least in the short term, as general economic health. Thus, some Third World producers of highly valued agricultural or mineral commodities have initially been able to pay rising prices for imported oil more easily than other developing countries whose economies, taken as a whole, are stronger or better balanced.

Ultimately, however, the general strength of a national economy determines its capacity to meet the total cost of all energy inputs and determines its capacity to invest in modifying the supply, conversion, or use of energy, thereby achieving the technical and economic components of energy congruence. The impact of financial factors on the relative power of different countries does not, therefore, depend only on the burden imposed by energy import costs on their external accounts, as Japan demonstrated between 1979 and 1981. To suppose that a country is necessarily at an economic disadvantage because it runs a current account deficit with, say, oil exporters would be simplistic, for to import oil from abroad may strengthen the economy more than committing larger resources to finding and extracting indigenous oil.[9] What matters is the relation between the cost of energy from various sources, the wealth-generating effect of using that energy, and the partially consequential ability to pay *inter alia* for further supplies of energy or an improvement in energy congruence.

Congruence

An assessment of the extent to which any country's energy system is congruent, as originally defined, must have two distinct bases. In the first place, we need to consider whether the pattern of energy conversion and use is well adapted to the current and prospective pattern of energy supply, quantitatively and qualitatively. A country without fossil fuels but with effectively unlimited access to hydroelectric power, but which made no use of electricity for rail transport, would, for example, be off balance. By the same token, a country whose labor force was heavily concentrated in the industrial sector but that used an unusually large amount of energy per unit of industrial production would lack the congruence needed either to withstand foreign competition or to safeguard domestic employment. Energy efficiency, in its broadest economic sense, is thus a major element of energy congruence.

One way of approaching the relationship of energy use to expected supply is to look at the apparent prudence with which different countries have committed themselves to long-term dependence on particular fuels, in light of their probable cost and availability. Within the OECD, for example, given its heavy dependence on expensive oil imports, it is alarming to find that more than 20 percent of all the electricity generated, including large amounts of base-load electricity, still comes from oil-fired plants, and that the highest proportions are not in countries with oil of their own but in others, such as Japan, Italy, and Ireland, which have to import all their supplies. It is also alarming to discover that highly industrial countries vary so considerably in the apparent efficiency with which their major industries make use of energy. In 1976, for example, each ton of oil equivalent consumed in the crude steel industry yielded 1.25 tons of steel in the United States, whereas the same amount of energy produced 2.97 tons of steel in Italy. Similarly, each ton of oil equivalent produced 2.98 tons of paper and board in West Germany, as compared with only 1.56 tons in Britain.[10]

A country's ability to maintain or improve energy congruence depends not only on access to resources, economic strength, and administrative skill, but also on whether its government can mobilize sufficient political and social support for prudent policies. There are countries at similar levels of economic development in which a domestic consensus on energy policy has recently been much easier to achieve than in others. Despite problems in particular cases, that has been the case in France and Japan, compared with West Germany and the United States.[11]

Energy congruence is essentially a compound of economic effectiveness and social harmony, secured and sustained within the framework of coherent policy, and is a national energy asset of incalculable value. Whether it continues to make a contribution to national power depends on whether the government and society concerned can recognize the national needs implied by a changing energy situation and can adjust their energy system to meet them without disrupting the domestic consensus. Facing the rest of the 1980s and the 1990s, few countries, if any, can afford merely to preserve their traditional patterns of energy supply and demand, however congruent they may have been in the past. Neither economic efficiency nor social consensus will be likely to stand the strain of insistent conservatism.

Winners and Losers

America and Russia

Energy factors have recently been related to a comparison between the superpowers. Their respective patterns of energy consumption are set out in Table 12.1. Each can be considered self-sufficient in all fuels except oil. In 1980, however, the United States had to import almost 40 percent of its oil supply (with almost 20 percent from the Middle East and North Africa alone), whereas the Soviet Union had a substantial surplus available for export. Their access to other fuels will change only slightly during the 1980s. The United States could become a somewhat more significant importer of gas, at least for a time, but should also become a larger exporter of coal. The Soviet Union will become a larger gas exporter. However, none of those changes will greatly affect their bilateral power relationship. Currently and prospectively, therefore, attention focuses mainly on the different situations of the superpowers in regard to dependence on foreign oil.

The United States has recently shown considerable success in reducing its oil imports, which declined by 20 percent between 1979 and 1980 and another 15 percent in 1981. Although a strong revival of economic activity could well reverse that trend, it may be able to hold its oil imports throughout the 1980s below the 8 million barrels a day reached in 1978. Soviet prospects are necessarily brighter, since the Soviet Union (as dis-

Table 12.1 The USA and USSR: A Basic Energy Comparison (1980)

	USA			USSR		
	mbdoe	% of total	% change from 1979	mbdoe	% of total	% change from 1979
CONSUMPTION:						
Oil	16.4	43	− 8	8.8	37	+2
Gas	10.2	27	− 1	6.6	28	+7
Coal	8.5	22	+ 4	6.9	29	0
Hydro	1.6	4	− 1	1.0	4	+5
Nuclear	1.5	4	− 1	.3	1	+8
TOTAL	38.2		− 3	23.6		+3
Oil production	10.2		0	12.2		+3
Implied oil imports	6.2		−20			
(exports)				(3.4)		(+5)
Oil imports as % of total oil		38			0	
Oil imports as % of primary energy		16			0	

Source: *BP Statistical Review of the World Oil Industry, 1980.*

tinct from Comecon as a group) is most unlikely to become a net oil importer at any time in the current decade. It may have to trim its exports, contain consumption, and shift some demand to other fuels, such as gas, to be sure of maintaining its self-sufficiency, but there is no reason to suppose it will fail to do so. In all probability, therefore, the two superpowers will remain much as they are today throughout this decade in terms of fuel import dependence.

In these circumstances, it is not surprising that the dependence of the United States on imported oil is interpreted, especially by Americans, as one of several indications that the balance of international power has shifted to the Soviet advantage. Nor is there any doubt that its need for foreign oil, especially from the Middle East, acts as a check on the freedom of international action that the United States otherwise enjoys. If OPEC exports were sharply reduced in the 1980s, it would also entail a considerable economic penalty, as supplies fell and prices rose. To that extent, the United States is more vulnerable than the Soviet Union. A common misunderstanding, however, is to suppose that American oil vulnerability is a new phenomenon, for the United States has, in reality, depended on imported oil for over a quarter of a century. The proportion of American oil demand covered by net imports has, of course, risen considerably in this period, from about 10 percent to about 40 percent. But the change has been longer and more gradual than many people now realize.[12] The sense of qualitative change arguably owes less to statistical measurements than to two quite different factors: the loss by largely American international companies of direct control over a large part of the world's oil reserves and oil production, and the progressive need to import more politically "unsafe" oil from the Gulf and Africa to supplement what had been assumed to be "safe" oil from Latin America.

But the United States is not crucially handicapped by its dependence on foreign oil. For it is spared one of the Soviet Union's major international burdens and responsibilities — as a supplier of fuel to its allies. Three quarters of Eastern Europe's oil has to come from Soviet sources; none of Western Europe's comes from the United States. Conversely, the fact that the United States has joined its major allies in depending on oil from the Middle East and Africa conveys an almost paradoxical political benefit, which is not enjoyed by the Soviet Union. Before the 1970s, American governments had no clear and obviously legitimate reason to be concerned directly with the flow of oil from those regions. Repeatedly, therefore, any effort to intervene politically in relation to oil was interpreted by both local regimes and U.S. allies either as merely gratuitous or, more probably, as an attempt to secure some commercial advantage for American companies. Now, however, the United States, exactly because of its recent energy dependence on the Gulf and Africa, has a generally acknowledged need to pursue national policies in regard to oil that will safeguard both its own energy interests and those of its closest allies.[13] In contrast, the Soviet Union can be seen in the same context only as a potential troublemaker.

None of that is to belittle the political importance of American oil imports or to deny that the political power of the United States in relation to a self-sufficient Soviet Union would be greater if the oil were not needed. The long rise in oil import dependence since the 1950s has increased the external vulnerability of the American energy system, and the United States is more exposed to some kinds of international pressure as a result. Still, over 80 percent of American energy supplies — an extraordinarily high proportion by the standards of the industrial world — comes from indigenous sources; and during the 1980s, that proportion is more likely to rise than to fall, provided the long-standing tendency of oil imports to grow can be held in check. More important, the United States retains enormous strength in other fuels such as coal and uranium, in energy technology of all kinds, and in its fundamental ability to generate capital for energy investments. Although it is far from clear that the obvious reduction in the relative international power of the United States over the last quarter of a century has, in reality, owed much to its growing dependence on foreign oil producers, Americans are naturally inclined to concentrate on oil imports as the obvious chink in their armor. But other industrial countries, most of which depend far more heavily on foreign fuel suppliers and none of which has nearly such impressive energy assets in the aggregate, would continue to regard the United States as an object of envy rather than pity.

The Soviet Union resembles the United States in having major indigenous fuel reserves, although its capacity to produce surpluses of oil, gas, and coal for export gives it an unquestioned current advantage. Also, like the United States, total energy productivity within its economy is relatively poor, while its heavy dependence on long-range internal transport and its climatic circumstances create greater problems. Further, the high unit energy consumption that persists in certain American industries is more than matched by the energy inefficiency of most Soviet industries. Overall energy productivity in the two countries is, in fact, strikingly similar: $1241 of GNP per ton of oil equivalent for the United States in 1979 as against $1236 for the Soviet Union.[14] But in the crucial area of access to energy technology and equipment and to the highly skilled personnel to exploit them, the United States is unrivaled, whereas the continued reliance of the Soviet Union on external suppliers is constantly apparent. The two countries differ no less strikingly in their inherent ability to pay for energy, and especially for timely energy investments — a field in which the United States also enjoys and will continue to enjoy a clear advantage, even if it does not always seem to be bent on exploiting it. From the point of view of the Soviet leaders, those deficiencies must rank as major weaknesses, weighing heavily against their indigenous fuel strength.

But the superpowers seem to differ most in the capacity of their governments and national institutions to shape and adapt energy policy and to mobilize a consensus in its support. In that respect, they appear to be almost at polar extremes of the international spectrum, ranging from American incoherence to Soviet regimentation. Again, however, the contrast is not a simple one. Frequently, the Soviet Union seems as much encumbered by the obscurantist complexity of its bureaucratic system as the United States is by the studied archaism of its political process. Nevertheless, the Soviet government, albeit inefficiently, is able to enforce its will domestically in the energy field, in spite of any popular preference, more easily than its American counterpart. Energy adaptation will thus remain a simpler executive process for the Soviets. Their handicap will rather be the continuing inadequacy of their administrative, technical, and, in some respects, economic resources.

Which nation has more to fear or hope from the decade? The answer is likely to depend less on any objective difference between their domestic energy situations than on their political and executive systems and on the course of events in the outside world. As long as the United States can continue to secure something close to its current level of oil and too many of its energy assets are not squandered through political incoherence, then its technological superiority, combined with its greater capacity to invest in energy adaptation and higher productivity, should be enough to enhance its relative power.

The effectiveness of domestic energy policy in the United States and of its relations with oil-exporting states in the Gulf, Africa, and Latin America represent two of the important international issues for the 1980s. Whatever the confusions of the 1970s, however, it can hardly be that the United States is incapable of a coherent domestic policy. It is not demonstrably inferior to the Soviet Union in its ability to influence international exporters of oil, and it is clearly superior in that respect to those other major industrial countries, such as Japan and West Germany, that have long depended more heavily on oil imports. That being so, the risks and uncertainties inevitably associated with a democratic internal system and with a measure of external fuel dependence constitute a challenge to American skill rather than an index of American weakness.

In America's case, the most important uncertainties concern energy imports, whereas they are linked to energy exports for the Soviets. East European dependence on Soviet energy resources and the resulting Soviet influence over East European governments has no parallel for the United States in its relations with the other members of the Atlantic alliance or the OECD.

The Soviet Union is actually a larger supplier of fuels to Western Europe than the United States itself, and it is likely to remain so at least until American coal exports begin to enter the European market on a large scale in the 1990s. Such Soviet fuel exports play a significant role in particular West European countries, as with oil in Finland and Iceland or with gas in Austria and West Germany. That fact prompts occasional alarm on the Western side that the Soviet Union will wield its energy assets to win political advantage. This concern was evident in 1980, when the West German and French governments decided that up to 30 percent of their natural gas supply might, in principle, come from Soviet sources. It became more widespread thereafter, as forecasts suggested that, by 1990, if all proposed agreements were completed, the European Community as a whole might be importing 20 percent of its gas from the Soviet Union. Concern is understandable; no country can be totally insensitive to the vulnerability of any significant share of its energy supply. This is yet another area, however, where the sense of danger has to be kept in perspective. It would be naive, for example, to suppose that dependence on Soviet fuels is more than a minor element in the constraints imposed on Finland and Austria to maintain their neutral status. But even if West Germany's and France's gas imports from the Soviet Union eventually reach the limit recently announced, they will amount to only about 5 percent of their energy supply. This will happen at a time when they are also gaining access to additional gas supplies from Western sources and are in any case using gas predominantly to meet less crucial parts of their energy demand.

On the other hand, the Soviet Union, which has hitherto been an impressively reliable energy supplier, will be struggling to sustain the level of its total fuel exports to Western Europe, not as an instrument of political leverage, however welcome that might be, but as a source of badly needed hard currency and as a means of securing access to some of the foreign technology and equipment on which it is itself dependent. Indeed, the currency, goods, and services exchanged for Soviet fuel exports to Western Europe (or Japan) should arguably be seen as a hostage to fortune no less substantial than the fuels that flow in the opposite direction. Those exports will still, of course, form one part of the intricate, if frail, economic network already linking the Soviet and European economies; and West European governments, before increasing their dependence on Soviet exports, will have to weigh the political implications carefully, including those of an alternatively heavier dependence on fuels from the Middle East and Africa. In doing so, they must also consider the overall pattern of their fuel imports; much may depend on whether their particular non-Soviet sources seem likely to offset (or alternatively to connive at) any Soviet attempt to manipulate the flow of oil or gas. On balance, however, it does not seem likely that the Soviet Union will be in a position during the 1980s to exploit the fuel dependence of any major West European country to its own strategic advantage. The stakes will be too high, and the Soviet share of West European energy markets too low, to make that sort of power play plausible.

In contrast, the link between Soviet energy strength and political influence in Eastern Europe is obvious and of long standing, but whether the Soviet Union can maintain its dominance of that market throughout the 1980s is nevertheless open to question, given its well-known difficulties in expanding, or even sustaining, its oil production.[15] Although those difficulties have sometimes been exaggerated, they may be sufficient for much of the 1980s to force Soviet leaders into a hard choice between restricting supplies to Eastern Europe, cutting back on hard currency sales outside Comecon, and reducing domestic consumption — while in any case increasing their export prices. All those options will probably have to be pursued to some extent, but the heaviest share of the burden may conceivably fall on other Comecon countries, which could find the real price of Soviet oil rising as the proportion of growing energy demand that Soviet oil can satisfy declines.

That scenario implies the possibility of both economic hardship and political discontent in Eastern Europe. If only for obvious reasons of self-interest, however, the Soviet Union will go to considerable lengths to minimize the potential impact of future limits on its oil exports to Comecon partners. Already, supplies of gas to them from the enormous Soviet reserves are rising, and while it is not yet clear that these supplies can fully offset a relative decline in Soviet oil sales, they will certainly make an important contribution to doing so, as will larger direct exports of Soviet electricity. In addition, the Soviet Union may seek to buy limited quantities of oil in the world market for transshipment to East European countries, as it has previously done in the case of Bulgaria. Thus, new restrictions on the flow of Soviet oil within Comecon will not necessarily reduce the overall importance of the Soveit Union as a regional energy supplier.

Nor are the East European countries themselves entirely helpless. Although the share of oil in their total energy supply rose from 21 percent in 1975 to 23 percent in 1980, they have recently begun to shift a larger share of their demand for oil to natural gas and nuclear power, and they will certainly make further progress in that direction during the 1980s. In 1979, East Europeans used 24 percent more energy per capita than West Europeans, but obtained only $1000 of GNP from each ton of oil equivalent, as against $2189.[16] They also have considerable scope, therefore, for improving the efficiency of their consumption, and thus their energy productivity, even without large capital investments. As an illustration, Hungary, by combining fuel substitution and improved energy efficiency, was able by 1980 to eliminate its previous need to supplement Comecon supplies by importing oil in return for hard currency.

The Warsaw Pact and NATO

Were Comecon to encounter much more serious energy supply problems in the 1980s than now seems probable, many in the West might fear a power play calculated to secure control of additional fuel resources. As it is, the Soviet Union, the only plausible author of such a power play, is clearly eager to have more gas from Iran and would doubtless be happy to gain preferential access to some oil from the Gulf as well. It is not currently under strong pressure, however, to import more fuel into Comecon than it can reasonably expect to obtain by normal commercial and diplomatic means: buying in the world market, entering into agreements on economic cooperation, and bartering arms. Only in the unlikely event of Comecon's facing a new and graver energy crisis, which required a dramatically expanding fuel supply, would a question of using Soviet military power to that end even arise. At such a juncture, however, for the reasons explored earlier, it is improbable that the Soviet government would, in fact, see military force as a rational option. Irrational action in the face of imminent

Table 12.2 NATO and the Warsaw Pact: A Basic Energy Comparison
(1980)

	NATO			The Warsaw Pact		
	mbdoe	% of total	% change from 1979	mbdoe	% of total	% change from 1979
CONSUMPTION:						
Oil	29.5	46	− 8	10.9	33	+2
Gas	14.8	23	− 1	8.0	24	+6
Coal	13.6	21	+ 3	12.6	38	+1
Hydro	4.1	6	0	1.1	3	+5
Nuclear	2.4	4	+ 3	.3	1	+6
TOTAL	64.4		− 3	32.9		+3
Oil production	14.2		0	12.6		+3
Implied oil imports (exports)	15.3		−15	(1.7)		(+7)
Oil imports as % of total oil		52			0	
Oil imports as % of primary energy		24			0	

Source: *BP Statistical Review of the World Oil Industry, 1980.*

energy "strangulation" is no more impossible in the Soviet case, of course, than in the Western. But the only circumstance, short of such desperation in which the risk of a Soviet military power play might grow, would be if both Soviet leaders and the governments of the fuel-producing countries concerned came to believe that the ability and determination to deter an adventure of that sort no longer existed outside Comecon.

Partly for that reason, there is a particular point in considering the impact of energy factors in the 1980s, not only on the relative positions of the superpowers, but also on the overall balance between the two great military alliances, NATO and the Warsaw Pact. The Warsaw Pact will have a clear advantage throughout the decade in terms of internal access to energy natural resources — oil, gas, coal, and uranium — in all of which its members are jointly self-sufficient. In contrast, as Table 12.2 shows, the NATO group has to import over half its oil as well as smaller shares of other fuels. NATO states will, however, retain an equally clear collective advantage in energy technology and in their capacity to pay for energy supplies and investments. In general, therefore, the balance of fundamental energy strengths and weaknesses between the two alliances mirrors that between the two superpowers. In this case, too, the decisive considerations in the 1980s will be the quality of policy on the two sides and the impact on them of wider international events.

If the international trade in fuels, and especially oil, were to suffer a major and prolonged interruption — as a result of international conflict or producer action affecting the equivalent of, say, most of Saudi Arabia's production, so that NATO oil imports were cut by up to 50 percent — the self-sufficient Warsaw Pact would have an ostensible advantage. But the NATO countries, by drawing on emergency oil stocks and indigenous production, should be able to operate without undue hardship for at least twelve months, even in the face of the most serious plausible disruption, much as was demonstrated in Chapter 5 for Japan in such a crunch. Conversely, several of the East European countries would be more swiftly vulnerable to any interdiction of oil and gas supplies from the Soviet Union. Initially, therefore, the balance between the two alliances would be closer than it might appear.

Obviously, the Warsaw Pact's position would be stronger if the disruption of international fuel trade persisted indefinitely. In circumstances of adequate international supply, however, the greater joint strength of the NATO countries in technology, productivity, and general economic performance, provided they are effectively exploited, should carry steadily increasing weight during the 1980s. The only way in which their collective position could then be undermined would be through a failure of political will or executive incompetence — in terms of national policies as well as in terms of energy cooperation. Just as energy problems could pose a threat in the 1980s to the economic welfare and political stability of certain Warsaw Pact countries, and thus to the effectiveness of the pact as a whole, so they could contribute to the serious economic difficulties that certain NATO countries, especially on the southern flank of Europe, are likely to face. Those difficulties can be either exacerbated or lessened by the energy and economic policies of other NATO states. In NATO as in the Warsaw Pact, therefore, much will depend on alliance cohesion and cooperation under economic pressure, including the pressure of energy factors.

Japan and China

Outside Europe, the major area of immediate demarcation between the Communist and non-Communist worlds is in East Asia, dominated by the physical volume of China and the economic mass of Japan. Japan has minimal natural resources but great technological capacity, together with a developed ability to generate wealth for making energy investments and paying energy bills and a high (but not unlimited) degree of social and political cohesion. China may be similar in that last respect, but is otherwise the mirror image of Japan.

In a heightened and protracted conflict, China's greater resource strength might tilt the balance, but in normal circumstances, the relative positions of the two countries are more complementary than competing, offering scope for more extensive energy collaboration in the 1980s. Whether that opportunity will be seized depends on politics as well as economics — on Chinese views about dependence on foreign cooperation and on Japanese sensitivity about the reaction of other countries, including the Soviet Union, to closer links with China. It also depends on social and political forces inside Japan as they bear, for example, on environmental issues, if only because a greater reliance on China for fuel would probably involve a shift from the use of oil to the use of coal. However, in East Asia, energy problems are more apt to encourage the pursuit of complementary interests than to cause conflicts.

The Third World

The world's need for energy will enhance the relative power of oil-exporters in the Third World — of OPEC members, but also of states such as Mexico, Oman, and Malaysia, which remain outside OPEC. It is they, or at least some of them, that have recently been able to expand their general economic and political influence by exploiting the dependence of others on their energy resources. Some, such as Nigeria, Indonesia, and Malaysia, may as a result begin to rise through the ranks of the so-called newly industrializing countries to positions of wider regional influence by the end of the 1980s.

The road to international power is not, however, a smooth one. The capacity of Third World oil exporters to exert leverage by supplying or withholding oil depends ultimately on the maintenance of demand for their product, and thus on the economic health of their customers. In the long run, the strength of their own economies depends upon industrial development, which requires an increasing reliance on foreign technology, goods, and markets. Most of them have to import growing quantities of capital and consumer goods, as well as foodstuffs, at a cost inevitably affected by the price they have charged for their own oil. Some of them, especially in the Gulf, keep a large part of their oil revenue within the industrial world, as deposits or investments. As a result, they are forced to take a progressively closer interest in the condition of the economies with which they trade. Exploiting their oil power without restraint might initially increase both their income and influence, but at the eventual expense not only of a depressed demand for oil but also of wider economic damage that would inevitably rebound on their own development, savings, and welfare.

To most people, Saudi Arabia seems to be the classic case of a country whose power is founded on the control of exportable oil (although its control of money is already an additional factor). It is also, however, a case of a country engaged in qualifying its power by committing itself to development policies that entail a growing degree of international interdependence. The social and political strain of such development, already demonstrated in Iran, is one of the reasons that some observers fear an abrupt and substantial drop in Saudi oil exports, by 50 percent or more, as a result of some dramatic change in either policy or regime.

If such a reduction simply tightened the international market without causing a major drop in demand, the result might paradoxically be to increase the influence of Saudi Arabia as well as the price commanded by the residue of Saudi oil production. A drastic cut, for whatever reason, would nevertheless reflect a decision to accept at least an equal probability of diminishing Saudi Arabia's relative international power, because other suppliers might take part of its market share in the short run or because wider economic repercussions might depress demand and reduce the value of Saudi reserves and investments in the long run. A smaller cut in a stable market would be an entirely different matter; both Saudi Arabia's internal stability and its influence on the policies of other OPEC members might well be enhanced if it reestablished a significant margin of reserve production capacity in such circumstances. Conversely, pressing Saudi leaders to maximize production may tend also to maximize the risk of domestic discord, leading directly to the dramatic disruption of supply that they and their customers are most anxious to avoid.

The rest of the 1980s could see changes of oil policy, or even leadership, in Saudi Arabia, but the resilience of the Saudi regime and the Saudi political and social system must not be underestimated. On balance, the probability is high that Saudi Arabia will continue to be both the largest oil supplier and the largest political beneficiary, in terms of relative power, from control of such a fuel resource.

Some of the other oil exporters — Algeria and Nigeria, for instance — may be unable or unwilling to maintain current production levels until 1990, and they may lose some of their potential power as a result when compared with more richly endowed colleagues; others, such as Iraq, may increase both market share and influence. More fundamentally, OPEC members, while continuing to support each other's international policies in many respects, have found great difficulty in regaining that capacity for collective action, especially on prices, that they demonstrated in the mid-1970s. One effect of their recurrent failure to do so in the future may be to encourage closer bilateral links between particular exporters and importers, with the implication that policy influence, on economic matters as well as on political issues such as the Arab-Israeli conflict, will come to be exerted as much in a bilateral as in a multilateral framework.

Certainly energy factors, including access to technology and finance as well as fuels, tend to increase the gap between average economic standards in the developed and developing worlds. Even setting OPEC members aside, however, Third World countries vary as much as their developed counterparts in the relative strength of their energy assets. Some have or will soon have indigenous access to significant fuel reserves. A few have a considerable mastery of energy technologies. A number have external trade balances to help them to pay for energy imports. In one or more of those respects, countries as different as Argentina, Cameroon, India, and Singapore will have an advantage over many of their neighbors, an advantage that will grow if the world energy situation becomes more stringent.

Many developing countries are nevertheless in danger of becoming economic, social, and political casualties of energy stringency. Their ability to escape that fate will depend even more than in the past on international assistance in developing indigenous energy sources and more efficient energy systems and, when necessary, in supporting increased indebtedness. In their case, the issue is not whether their relative power will increase but whether they will survive.

The Crucible of Adaptation

Some states entered the 1980s in favored energy circumstances because they controlled fuel resources, because they had technological strength, or because their economies were particularly robust. But the eventual impact of energy on relative international power in a period of such change and uncertainty will depend less on the assets with which countries entered the decade than on their capacity to adapt to a changing national and international energy situation. Some nations will enjoy relative self-sufficiency — in fuels, technology, financial capacity, or in all three at once. Others, without moving closer to autarky, will use their international bargaining strength to achieve effective energy security. And yet others, unable or unwilling to move in either direction, will find their potential power slipping away. Moreover, the ranking of states, in terms of how energy will affect their power, will change repeatedly as energy supply, demand, and price all shift and as national capacities for dealing with the evolving energy situation are more clearly revealed.

Because it entails additional investment in alternative options, energy uncertainty must itself be regarded as a cost to the international system, just as an increase in real price or a divergence of notional and effective demand implies a cost, and all such costs have to be borne by someone, eventually, as a charge on economic welfare.

The fundamental issue for the 1980s is not whether such costs can be avoided, because they cannot. It is rather how they are to be distributed, internationally and within national economies. The pattern of distribution, rather than the aggregate cost, will determine the impact of energy on the relative power of nations. But it is the total cost that will ultimately measure the impact of energy on the overall prosperity of the world economy, from which no nation can entirely divorce itself.

From that flows the basic dilemma for everyone concerned with energy policy: whether to maximize national advantage (or minimize national loss) by exploiting particular energy assets separately and to the utmost, without regard to regional or global repercussions, or to assume some larger share of an irreducible total cost in the interest of wider security and stability. Energy stringency and uncertainty will make patterns of power, as well as profit and loss, more fluid in the 1980s than in the recent past. And that situation will give some nations an opportunity to increase their relative power, as well as their economic return, by exploiting their energy assets to the fullest. Some, alternatively, will be able to use strength in other fields to obtain a disproportionate energy advantage. Relative power can be used to influence energy circumstances, just as energy factors can influence relative power. In neither case, however, will those concerned be able to act entirely independently without putting a strain on an already fragile regional or global system.

That has already been much advertised as a problem for OPEC states. But it is also a difficult dilemma for the industrial nations of the West. They may be able to cope with energy stringency by transferring the whole cost to weaker nations, in the Third World or within their own ranks, in the form of depressed economic standards. To do so could enhance their own relative power and minimize their energy burden, but would risk more active conflict between richer and poorer countries and still larger economic and security penalities, whose impact they would not easily avoid. Alternatively, by adopting a different approach, they might accept a larger share of the burden themselves at the expense of a lower rate of economic growth with all that would imply for their own economies and for the international economy.

Let there be no mistake — each of those courses would be costly to the world economy. Yet the total cost at issue cannot in any case be avoided. Those responsible for policy in both the industrial nations and the oil-exporting countries must consider whether the impact of energy on the strength and stability of the international system as a whole may not be even more important in the end to their own national security and welfare than its impact on the relative power of individual states.

Appendix
Notes
Index
Contributors

Appendix   The Economic Performance of the "Big Seven"
(percent)

	Pre-embargo									Post-embargo							
	1965	1966	1967	1968	1969	1970	1971	1972	1973	1974	1975	1976	1977	1978	1979	1980	1981
United States																	
GDP Growth	6.2	6.0	2.6	4.7	2.8	-0.2	3.4	5.5	5.4	-0.6	-0.9	5.4	5.4	4.4	2.8	-0.2	1.8
Consumer Prices	1.7	2.9	2.8	4.2	5.4	5.9	4.3	2.3	6.2	11.0	9.1	5.8	6.5	7.7	11.3	13.5	10.4
Unemployment	4.4	3.6	3.7	3.4	3.4	4.8	5.7	5.4	4.7	5.4	8.3	7.5	6.9	5.9	5.7	7.0	7.5
Japan																	
GDP Growth	5.1	10.6	10.8	12.8	12.3	9.8	4.6	8.8	8.8	-1.0	2.3	5.3	5.3	5.0	5.5	4.2	3.8
Consumer Prices	6.6	5.1	4.0	5.3	5.2	7.7	6.1	4.5	11.7	24.5	11.8	9.3	8.1	3.8	3.6	8.0	4.9
Unemployment	1.2	1.3	1.3	1.2	1.1	1.1	1.2	1.4	1.3	1.4	1.9	2.0	2.0	2.2	2.1	2.0	2.2
Germany																	
GDP Growth	5.6	2.5	-0.2	6.3	7.8	6.0	3.2	3.7	4.9	0.5	-1.8	5.2	3.0	3.3	4.6	1.8	-1.0
Consumer Prices	3.4	3.5	1.4	2.9	1.9	3.4	5.3	5.5	6.9	7.0	6.0	4.5	3.7	2.7	4.1	5.5	6.0
Unemployment	0.3	0.2	1.3	1.5	0.9	0.8	0.9	0.8	0.9	1.6	3.7	3.7	3.7	3.5	3.2	3.1	5.0
France																	
GDP Growth	4.8	5.2	4.7	4.3	7.0	5.7	5.4	5.9	5.4	3.2	0.2	5.2	3.1	3.7	3.5	1.3	0.5
Consumer Prices	2.5	2.7	2.7	4.5	6.4	5.2	5.5	6.2	7.3	13.7	11.8	9.6	9.4	9.1	10.8	13.6	13.3
Unemployment	1.5	1.8	1.9	2.6	2.3	2.4	2.6	2.7	2.6	2.8	4.1	4.4	4.7	5.2	5.9	6.3	7.5

Appendix   The Economic Performance of the "Big Seven" (continued)
(percent)

| | Pre-embargo | | | | | | | | | Post-embargo | | | | | | | |
	1965	1966	1967	1968	1969	1970	1971	1972	1973	1974	1975	1976	1977	1978	1979	1980	1981
United Kingdom																	
GDP Growth	2.3	2.0	2.6	4.1	1.5	2.2	2.7	2.2	7.5	-1.2	-0.6	3.6	1.3	3.3	1.4	-1.8	-2.0
Consumer Prices	4.8	3.9	2.5	4.7	5.4	6.4	9.4	7.1	9.2	16.0	24.2	16.5	15.8	8.3	13.4	18.0	11.9
Unemployment	2.3	2.3	3.4	3.4	3.1	3.1	3.7	4.1	3.0	2.9	3.9	5.5	6.2	6.1	5.7	7.4	10.5
Italy																	
GDP Growth	3.3	6.0	7.2	6.5	6.1	5.3	1.6	3.2	7.0	4.1	-3.6	5.9	1.9	2.7	4.9	4.0	0.0
Consumer Prices	4.6	2.3	3.7	1.4	2.6	5.0	4.8	5.7	10.8	19.1	17.0	16.8	18.4	12.1	14.8	21.2	19.3
Unemployment	5.3	5.7	5.3	5.6	5.6	5.3	5.3	6.3	6.2	5.3	5.8	6.6	7.0	7.1	7.5	7.4	8.2
Canada																	
GDP Growth	6.8	7.0	3.4	5.6	5.2	2.6	7.0	5.8	7.5	3.5	1.1	5.8	2.4	4.0	3.2	-0.1	3.0
Consumer Prices	2.4	3.7	3.6	4.0	4.6	3.3	2.9	4.8	7.6	10.8	10.8	7.5	8.0	9.0	9.1	10.1	12.5
Unemployment	3.6	3.3	3.8	4.4	4.4	5.6	6.1	6.2	5.5	5.3	6.9	7.1	8.0	8.3	7.4	7.5	7.5

Source: *OECD Economic Outlook*, December 1981, pp. 12, 19, 131, 140, 142.

Notes

Chapter 1　Crisis and Adjustment: An Overview

1. See Inguar Svennilson, *Growth and Stagnation in the European Economy* (Geneva: United Nations, 1954); Donald Winch, *Economics and Policy: An Historical Study* (London: Fontana, 1972); "An Economic World Restored" in Daniel Yergin, "The Rise of the National Security State" (Ph.D. diss., Cambridge University, 1974).

2. Reserves from American Petroleum Institute, *Basic Petroleum Databook*, II-1, IV-3, I-6; on price, Hollis Chenery, "Restructuring the World Economy: Round Two," *Foreign Affairs*, Summer 1981, p. 1105; on consumption and shares, United Nations, *World Energy Supplies, 1950–1974*, ser. 5, no. 19 (New York: 1976).

3. For the two oil shocks, see Robert Stobaugh and Daniel Yergin, *Energy Future: Report of the Energy Project at the Harvard Business School* (New York: Ballantine, 1980), chap. 2, app. 2. For the alteration in oil market, see Ian M. Torrens, *Changing Structures in the World Oil Market* (Paris: Atlantic Institute, 1980). "Almost universal boom" and "flamboyant growthmanship" are both phrases of Herbert Block in, respectively, *The Planetary Product in 1980: A Creative Pause* (Washington, D.C.: Department of State, 1981), p. 23, and *The Planetary Product in 1978* (Washington, D.C.: Department of State, 1979), p. 1.

4. *OECD Economic Outlook,* July 1981, p. 8.

5. Richard Nixon, *R.N.: The Memoirs of Richard Nixon* (New York: Warner, 1979), p. 52.

6. International Monetary Fund, *World Economic Outlook* (Washington, D.C.: 1981), p. 7.

7. Robert Hall and Robert Pyndyck, "What to Do When Energy Prices Rise Again," *The Public Interest*, no. 65 (Fall 1981), p. 65; Congressional Budget Office, *The Effect of OPEC Oil Pricing on Output Prices* and *Exchange Rate in the United States and Other Industrialized Countries* (Washington, D.C.: February 1981); Dale Jorgenson, "Energy Prices and Productivity Growth," *Scandinavian Journal of Economics* 83 (1981): 165–79.

8. *OECD Economic Outlook;* Robert Stobaugh, Harvard Business School Energy Project.

9. LDC data from IMF, *World Economic Outlook,* pp. 122, 132: World Book, *World Development Report, 1981* (New York: Oxford University Press, 1981), pp. 1–3, 35, 42–43.

10. Chenery, "Restructuring the World Economy," p. 1107.

11. IMF, *World Economic Outlook,* pp. 97–98; IEA data.

12. International Energy Agency, *Energy Conservation: The Role of Demand Management in the 1980s* (Paris: 1981), p. 19.
13. See Sheikh Yamani speech in William B. Quandt, *Saudi Arabia in the 1980s: Foreign Policy, Security, and Oil* (Washington: Brookings, 1981), pp. 164–65.
14. European Community, *In Favor of an Energy Efficient Society*, Energy Ser. no. 4 (Brussels: 1979), chap. 3–4.
15. Ibid., p. 72.
16. Winston Churchill, *The World Crisis*, pp. 135–40.
17. Karl Kaiser, Winston Lord, Thierry de Montbrial, and David Watt, *Western Security: What Has Changed? What Should Be Done? A Report Prepared by the Directors of Forschungsinstitut der Deutsche Gesellschaft für Auswärtige Politik, Council on Foreign Relations, Institut Français des Relations Internationales, and Royal Institute of International Affairs* (New York: Council on Foreign Relations, 1981).
18. For different views on the security question, see Joshua Epstein, "Soviet Vulnerability in Iran and the RDF Deterrent," Dennis Ross, "Considering Soviet Threats to the Persian Gulf," *International Security*, Fall 1981, pp. 126–80; Robert W. Tucker, *The Purposes of American Power: An Essay on National Security*, a Lehrman Institute Book (New York: Praeger, 1981); Quandt, Saudi Arabia; Christopher Van Hollen, "Don't Engulf the Gulf," *Foreign Affairs*, Summer 1981, pp. 1064–78.
19. See Angela Stent, *Soviet Energy and Western Europe*, Washington Paper no. 90 (New York: Praeger, 1982); Hans W. Maull, *Natural Gas and Economic Security: New Problems for the West*, Atlantic paper no. 43 (Paris: Atlantic Institute, 1981).
20. Ulf Lantzke, "Energy Vulnerability and the Industrial World," paper presented at the 20th anniversary meeting of the Atlantic Institute, October 1981, p. 9.
21. This bias results from such factors as the well-developed mechanisms that exist for mobilizing capital for energy supply and the tax code. For the latter, see Alliance to Save Energy, "The Administration's FY 1983 Energy Policy: An Unbalanced Budget," March 1982.
22. On planning for energy emergencies, see David A. Deese and Joseph S. Nye, eds., *Energy and Security* (Cambridge, Mass.: Ballinger, 1980); William Hogan, "Oil Stockpiling: Help Thy Neighbor," Harvard Energy and Environmental Center Discussion Paper.

Chapter 2   World Energy to the Year 2000

1. For a review of systematic problems appearing in some of the most important models to appear shortly after the first oil shock, see Sergio Koreisha and Robert Stobaugh, "Appendix I: Limits to Models," in *Energy Future: Report of the Energy Project and the Harvard Business School*, ed. Robert Stobaugh and Daniel Yergin (New York: Ballantine, 1980).
2. Other factors are listed in Koreisha and Stobaugh, ibid.
3. International Energy Agency, "Assessment," May 7, 1979; Walter Levy, "The Years That the Locust Hath Eaten: Oil Policy and OPEC Development Prospects," *Foreign Affairs*, Winter 1978–79.
4. Bijan Mossavar-Rahmani, "Tilting toward the Arabian/Persian Gulf," Harvard Energy Security Program, March 1982; *Oil and Gas Journal*, January 4, 1982, p. 51.
5. Exxon Company, U.S.A.'s "Energy Outlook," December 1979, December 1980. There is an abundance of oil forecasts; in the last two years, I have reviewed over one hundred. Some of the better known, in addition to Exxon's, include those of

the International Energy Agency, Royal/Dutch Shell Group, and Petroleum Industry Research Foundation, to name just a few.

6. Angela E. Stent, *Soviet Energy and Western Europe,* Washington Paper no. 90 (New York: Praeger, 1982).

7. The following discussion of obstacles is drawn from I. C. Bupp and Frank Schuller, "Natural Gas: How to Slice a Shrinking Pie," in *Energy Future.* Also, for more information on natural gas, see Jensen Associates, *Imported LNG Project, Supply and Consumption,* Report to Congress of the United States Office of Technology Assessment, September 1979; U.S. General Accounting Office, "Implications of U.S.-Algerian Liquefied Natural Gas Price Dispute and LNG Imports"; and IEA, "World Oil Demand and Supply," natural gas section, April 24, 1981; and Joseph Riva, *U.S. Domestic Natural Gas Production during the 1980s,* Congressional Research Service, October 24, 1980.

8. Exxon, "Energy Outlook."

9. Guy Elliott Mitchell, "Billions of Barrels of Oil Locked Up in Rocks," *National Geographic,* February 1918, p. 205.

10. Carroll Wilson, *Coal — Bridge to the Future* (Cambridge, Mass.: Ballinger, 1980); IEA, *Coal Production,* November 6, 1980; and Leonard Waverman and Arthur Donner with Dianne Groome, "Investments in Energy Supply Industries and the Economy of the OECD," Institute for Policy Analysis, University of Toronto, prepared for the IEA, March 1981; Commission of the European Community, Substitution of Coal for Oil in "Other Industry," COM(81) 229, May 6, 1981.

11. This section draws heavily on I. C. Bupp, "The Nuclear Stalemate," in *Energy Future;* I. C. Bupp and J. C. Derian, *The Failed Promise of Nuclear Power* (New York: Basic Books, 1980); and I. C. Bupp, "The French Nuclear Harvest," *Technology Review,* November–December 1980, pp. 30–40.

12. Exxon president Howard Kauffmann was quoted in the *Wall Street Journal,* February 24, 1982, p. 2. One international organization assumed that world oil prices would increase 1.0 percent annually between 1980 and 1985, 2 percent annually between 1985 and 1990, and 2.5 percent annually between 1990 and 2000, while recognizing that their best econometric models indicated a 3 to 3.5 percent increase between 1980 and 1990 and even more between 1990 and 2000. A study released in mid-May 1980 by the Petroleum Industry Research Foundation in the United States assumed that world oil prices would rise several percentage points faster than the inflation rates in the industrial nations. A major international oil company estimated that oil prices would rise 1.5 times the general inflation rate. Walter Levy warned of a possible "takeoff" in future oil prices in "Oil and the Decline of the West," *Foreign Affairs,* Summer 1980. By early 1982, many observers were predicting relatively level oil prices for the next several years.

13. For a discussion of the situation of the developing countries, see William Martin and Frank Pinto, "Energy for the Third World," *Technology Review,* June–July 1978; Joy Dunkerley, William Ramsay, Lincoln Gordon, and Elizabeth Cecelski, *Energy Strategies for Developing Nations* (Baltimore: Johns Hopkins University Press, 1981); as well as Chapter 9 in this book.

Chapter 3 The Bedeviled American Economy

1. Petroleum conservation regulations were enacted in the United States after massive oil discoveries in the 1930s. The conservation system was a "pro-rationing" system, limiting the output of domestic oil wells to a percentage of their capacity in order to balance supply and demand. Oil discoveries in the 1950s ex-

tended excess capacity into the 1960s, but the margin of excess capacity diminished through the decade. As long as excess capacity existed, the United States could balance out shortfalls in the world oil market by increasing domestic supply and petroleum exports, as was done in the Suez crises of 1956 and 1967. When the margin of surplus capacity was exhausted in the early 1970s, the United States could no longer compensate for production cutbacks elsewhere. For further discussion of the pro-rationing system, see Wallace Lovejoy and Paul Homan, *Economic Aspects of Oil Conservation Regulation* (Baltimore: Johns Hopkins University Press, 1967).

2. U.S. Department of Energy, Energy Information Administration, *Annual Report of Congress 1980,* vol. 2, p. 49. The breakdown of the pro-rationing system was not the only factor in the U.S. turn toward imported oil. Shortages of natural gas, the product of almost two decades of price control, led to a shift in energy demand toward petroleum during these years.

3. For the data on the OECD countries, see OECD, *Energy Balances of OECD Countries, 1960–74,* p. 23. For the effect on oil production capacity, see William Nordhaus, "Oil and Economic Performance," *Brookings Papers on Economic Activity* 2:1980, fig. 4, p. 369.

4. For a further discussion of the effects of oil price increases on aggregate demand, see Robert Dohner, "Energy Prices, Economic Activity and Inflation: A Survey of Issues and Results," in *Energy Prices, Inflation, and Economic Activity,* ed. Knut Mork (Cambridge, Mass.: Ballinger, 1981), pp. 17–23.

5. This is the result of multiplying 2.23 billion barrels of crude oil and petroleum product imports in 1974 by $7.37, the change in crude oil import prices from Customs data, adjusted for the intervening rise in the general price level.

6. Personal income and personal disposable income from the national accounts divided by the deflator for personal consumption expenditures, from U.S. Council of Economic Advisers, *Economic Report of the President, 1980,* tables B-3, B-20.

7. GATT, *International Trade, 1975/76,* p. 5.

8. The real supply of money is defined here as M2 divided by the Consumer Price Index.

9. The Tax Reduction Act of 1975 was a temporary tax cut, achieved mainly through a rebate of 1974 taxes. The rebate appears to have been largely absorbed in additional savings in 1975, with very little effect on consumption demand. See Alan Blinder, *Economic Policy and the Great Stagflation* (New York: Academic Press, 1979), pp. 150–51.

10. U.S. Department of Energy, *Monthly Energy Review,* May 1981. The price referred to is the refiner acquisition price of imported crude oil.

11. Council of Economic Advisers, *Economic Report, 1980,* p. 65.

12. Ibid., pp. 49–53.

13. Council of Economic Advisers, *Economic Report, 1981,* p. 136.

14. Part of the surge in 1974 was due to the lapse of the price control program. On this, see Blinder, *Economic Policy,* chap. 6.

15. Jeffrey Sachs, "Wages, Profits, and Macroeconomic Adjustment in the 1970's: A Comparative Study," *Brookings Papers* 2:1979, pp. 269–319.

16. Japan and Germany are good examples. See Yoichi Shinkai, "Oil Crises and Stagflation (or Its Absence) in Japan," Discussion Paper 110, Institute of Social and Economic Research, Osaka Univ., November 1980. For Germany, see *OECD Economic Survey: Germany* (Paris: OECD, May 1980), p. 20. On this point, see also Council of Economic Advisers, *Economic Report, 1981,* pp. 184–85.

17. Blinder, *Economic Policy,* pp. 146–49, 185–94.

18. Council of Economic Advisers, *Economic Report, 1980,* p. 61. See also p. 79.

Similar rationales are expressed in *Economic Report, 1975,* p. 42, and *Economic Report, 1981,* p. 156.

19. The share of energy in total costs has been increasing throughout the 1970s as the relative price of energy has risen. The 5 percent figure is an approximation to the mid-decade share. See fn. 43.

20. For studies on the decline in productivity in the United States, see Edward Denison, "Explanations of Declining Productivity Growth," *Survey of Current Business* 59:8, pt. 2 (August 1979): 1–24; J. Randolph Norsworthy et al., "The Slowdown in Productivity Growth: Analysis of Some Contributing Factors," *Brookings Papers* 2:1979, pp. 387–421; and Ernst Berndt, "Energy Price Increases and the Productivity Slowdown in U.S. Manufacturing," in *The Decline in Productivity Growth,* Federal Reserve Bank of Boston, Conference Series 22 (1980), pp. 60–89.

The difficulty of measuring energy inputs hampers studies of its effects on productivity. This is particularly true of the Berndt study, where the data used indicate that energy use per unit of production in manufacturing increased after 1973. In a recent article, Dale Jorgenson argues that higher energy prices do explain the productivity slowdown. In a sectoral model that relates productivity growth rates to relative input prices, he finds energy price effects large enough to account for much of the retardation of productivity growth after 1973. See "Energy Prices and Productivity Growth," *Scandinavian Journal of Economics* 83 (1981):165–79.

21. A similar approach by Jack Alterman finds energy inputs in manufacturing reduced by 15 percent between 1971 and 1976. See Joy Dunkerley, Jack Alterman, and John Schanz, *Trends in Energy Use in Industrial Societies,* Report EA-1471 (Palo Alto: Electric Power Research Institute, 1980), table E-18.

22. Denison, "Explanations of Declining Productivity Growth," table 3, p. 14.

23. This is a short summary of a complicated debate on energy-capital complementarity. For further discussion, see Ernst Berndt and David Wood, "Engineering and Econometric Interpretations of Energy-Capital Complementarity," *American Economic Review* 69:3 (1979):342–54, and William Hogan, "Capital-Energy Complementarity in Aggregate Energy-Economic Analysis," *Resources and Energy* 2:3 (1979).

24. Council of Economic Advisers, *Economic Report, 1981,* table 9, p. 71.

25. Bankers Trust Company, *U.S. Energy and Capital* (New York: Bankers Trust, 1980), p. 23.

26. "Energy and the Future U.S. Economy," *The Wharton Magazine* 3:4 (Summer 1979):15–21. See also Edward Hudson and Dale Jorgenson, "Energy Policy and U.S. Economic Growth," *American Economic Review* 68:2 (May 1978):118–22.

27. A complete accounting of changes in the energy-GNP ratio is done in Dunkerley et al., *Trends in Energy Use,* app. E, table E-18, for 1973–76 and three previous periods.

28. The data on GNP in 1972 dollars are from *Economic Report of the President, 1981.* The data on electricity consumption are from Department of Energy, *Annual Report to Congress, 1980,* vol. 2, table 64, p. 153.

29. Dunkerley et al., *Trends,* table E-18.

30. U.S. Department of Transportation, *Highway Statistics,* annual.

31. See Hudson and Jorgenson, "Energy Prices," p. 882.

32. The composition of consumption demand is sensitive to the business cycle. The years chosen are at or near cyclical peaks.

33. The growing share of the service sector in the advanced economies has been widely noted and discussed. This would contribute to a reduction over time in the energy content of GNP in those economies. One must be somewhat careful, how-

ever, for services as generally measured include transportation and several direct purchases of energy. For recent discussions of the increasing share of the service sector, see Angus Maddison, "Economic Growth and Structural Change in the Advanced Countries" and Maurice Lengelle, "Development of the Service Sector in the OECD Countries: Economic Implications," both in *Western Economies in Transition,* ed. Irving Leveson and Jimmy Wheeler (Boulder, Colo.: Westview Press, 1980).

34. GNP in 1972 dollars from Council of Economic Advisers, *Economic Report, 1981,* table B-2. Manufacturing output: Gross product in manufacturing less gross product in petroleum refining (SIC 29) (see fn. 35) in 1972 dollars from Bureau of Economic Analysis, unpublished data. The growth rates for GNP and manufacturing output for 1956 and 1973 were 3.9 and 4.2 percent per year, respectively. The years 1956, 1973, and 1978 were chosen for comparison since manufacturing output is sensitive to the business cycle. The three years are peak or near peak years.

35. The treatment of petroleum refining is a matter of convention. In the industrial statistics and national account statistics of the United States, petroleum refining is a manufacturing industry. In the energy statistics of the Department of Energy, petroleum refining is an energy conversion industry. Petroleum consumption by energy-using sectors is defined as the consumption of refined petroleum products (i.e., it does not include the energy consumed in petroleum refining). We use this convention here to make the results comparable with the industrial sector energy data. However, petroleum refining is included as an addendum to Table 3.5.

36. Since only limited data are available for this type of calculation, the change in energy intensity due to production shifts only include changes in output shares among two-digit Standard Industrial Classification (SIC) industries. Thus they do *not* capture, for example, changes in the production of aluminum relative to steel (both SIC 33) or changes in the production of different types of chemicals (SIC 28). Any changes of the mix within two-digit SIC industries is counted as a change in efficiency in Table 3.5. Finally, the sensitivity of this calculation to the steel industry should be noted. The year 1974 was a peak year for steel shipments, and this fact accounts for the shift in output shares toward greater energy use between 1971 and 1974 and for part of the shift in output shares toward lower energy use between 1974 and 1977.

37. The data in this paragraph are drawn from Richard Greene, "Employment Trends in Energy Extraction," *Monthly Labor Review,* May 1981, p. 5, and from U.S. Department of Labor, *Employment and Earnings,* various issues.

38. The rapid expansion of energy production and transportation in the western states has become a heated political issue. What is involved is rapid expansion in underpopulated areas and the "boomtowns" that are created, the allocation of scarce water rights for energy and nonenergy uses, and the feeling that energy policy is being dictated from Washington. "Montana does not intend to become a boiler room for the nation," declared Bill Christiansen, its lieutenant governor. Quoted in Lynton Hayes, *Energy, Economic Growth, and Regionalism in the West* (Albuquerque: University of New Mexico Press, 1980), p. 3.

39. U.S. Department of Labor, *Employment and Earnings,* various issues. As Daniel Yergin points out in Chapter 4, consumer demand shifted toward small cars in 1974, only to return to larger cars and personal trucks in the next few years. By early 1979, substantial dealer inventories of small cars had built up. The 1979 oil price shock had what appears now to be a more permanent effect, greatly increasing the demand for small, fuel-efficient cars. The intervening resumption of the demand for large cars in the United States has been blamed on the price control

program, which kept domestic petroleum prices below world market levels. This conclusion is too strong, for the effects of oil price decontrol on gasoline prices would have been relatively small. As Yergin points out, the change in perceptions after the second oil price shock was probably more important in accounting for the extent of the shift toward small car demand.

40. U.S. Interagency Task Force on the Steel Industry, *Report to the President: A Comprehensive Program for the Steel Industry* (Solomon Report) 6 December 1977, pp. 3, 29.

41. U.S. Office of Technology Assessment, *Technology and Steel Industry Competitiveness,* 1980, table 42, p. 142.

42. A survey by McGraw-Hill found 26 percent of the American steel industry's plants and equipment to be outmoded, compared to an average of 12 percent for U.S. durable goods industries as a whole. See McGraw-Hill, "How Modern Is American Industry?" November 1979, quoted in ibid., p. 129.

43. Sam Schurr et al., *Energy in America's Future: The Choices Before Us* (Baltimore: Johns Hopkins University Press, 1979), p. 80. The authors estimate the share of energy in total expenditure as 5.2 percent in 1976 (p. 79).

44. However, this need not be the case during the transition period as the capital stock adjusts to higher energy prices. As Robert Leone pointed out in comments on a draft of this chapter, during an energy crisis, the United States could use more intensively the most energy-efficient units of the capital stock. If wide differences in energy efficiency exist (as is the case in the current U.S. automobile stock), this could result in significant short-run energy savings. There would be problems, however, in coordinating this kind of response, problems that, for instance, would be more easily solved in two-car families than in one-car families. In the production sphere, this is the response the market would generate if left to allocate activity. However, in a true crisis, policy-makers may be unwilling to allow the market to assign high profits and activity to some plants or firms and high unemployment to other installations. This poses one of the most difficult problems of policy, requiring a choice between adjustments to reduce energy use and notions of fairness that demand that sacrifice be shared during a crisis.

A second possibility is that, due to the underlying uncertainty in the energy market, the capital stock is replaced by units that are more flexible in their choice of fuel. Manufacturers of industrial boilers now report a strong demand for units with fuel-switching capacity.

45. OECD, *Energy Balances of OECD Countries, 1960-74, 1974-78.*

46. This was true of both the 1974 and the 1979 oil price rises. See Nordhaus, "Oil and Economic Performance," pp. 368–69. In contrast, the Iran-Iraq war, which shut down part of Middle East oil production, came at a time of low capacity use and excess supplies. Despite early fears, the war produced no jump in oil prices. Recent models of OPEC price decisions have stressed the role of slack or tightness in international oil markets. See U.S. Department of Energy, Energy Information Administration, *Annual Report to Congress, 1979,* vol. 3, pp. 9–11, and Jeffrey Sachs, "Discussion," *Brookings Papers* 2:1980, pp. 393–94.

47. National Academy of Sciences, Committee on Nuclear and Alternative Energy Systems (CONAES), *Alternative Energy Demand Futures to 2010* (Washington: NAS, 1979), pp. 14, 16. The range quoted is that from Scenario A to Scenario D and compares energy use in 2010 to that of 1975. The assumptions incorporate a wide range of energy prices and policies.

48. James Just and Lester Lave, "Review of Scenarios of Future U.S. Energy Use," *Annual Review of Energy* 4 (1979): 501–36.

49. For earlier years, see Joel Darmstadter, *Energy in the World Economy* (Baltimore: Johns Hopkins University Press, 1972), table 1.

50. OECD, *National Accounts of OECD Countries, 1950–78,* vol. 1, *Main Aggregates* (Paris: OECD, 1980).

51. Joel Darmstadter, Joy Dunkerley, and Jack Alterman, *How Industrial Societies Use Energy* (Baltimore: Johns Hopkins University Press, 1977), p. 189.

52. Ibid., table 6-13, p. 133.

53. The attempt of Japan to diversify out of energy-intensive manufacturing industries is discussed in Chapter 5 by Teruyasu Murakami.

54. H. N. Fullerton and P. O. Flaim, "New Labor Force Projections to 1990," *Monthly Labor Review,* December 1976, pp. 3–13. The 1.3 percent growth rate applies to the Series II population projections.

55. Council of Economic Advisers, *Economic Report, 1980,* p. 88.

56. CONAES, *Alternative Energy Demand Futures,* pp. 135, 162.

57. U.S. Department of Energy, *Annual Report to Congress, 1980,* vol. 3, table 3-5, p. 53, and CONAES, *Alternative Energy Demand Futures,* tables 46–47 (pp. 151–52), Scenario B. Transportation Energy Use accounts for 26 percent of total energy use. If transportation energy demand remained constant and all other energy demands and GNP grew at 2.5 percent per year, then in twenty-two years (1978–2000), the energy-GNP ratio would fall by 11 percent.

58. CONAES, *Alternative Energy Demand Futures,* p. 51.

59. The most persuasive studies are those for industries with roughly homogeneous production. See Battelle Columbus Laboratories, *Potential for Energy Conservation in the Steel Industry* (Columbus, Ohio: Battelle, 1975), and Gordian Associates, Inc., "Industrial International Data Base Pilot Study: The Cement Industry" (New York: Committee on the Challenges of Modern Society, North Atlantic Council, 1976). See also Stanford Research Institute, *Comparison of Energy Consumption between West Germany and the United States* (Menlo Park, Calif.: SRI, 1975).

60. CONAES, *Alternative Energy Demand Futures,* pp. 105–6. The study points out that this energy savings comes through investment in new facilities and could not be achieved by refitting the existing capital stock. But it assumes no more rapid investment than the historical rate of turnover of the capital stock.

61. Air Transport Association of America, *Air Transport, 1981* (Washington: ATA, 1981), p. 12.

62. A few forecasts have produced conditions for zero-growth energy paths. See Dale Jorgenson and Edward Hudson, "Economic Analysis of Alternative Energy Growth Patterns, 1975–2000," Appendix F of S. David Freeman et al., *A Time to Choose* (Cambridge, Mass.: Ballinger, 1974); Hudson and Jorgenson, "Energy Prices and U.S. Economic Growth," *American Economic Review* 68:2 (May 1978); and Scenario A of CONAES, *Alternative Energy Demand Futures,* pp. 10, 32–33. These scenarios have attracted more controversy than any other aspect of energy demand modeling. The zero-energy-growth paths in the Hudson and Jorgenson studies are accompanied by significantly reduced economic growth rates. The CONAES study imposed a given economic growth rate, but assumes very energetic demand reduction policies and "substantial changes in consumer behavior and purchasing patterns."

63. The measure that economists use for economic flexibility is the "elasticity of substitution," the percentage change in input proportions with a given percentage change in relative input prices. An elasticity of zero implies that no substitution is possible, while an elasticity of 1 or more implies very great substitution possibilities. Estimates of this elasticity for the U.S. economy range between 0.2 and 0.6. For further discussion, see William Hogan and Alan Manne, "Energy-Economy

Interactions — The Fable of the Elephant and the Rabbit?" in *Modelling Energy-Economy Interactions: Five Approaches,* ed. Charles Hitch, Research Paper 5 (Washington: Resources for the Future, 1977); and CONAES, *Energy Modelling for an Uncertain Future* (Washington: National Academy of Sciences, 1978), pp. 36–46.

64. CONAES, *Energy Modelling,* pp. 46–47, 106–10.

65. Hudson and Jorgenson, "Energy Policy and Economic Growth," pp. 118–22. One must be careful in grafting one study onto another. Their basic economic growth rate is 3.2 percent per year, which is higher than what we have been discussing, and this makes an actual reduction in energy use harder to achieve.

66. Council of Economic Advisers, *Economic Report, 1974,* p. 211.

67. See, for instance, the discussion in Council of Economic Advisers, *Economic Report, 1981,* pp. 96–97. Natural gas also presents a problem for the United States. The gradual decontrol of prices under the Natural Gas Policy Act will not bring prices up to world levels when controls expire in 1987, and it may produce a sharp rise in prices. In such a circumstance, it may be difficult to avoid additional control legislation.

68. This shows up in a variety of "sensitivity tests" in energy demand forecasts. For example, in the most recent Department of Energy forecast, an 8.8 percent change in the level of output with constant energy prices results in an 8.3 percent change in primary energy use and an 8.8 percent change in net energy use (Department of Energy, *Annual Report to Congress, 1980,* vol. 3, p. 65). In the CONAES study, Scenarios B and B' differ only in the assumed economic growth rate. Output in B' is 41 percent higher and energy use 43 percent higher at the end of the period than in Scenario B (CONAES, *Alternative Energy Demand Futures,* pp. 18, 31). See also Morris Adelman, "Energy-Income Coefficients and Ratios: Their Use and Abuse," *Energy Economics* 2:1 (1980):2–4.

69. Nathan Rosenberg, "Historical Relations between Energy and Economic Growth," in *International Energy Strategies,* ed. Joy Dunkerley (Cambridge, Mass.: Oelgeschlager, Gunn and Hain, 1980), pp. 55–70. See also his *Technology and American Economic Growth* (White Plains, N.Y.: Sharpe, 1972), pp. 55–58.

70. Sam Schurr, "Energy, Economic Growth, and Human Welfare," *EPRI Journal* 3:4 (May 1978):14–18.

71. Bankers Trust, *U.S. Energy and Capital,* p. 23.

72. Council of Economic Advisers, *Economic Report, 1980,* p. 115.

73. Bankers Trust, *U.S. Energy and Capital,* p. 23.

74. Council of Economic Advisers, *Economic Report, 1981,* pp. 70–72.

75. Council of Economic Advisers, *Economic Report, 1979,* p. 125. The measure used here is real nonresidential fixed investment as a percent of real GNP.

76. Council of Economic Advisers, *Economic Report, 1981,* table B-21, p. 258, personal savings as a percentage of personal disposable income.

77. Jeffrey Sachs, "The Current Account and Macroeconomic Adjustment in the 1970s," *Brookings Papers on Economic Activity* 1:1981, pp. 201–268.

78. See Peter K. Clark, "Investment in the 1970's: Theory, Performance and Prediction," *Brookings Papers* 1:1979, pp. 95–97, and Robert Tannenwald, "The Changing Mix of Business Investment," *Data Resources Review,* December 1980.

79. William Fellner, "Corporate Asset-Liability Decisions in View of the Low Market Valuation of Equity," in *Contemporary Economic Problems 1980,* ed. William Fellner, (Washington: American Enterprise Institute, 1980), pp. 79–85.

80. Assar Lindbeck, "Can the Rich Nations Adapt? Needs and Difficulties," *The OECD Observer,* January 1981, pp. 1–6.

81. See OECD Development Center, *Adjustment for Trade* (Paris: OECD, 1975).

Chapter 4 America in the Strait of Stringency

1. Daniel Bell, "The Public Household," *The Public Interest,* Fall 1974, pp. 42–43.
2. Adolph Berle, *Navigating the Rapids* (New York: Harcourt Brace Jovanovich, 1973), p. 575.
3. Lewis J. Perelman, "Speculation on the Transition to Sustainable Energy," *Ethics* 90 (April 1980), p. 394.
4. Frederick Jackson Turner, *The Frontier in American History* (New York: Henry Holt, 1920), p. 293; David M. Potter, *People of Plenty: Economic Abundance and the American Character* (Chicago: University of Chicago Press, 1954), pp. 84, 85, 89.
5. Sam Schurr and Bruce Netschert, *Energy in the American Economy, 1850–1975* (Baltimore: Johns Hopkins University Press, 1960), p. 511; *Monthly Energy Review;* Potter, *People of Plenty,* p. 89.
6. Quoted in Schurr and Netschert, *Energy in the American Economy,* pp. 35, 49–51.
7. Harold Williamson, Ralph Andreano, Arnold Daum, and Gilbert Klose, *The American Petroleum Industry: The Age of Energy* (Evanston, Ill.: Northwestern University Press, 1963), p. 509. Michael Stoff, *Oil, War, and American Security* (New Haven: Yale University Press, 1980).
8. On fears at the end of World War II, see U.S. Congress, Senate Special Committee Investigating Petroleum Resources, *American Petroleum Interests in Foreign Countries: Hearings,* 79th Cong., 1st sess., 1945; *Wartime Petroleum Policy: Hearings,* 79th Cong., 1st sess., 1945. On overproduction and hot oil, see Williamson et. al., *Age of Energy,* pp. 540–56. In addition to balancing supply against demand, the production controls were also meant to curtail waste and increase the ultimate recovery of oil.
9. Energy Information Agency, *Report to Congress, 1979,* vol. 2, p. 15; *Fortune,* August 1959, pp. 95–96. On subsidies to energy production and consumption, see Batelle Memorial Institute, *An Analysis of Federal Incentives Used to Stimulate Energy Production* (Springfield, Va.: NTIS, 1978).
10. Motor Vehicle Manufacturers Association, *Motor Vehicle Facts and Figures;* Alan Altschuler, *The Urban Transportation System: Politics and Policy Innovation* (Cambridge, Mass.: MIT Press, 1981).
11. Dorothy K. Newman and Dawn Day, *The American Energy Consumer* (Cambridge, Mass.: Ballinger, 1975), p. 82; Richard Stein, *Architecture and Energy* (New York: Doubleday, 1977), pp. 60–61.
12. On industrial efficiency, see John Myers and Leonard Nakamura, *Saving Energy in Industry: The Post Embargo Record* (Cambridge, Mass.: Ballinger, 1978).
13. Energy Information Agency, *Report to Congress, 1979,* pp. 58, 7.
14. For Dow, see Robert Stobaugh and Daniel Yergin, eds., *Energy Future: Report of the Energy Project at the Harvard Business School* (New York: Ballantine, 1980), p. 205; for Reagan, *Wall Street Journal,* 5 September 1980.
15. For constraints on U.S. oil production, see Richard Nehring with E. Reginald Van Driest II, *The Discovery of Significant Oil and Gas Fields in the United States* (Santa Monica, Calif.: Rand Corporation, 1981), R-2654/1/USGS/DOE. The changes in the world oil market in these years are analyzed in Chapter 2 of *Energy Future.*
16. Crauford Goodwin, ed., *Energy Policy in Perspective: Today's Problems, Yesterday's Solutions* (Washington: Brookings Institution, 1981), p. 399. The discussion that follows on the internal administration struggles in the Nixon and Ford years draws on this book.

17. U.S. Congress, Senate Energy Committee, *Executive Energy Messages,* 95th Congress, 2nd sess., 1978.

18. Ibid.

19. I. C. Bupp and J. C. Derian, *The Failed Promise of Nuclear Power: The Story of Light Water* (New York: Harper Colophon Books, 1981), pp. 132–35, 153–69; Daniel Yergin, "The Terrifying Prospect: Atomic Bombs Everywhere," *Atlantic Monthly,* April 1977.

20. Quoted in John H. Gibbons and William U. Chandler, *Energy: The Conservation Revolution* (New York: Plenum Press, 1981), p. 28. Also see Marc Ross and Robert Williams, *Our Energy: Regaining Control* (New York: McGraw-Hill, 1981).

The United States consumed about 75 quads of energy in 1973 and in 1981. Following is a sample of some forecasts from the period from Gibbons and Chandler, *The Conservation Revolution,* p. 22, and Ross and Williams, *Our Energy,* p. 19.

	1990	2000
Federal Power Commission (1970)	140	
Atomic Energy Commission (1973)		150–200
National Petroleum Council (1973)		300
U.S. Bureau of Mines (1973)		163
Edison Electric Institute (1970) moderate growth		161
Stanford Research Institute (1977) base case		141

21. Goodwin, *Energy Policy;* Ford Foundation, Energy Policy Project, *A Time to Choose: America's Energy Future* (Cambridge, Mass.: Ballinger, 1974).

22. Jack M. Hollander, *U.S. Energy Demand and Supply Scenarios: A Retrospective Appraisal of CONAES* (Berkeley: Energy and Resources Group, July 1980), p. 26. CONAES developed four demand scenarios, all of which would be compatible with a doubling to tripling of real GNP by the year 2010. Thus demand could range from 58 to 133 quads. Committee on Nuclear and Alternative Energy Systems (CONAES), *Energy in Transition: 1985–2010* (San Francisco: W. H. Freeman, 1980).

23. U.S. Congress, Senate Energy Committee, *The President's Energy Program: A Compilation of Documents,* 95th Cong., 1st sess., 1977, pp. 2–10.

24. The developments of 1979 are treated in detail in Benjamin Brown and Daniel Yergin, "Synfuels 1979," Kennedy School of Government case.

25. *Wall Street Journal,* 1 October 1978.

26. Report of the Energy Policy Task Force, 5 November, 1980; Reagan in *National Journal,* 6 December 1980, p. 2075; Daniel Boggs session of the Harvard Energy and Security Seminar, 2 November 1981.

27. Weinberger in *New York Times,* 6 March 1981. Statement by James B. Edwards, Secretary of Energy, before the Joint Economic Committee, 26 February 1981. The lessened concern about oil imports was reflected in the Reagan administration's National Energy Plan, *Securing America's Energy Future: A Report to Congress,* July 1981.

28. "Temporary inconvenience" from Neil diMarchi in Goodwin, *Energy Policy,* p. 513; O'Neill in *Congressional Quarterly Almanac: 1977,* p. 713.

29. On energy prices and consumers in the OECD, see International Energy Agency, *Energy Conservation: The Role of Demand Management in the 1980s* (Paris: OECD, 1981).

30. Thomas Schelling, *Thinking Through the Energy Problem* (Washington: Committee for Economic Development, 1979), p. 112.

31. As William Schneider, a public affairs analyst, has observed: "There is an enormous sensitivity on the part of the American public to price. This shows up in any poll you take on the subject . . . As far as the public is concerned, *right now,* the energy crisis is not the most important problem facing the country. It hasn't really ever been. The most important problem is inflation. Any energy solution that involves any direct increase in the price of energy, in the price of anything, is rejected out of hand." In Daniel Yergin, ed., *The Dependence Dilemma* (Cambridge, Mass.: Center for International Affairs, 1980), p. 81. For public opinion, Barbara C. Farhar, Charles T. Unseld, Rebecca Varies, and Robin Crews, "Public Opinion about Energy," *Annual Review of Energy,* vol. 5 (Palo Alto: Annual Reviews, 1980), pp. 143, 149, 155–56. For worries in the Ford administration, see Goodwin, *Energy Policy in Perspective,* p. 507.

32. Stobaugh and Yergin, *Energy Future,* pp. 273–74.

33. For concern about the possible "crowding out" of energy efficiency investment by energy production investment, see remarks by William Sneath, the chairman of Union Carbide, in *Energy User News,* 8 October 1979, and Bankers Trust Company, *U.S. Energy and Capital: A Forecast, 1980–1990,* pp. 23–24.

34. The issue is analyzed in Harold Beebout, Gerald Peabody, and Pat Doyle, "The Distribution of Household Energy Expenditures and the Impact of Higher Prices" in *High Energy Costs: Assessing the Burden,* ed. Hans Lansburg (Baltimore: Johns Hopkins University Press, 1981); Congressional Budget Office, *Low Energy Income Assistance* (Washington: GPO, 1981); Fuel Oil Marketing Advisory Committee of the U.S. Department of Energy, *Low Income Energy Assistance Program: A Profile on Need and Policy Options* (Washington, 1981); Kenneth J. Arrow and Joseph P. Kalt, *Petroleum Price Regulation: Should We Decontrol?* (Washington: American Enterprise Institute, 1979); and R. Herenden and J. Tanaka, "Energy Cost of Living," *Energy* 1 (June 1976). Not surprisingly, given the uncertainties of the estimates, there is a good deal of disagreement as to the regressiveness when indirect costs are included as well.

35. David Schooler in Yergin, *Dependence Dilemma,* p. 80.

36. For state tax revenues, see *National Journal,* 27 March 1980, pp. 47–71. For New England, Congressional Budget Office, *Low Income Energy Assistance,* p. 9.

37. *New York Times,* 6 November 1979; Barbara Farhar et al., "Public Opinion about Energy," *Annual Review of Energy,* vol. 5, pp. 143–45, 149; Russell Long in *National Journal,* 5 November 1977, p. 1717; research director in private letter.

38. John J. McCloy, *The Great Oil Spill* (New York: Chelsea House, 1976); Anthony Sampson, *The Seven Sisters* (New York: Viking, 1975), pp. 205–7, 265–75. William Schneider, "Public Opinion and the Energy Crisis," in Yergin, *Dependence Dilemma,* pp. 154–55. On advising Ford, see Goodwin, *Energy Policy in Perspective,* p. 507.

39. B. Bruce-Briggs, "Gasoline Prices and the Suburban Way of Life," *Public Interest,* Fall 1974, pp. 143–45.

40. Schlesinger quoted in *National Journal,* 13 May 1978, p. 767.

41. See Arrow and Kalt, *Petroleum Price Regulation.* "Darts and arrows" from Brown and Yergin, "Synfuels 1979."

42. When Stobaugh first presented his innovative analysis of the social cost of imported oil in 1978, many other specialists refused to take it seriously. Within two years, the notion had become almost commonplace. For the development of his argument, and refinements in the calculation of the premium, compare his Chapter 2 in the 1979 and 1980 editions of *Energy Future.* Thomas Schelling also empha-

sized the social costs in his work at about this time. For a sophisticated analysis of an oil import premium, see William Hogan's chapter and appendix in David A. Deese and Joseph S. Nye, eds., *Energy and Security* (Cambridge, Mass.: Ballinger, 1981).

43. Schelling has described as a "shotgun effect" the use of price controls as a de facto welfare system; see *Thinking Through the Energy Problem.*

44. *Monthly Energy Review.*

45. *Business Week,* 6 April 1981, p. 58.

46. Two of the leading sources were the research programs at the Lawrence Berkeley Laboratory and at Princeton's Center for Energy and Environmental Studies. *Energy Future* suggested a 30 to 40 percent reduction per unit of GNP as a practical midterm goal. By 1981, the United States was about 17 percent more energy efficient than in 1973.

47. Alliance to Save Energy/Cambridge Survey Report; 1981 surveys by Yankelovich, Skelly and White; Castle cited in Altshuler, *Urban Transportation System,* p. xxix.

48. *Roper Reports,* February, April 1980.

49. *Los Angeles Times,* 21 January 1979; Henry Ford, speech to White House Conference on Balanced National Growth and Economic Development, 30 January 1978.

50. "Automobiles — Turning Around on a Dime — Interview with Marina V. N. Whitman," *Challenge,* May–June 1981, pp. 37–39.

51. Sam Schurr, ed., *Energy in America's Future: The Choice before Us* (Baltimore: Johns Hopkins University Press, 1979), pp. 21–22.

52. Fred Hirsch, *Social Limits to Growth* (Cambridge, Mass.: Harvard University Press, 1979), p. 2; Lester Thurow, *The Zero-Sum Society* (New York: Basic Books, 1980), pp. 11–12, 28.

53. Data from Steven Barnett.

54. Neil J. Smelser, "Energy Restriction, Consumption and Social Stratification," in Charles T. Unseld, Denton E. Morrison, David L. Sills, and C. P. Worl, eds., *Sociopolitical Effects of Energy Use as Policy,* Supporting Paper Number Five, Study of Nuclear and Alternative Energy Systems (Washington: National Academy of Science, 1979), pp. 225–26.

55. Ibid., p. 221.

56. Michael R. Beschloss, *Kennedy and Roosevelt: The Uneasy Alliance* (New York: Norton, 1980), p. 274.

57. Kenneth Boulding, "The Anxieties of Uncertainties in the Energy Problem," in *Prospects for Growth: Changing Expectations for the Future,* ed. Kenneth D. Wilson (New York: Praeger, 1977).

58. See Solar Energy Research Institute, *A New Prosperity: Building a Renewable Energy Future* (Andover, Mass.: Brick House, 1981); Ross and Williams, *Our Energy;* and CONAES, *Energy in Transition.*

59. *Energy Efficient Housing: A Prairie Approach* (Provinces of Alberta, Saskatchewan, and Manitoba, 1981); Office of Energy Conservation, Saskatchewan Mineral Resources, *Saskatchewan Conservation Energy Seminars: 1980.*

60. Charles L. Gray, Jr., and Frank von Hippel ("The Fuel Economy of Light Vehicles," *Scientific American,* May 1981, pp. 48–59) argue that a fleet average of 60 miles per gallon is possible by 1995. This would result from the intersection of five paths: weight reduction, reduction in aerodynamic drag, reduction in rolling resistance, more efficient engines, and more efficient transmissions.

61. Solar Energy Research Institute, *A New Prosperity;* Roger Seasonwein Associ-

ates, "The Conservation Decision"; Bankers Trust Company, *Energy Viewpoint,* May 1981; communication from the Alliance to Save Energy.
62. Daniel Yankelovich and Bernard Lefkowitz, "National Growth: The Question of the Eighties," *Public Opinion,* December–January 1980, pp. 48–49; surveys by Yankelovich, Skelly and White.

Chapter 5   The Remarkable Adaptation of Japan's Economy

1. Economic Planning Agency, *Economic Survey of Japan, 1978–1979* (Tokyo: The Japan Times, 1980), p. 143.
2. The structurally depressed industries include: open-hearth furnace, aluminum smelting, aluminum rolling, shipbuilding, chemical fertilizer, polyvinyl chloride resin, corrugated cardboard, textiles (spinning, weaving), plywood, and sugar refining.
3. In October 1980, the New Energy Development Organization (NEDO) was established, aiming at accelerating the commercial development of new sources of energy, rather than the basic research that is the primary focus of the Sunshine Project.
4. For example: Yoshio Suzuki, *Supply Shock No Keizaigaku (Economics of Supply Shock)* (Tokyo: Shukan Toyo Keizai, 23 February 1980; Takashi Tamaki, *Sekai-teki Inflation Kasoku To Dou Tatakau Ka (How to Cope with World Inflation)* (Tokyo: Japanese Economist, 22 January 1980); Masaru Yoshitomi, *Sekiyu To Dollar Ni Semerareru Nippon (Japan Attacked by Oil and Dollar)* (Tokyo: Chuo-kouron, February 1980). The titles of Japanese books have been translated into English by the author except where an English title was already given.
5. Japan took a significant step forward by implementing the New Foreign Exchange Law on 1 December 1980. This law introduced a fundamental change in policy toward exchange control because the principle of freedom is laid down, with provisions for control only in emergencies. It will encourage more active participation of foreign capital in the Japanese capital market.
6. MITI, *80 Nendai Tsusho Sangyo Seisaku No Vision (Visions for International Trade and Industry Policy in the 1980s)* (Tokyo: 1980).
7. Energy Sogo Suishin Iinkai (Energy Policy Development Committee), *Sogo Energy Seisaku Suishin No Tameno Shikin Kakuho No Kinyousei Ni Tsuite (Toward Earmarking Capital for Implementing the Comprehensive Energy Policy),* a policy recommendations leaflet issued on 20 December 1979.

Chapter 6   Japanese Society and the Limits of Growth

1. For more detailed argument on this, see Michel Crozier, Samuel P. Huntington, and Joji Watanuki, *The Crisis of Democracy: Report on the Governability of Democracies to the Trilateral Commission* (New York: New York University Press, 1975).
2. For an analysis of the Japanese budget-making process, see John Craigton Campbell, *Contemporary Japanese Budget Politics* (Berkeley: University of California Press, 1977).
3. Nobuyoshi Namiki, *Nihon Keizai Ittō Ryōdan (Japanese Economy: An Incisive Analysis)* (Tokyo: Nihon Keizai Shimbun Sha, 1980), p. 69. Recent figures of government expenditure/GNP are: Japan (1973) 30.5 percent, U.S. (1972) 34.5 percent, U.K. (1973) 43.9 percent, West Germany (1973) 44.6 percent, and France (1973) 44.0 percent. Okurasho Shukeikyoku (Budget Bureau, Ministry of Finance), *Saishutsu Hyakka (A Handbook on Government Expenditure)* (Tokyo: Okurasho Insatsukyoku, 1980), p. 7.

4. Campbell, *Japanese Budget Politics.*

5. *Prospects for Trade and Industrial Policy in the 1970s: An Interim Report of the Advisory Council for Industrial Structure, MITI* (in Japanese).

6. Types and examples of "knowledge-intensive industry" mentioned in the above report are as follows: (1) research and development industry: computer, airplane, electric car, industrial robot, nuclear related, IC, fine chemical, new synthetic chemistry, new metal, special ceramic, and ocean development; (2) Sophisticated assembling industry: communication instruments, business machine, computerized machine tool, pollution-control instruments, educational instruments, factory-produced housing, automated storage systems, large construction machinery; (3) Fashion-type industry: High-quality cloth and furniture, electronic musical instruments; (4) Knowledge industry: information processing service, information supply service, software, systems engineering, consulting.

7. Noriaki Sasaki, *Gendai Enerugii Kikiron (Present Energy Crisis)* (Tokyo: Shin Nippon Shuppansha, 1978), p. 96.

8. Namiki, *Japanese Economy,* pp. 63–64.

9. MITI, *21 Seiki eno Enerugii Senryaku (Energy Strategy for the 21st Century)* (Tokyo: Tsusho Sangyō Chōsakai, 1979), p. 160.

10. Naikaku Sōridaijin Kanbo Kōhōshitsu (Public Relations Section, Prime Minister's Chamber of the Cabinet), "Shō Enerugii, Shō Shigen ni kansuru Yoron Chōsa" ("Public Opinion Survey on Conservation of Energy and Resources") (mimeo: 1978).

11. Naikaku Sōridaijin Kanbo Kōhōshitsu (Public Relations Section, Prime Minister's Chamber of the Cabinet), "Kokumin Seikatsuni Kansuru Yoron Chōsa," ("Public Opinion Survey on People's Living") (mimeo: 1979).

12. Unyushō Daijinkanbo Jōhōkanribu (Information Keeping Section, Prime Minister's Chamber of the Cabinet, Ministry of Transportation), ed., *Unyu Keizai Zusetsu Shōwa 55nenban (An Illustrated Analysis of Transportation and Economy)* (Tokyo: Unyu Keizai Kenkyu Senta, 1980), p. 83.

13. *Gekkan Yoron Chōsa (Monthly Public Opinion Survey),* July 1980. According to that survey, 56 percent of the respondents said that they are keeping the temperature of the room below 64 degrees Fahrenheit (among those 56 percent, 30 percent said that they are keeping the room temperature below 60 degrees!).

14. Keizai Kikaku Chō (Economic Planning Agency), *Shōwa 54nenban Kokumin Seikatsu Hakusho (White Paper on People's Living)* (Tokyo: Okurasho Insatsu-kyoku, 1979), pp. 43–46.

15. PHP Kenkyūjo, ed., *Suji demiru Nihon no Ayumi (Change of Japan Shown by Figures)* (Kyoto: PHP Kenkyujo, 1980), p. 344.

16. As for advisory commissions in Japan, see Ezra F. Vogel, ed., *Modern Japanese Organization and Decision-making* (Berkeley: University of California Press, 1975), pp. 43–44, 267–72.

17. When a think tank conducted a survey on the energy policy of six political parties in Japan concerning the basic long-range perspective on the supply and consumption of energy in Japan, LDP answered that basically it agrees with government's perspective, leaving that part of the questionnaire blank. See Shakai Keizai Kokumin Kaigi (Socioeconomic Council of Japan), *Shindankai no Enerugii Seisaku (Energy Policies in a New Stage)* (Tokyo: SECJ, 1978), p. 39.

18. Toyoaki Ikuta, "Energy Problem of Japan," mimeo dated 25 May 1980, pp. 5–6.

19. The share of oil imported to Japan through the eight major oil companies has dropped from 65.8 percent in 1978 to 51.9 percent in 1979. Sekiyubu, Shigen Enerugii Chō (Oil Bureau, Agency of Resources and Energy), *Sekiyu Kankei Shiryo*

(*Data Related to Oil*), 16 July 1980. Replacement came from increase of DD or GG oil, which is handled mainly by big Japanese trading companies. See Keitaro Hasegawa, *Nihon wa Sekiyu ni Kateru! (Japan Can Beat the Oil)* (Tokyo: KK Best Sellers, 1980), pp. 65–68. Hasewaga argues that supply through Japanese trading companies will be more stable than the supply through major oil companies, which have cut the supply to Japan on short notice several times since 1973. However, Ikuta's argument suggests that this shifting of oil importers from major oil companies to Japanese trading companies can cause a rise in the import price of oil.

20. *Shinano Mainichi Shimbun,* 28 December 1979.

21. *Energy in Japan,* Quarterly Report, no. 54, September 1981; MITI, *Energy in Japan: Facts and Figures,* November 1981; Institute of Energy Economics, *Introduction to Energy and Japan,* October 1981; Shigen Enerugii Chō, Tsusansho (Agency for Resources and Energy, MITI), *55-nendo no Sekiyu Jukyū Mitōshi: Jisseki (Estimation of the Supply and Demand of Oil in the 1980 Fiscal Year: Actual Record),* as reported in *Yomiuri Shimbun,* 31 January 1981.

22. An English edition is available as Economic Planning Agency, *New Economic and Social Seven-Year Plan* (Tokyo: Printing Bureau, Ministry of Finance, 1979).

23. It took the form of a report, by the Demand and Supply Section, Comprehensive Investigation Council on Energy, which is a deliberation council in the Ministry of International Trade and Industry.

24. Ikuta, *Energy Problem of Japan,* pp. 10–11.

25. Ibid., pp. 18–19.

26. Kōreisha Koyō Kaihatsu Kyōkai (Association for Development of Employment for the Aged), ed., *Kōreikashakai eno Chōsen (A Challenge to Aging Society)* (Tokyo: Koreisha Koyo Kaihatsu Kyokai, 1979), p. 68.

27. Economic Planning Agency, *New Economic and Social Seven-Year Plan,* p. 13.

28. Ibid., p. 7.

29. PR Section, "Public Opinion Survey on People's Living."

30. Kōgai Kenkyu Kai (Research Group on Pollution), ed., *Zenkoku Shimin Undodantai Meibo (List of Citizens' Movements Groups in Japan)* (Tokyo: Kogai Mondai Kenkyukai, 1979).

31. *Nihon Keizai Shimbun,* 4 and 5 July 1980.

32. Ibid.

33. Ryukichi Imai, *Kaku Shinjidai to Enerugii Senryaku (New Era of Nuclear and Energy Strategy)* (Tokyo: Denryoku Shimpō Sha, 1980), p. 206.

34. See Hiroyuki Saitō, "Chiho jichitai to shin enerugii" ("Local Bodies and New Energy") *Sekiyu to Sekiyukagaku* 24:7 (June 1980). Also see *Todofuken Tenbo,* February–March 1980.

35. PR Section, "Public Opinion Survey on People's Living."

36. *Asahi Shimbun,* 13 August 1980.

37. In April 1980, Amory Lovins, the author of *Soft Energy Paths,* visited Japan and had opportunities to talk with Japanese energy specialists. On one such occasion, Toyoaki Ikuta argued that Japan's energy policy should be, and actually is, oriented to neither a hard path nor a soft path. It seeks both paths, and in that sense it should be called a "flexible energy path" (*Asahi Shimbun,* 5 April 1980). Actually, "Diversification of electric supply sources" has become one of the key targets of electricity policy in Japan.

38. Robert Stobaugh and Daniel Yergin, eds., *Energy Future: Report of the Energy Project at the Harvard Business School* (New York: Ballantine, 1980), chap. 5.

39. Sakaiya Taichi, *Yudan.*

40. Religious homogeneity needs some comments. First, it means that the proportion of believers in some established religions is low, about one third of the popula-

tion. Two thirds are believers in no particular religion. Second, religiosity — respect for the religious mind — is high. In the 1978 survey, 74 percent of the respondents answered that they respect the religious mind. In other words, the majority of the Japanese do not care about the content of religious doctrine, but they respect such values as devotion, self-discipline, brotherhood, harmony, etc., in any religion. These features give Japanese society a high degree of religious tolerance and a high degree of moral integration not based on any particular religion.

Chapter 7 Europe's Farewell to Full Employment?

1. G. F. Ray and G. M. Walsh, "The European Energy Outlook to 1985," *National Institute Economic Review,* no. 86, November 1978; and G. Ray and C. Robinson, *The European Energy Outlook to 1990* (London: Staniland Hall, 1982).

2. If the Danish experiments aimed at harnessing the wind materialize, a small contribution may come from that source, but it is unlikely to be significant in this decade. Similar is the case of attempts in other countries in the area of energy types that are considered noncommercial (or nonconventional) as yet.

3. The growth of electricity consumption in Western Europe has been much slower since 1973 than before. The annual average growth rates (percents) were: 1950–60, 8.2; 1960–70, 7.3; 1970–73, 7.1; 1973–78, 3.1. From United Nations, *World Energy Supplies.*

4. For a renewable energy scenario for the U.K., see National Centre for Alternative Technology, paper delivered at an energy seminar of the Social Science Research Council, London, 1978; Gerald Leach et al., *A Low Energy Strategy for the United Kingdom* (London: Science Reviews, 1979).

5. *OECD Economic Outlook,* December 1979, p. 37.

6. Ibid. Around June 1979, the average OPEC price was raised by about $2 a barrel.

7. See, for example, the years 1974 and 1977 in Table 7.5.

8. *In Favor of an Energy-Efficient Society.* Energy Series no. 4 (Brussels: European Community, 1980).

9. Growing shortages of wood resources disturbed the Athenian economy in the fifth century B.C., but the first well-documented fuel crisis hit England in the sixteenth century. Growing industrial activity — shipbuilding, construction, iron-making (for which wood was used in the form of charcoal), glass, and other manufacturing — was a heavy drain on limited forest resources. The primitive coal-getting operations produced only small quantities and coal's use was objected to, causing "clergy and nobility to complain of danger of contagion from the stench of burning coal." The situation came to a head in the winter of 1542–43, a critical year, when the London city fathers imposed a levy on all citizens to finance the holding of a safety stock of "seacoal." This plan came to nothing much. Coal-burning spread, but, nevertheless, in 1615 a strict act wholly prohibited the use of wood as a domestic fuel. Wood continued to be used in shipbuilding and in industry, where the greatest demand came from charcoal-based ironmaking. The outcome was a wandering iron industry, moving from one deforested area to the next forest, eventually harvesting most of the British forests. The technological solution that saved the country from the final impasse came from two sides at about the same time in the early eighteenth century: the replacement of charcoal by coking coal (Abraham Darby, 1709) and the introduction of steam engines for the elimination of water from the mines (Savery, 1698, and Newcomen, 1705), after which coal production rose rapidly. Despite large forests, the eastern states of the United

States also struggled with local shortages of fuelwood; this is clear from the writings of Benjamin Franklin, who even invented, among many other things, a new type of fireplace with very low fuel consumption. For more details, see G. F. Ray, "Energy Economics: A Random Walk in History," *Energy Economics,* July 1979.
10. This statement is based on the OECD's definition of the savings of all the residents and institutions in the countries listed and does not contradict the fact that, especially in some countries, personal savings alone might have increased. This turn in savings bears a direct relevance to investment. See OECD, *National Accounts of OECD Countries.*
11. An ingenious analytical and detailed description of the possible emergence of an oil shortage on the scale envisaged in Scenario C was given for one country only, France, in a monograph: H. Aujac and J. Rouville, *La France sans Pétrole* (Paris, Calmann-Levy, 1979). Written in 1978, it analyzes the consequences of a very major cutback of OPEC supplies on the French economy (and, in less detail, on other countries) and the severe measures the authorities would have to contemplate.
12. Sir Alec Cairncross, "Farewell to Full Employment," Ellis Hunter Memorial Lecture, University of York, 1979.
13. Wartime shortages led to the nitrogen fixation process, to synthetic rubber and detergents, etc. Technological advance responding to market pressure, or scientific "push" resulted in manmade fibers, in the large variety of synthetic resins and plastic materials, and so forth. More recent examples are synthetic (industrial) diamonds and quartz for electronic use. For details, see G. F. Ray, "The Contribution of Science and Technology to the Supply of Industrial Materials." *National Institute Economic Review,* no. 92, May 1980.
14. Asa Briggs, *Technology and Economic Development* (Penguin-Pelican, London-Harmondsworth, 1965), p. 16.

Chapter 8   The Social Contract under Stress in Western Europe

1. European Community, *Investment and Jobs in an Energy-Efficient Society* (Brussels: May 1981).
2. European Community, *In Favor of an Energy-Efficient Society,* Energy Series (Brussels: 1979), pp. 27–38.
3. Ibid., p. 42.
4. European Community, *Investment and Jobs,* p. 72.
5. Poll results from *Corporate Strategy Guide, 1979–80* (United Kingdom); Institut für Demoskopic Allensbock et Institut Sample: SOFRES and IFOP.
6. Energy Commission of the Eighth French Plan, "Energy and Raw Materials," Report, French Documentation, 1980.
7. Report 5, the Energy Commission, Utility Plan, General Documentation, p. 80.
8. In November 1978, in the same countries of the original European Community, the proportion of households owning washing machines was between 72 and 87 percent, between 63 and 76 percent for automobiles, and 87 to 97 percent for refrigerators. From *Economic Tables for France.*
9. Bernard Cathelat, *The Lifestyles of the French, 1978–1988* (Paris: Editions Internationales Alain Stanké, 1977).
10. COFREMCA (French firm of Applied Psychology and Sociology); also see OECD, *Interfutures: Facing the Future* (Paris: OECD, 1979), pp. 97–112.
11. For instance, unpublished surveys by the French polling firm SOFRES.
12. Leon N. Lindberg, *The Energy Syndrome: Comparing National Responses to the Energy Crisis* (Lexington, Mass.: Lexington Books, 1977).

13. For a comparative study of the interaction of governments, electricity-generating authorities, and the public in France, Sweden, and the United States, see I. C. Bupp and J. C. Derrau, *The Failed Promise of Nuclear Power: The Story of Light Water* (New York: Harper Colophon Books, 1981).

14. European Community, *The European Community and the Energy Problem* (Brussels: 1980).

15. H. Aujac and J. de Rouville, *France Without Oil* (Paris: Calman-Levy, 1979).

16. Ibid.

17. European Community, *A Blueprint for Europe: Report of a Study Group* (Brussels: 1977).

18. Ibid.

Chapter 9   The Global Poor

1. World Bank, *World Development Report, 1980* (Washington: June 1980).

2. World Bank, *World Development Report, 1981* (Washington: June 1981).

3. J. K. Paul, ed., *Ethyl Alcohol Production and Use as a Motor Fuel* (New Jersey: Noyes Data Corporation, 1979); World Bank, unpublished report.

4. See Vaclac Smil and William E. Knowland, *Energy in the Developing World: The Real Energy Crisis* (New York: Oxford University Press, 1980), pp. 227–39. Also on the nuclear option, National Center for Analysis of Energy Systems, *Energy Needs, Uses and Resources in Developing Countries* (Brookhaven National Laboratory, 1978), and D. G. Fallen Bailey and T. A. Byer, *Energy Options and Policy Issues in Developing Countries (Washington: World Bank, 1979).*

5. For a useful survey of the potential and technological options involved in renewable resource use in LDCs, see Philip F. Palmedo et al., *The Contribution of Renewable Resources and Energy Conservation as Alternatives to Imported Oil in Developing Countries* (Port Jefferson, N.Y.: Energy/Development International, 1980). For two case studies, see *Renewable Energy in Egypt: An Analysis of Options* (McLean, Va.: The Mitre Corporation, 1980), and Ranvir K. Trehan et al., eds., *Potential for Energy Farms in the Dominican Republic: A Preliminary Analysis* (McLean, Va.: The Mitre Corporation, 1980).

6. John S. Spears, "Wood as an Energy Source: The Situation in the Developing World," paper presented to the 103rd annual meeting of the American Forestry Association; Erik Eckholm, *Planting for the Future: Forestry for Human Needs*, Worldwatch Paper 26 (Washington, D.C.: Worldwatch Institute, 1979).

7. John Gladhill and Calvin Warwick, *Low Head Hydro: An Examination of an Alternative Energy Source* (Boise: Idaho Water Resources Institute, 1979); "The Economics of Small Hydro," *Water and Dam Construction*, January 1979; Sharat K. Tewani, "Economics of Wind Energy Use for Irrigation in India," *Science*, 3 November 1978. Henry R. Bungay, *Energy: The Biomass Options* (New York: John Wiley, 1981).

8. *Oil and Gas Journal*, 22 February 1980, 28 December 1981.

9. World Bank, *Energy in the Developing Countries*.

10. Paula York, ed., *International Petroleum Review* (New York: McGraw-Hill, 1980).

11. It is interesting to note that foreign oil companies have been more than willing to explore the possibilities of operating in LDCs through national subsidiaries functioning in a minority share concession arrangement or through the national companies directly. Similarly, "barter-investment arrangements" have been floated, where diversified Western oil companies would offer technologies and management know-how in non-oil activities (agriculture, consumer good manufacturing, construction, transportation systems, etc.) in exchange for access to oil

exploration and production activities. An interesting experiment is the creation of a new internationally financed (with Swiss, Kuwaiti, Swedish, and Canadian capital) firm, International Energy Development Corporation, led by former PetroCanada head Maurice Strong. It basically aims at exploiting opportunities commercially in marginal potential LDCs, where other private sector interest in exploration and production activities is lacking. IEDC has to survive in a competitive market on the basis of viable economic returns; hence it will not only represent an important test case for private sector involvement in such activities but it will also test the viability of project financing and lending in marginal energy projects.

For a discussion of the future market role of national oil companies, see *OPEC and Future Energy Markets* (London: The McMillan Press, 1980) and Fadhil J. Al-Chalabi, *OPEC and the International Oil Industry: A Changing Structure* (London: Oxford University Press, 1980).

12. "West Africa: A Special Report," *Oil and Gas Journal,* January 1981.

13. T. Hoffman and B. Johnson, "Bypassing Oil and the Atom: The Politics of Oil and World Energy," *Energy Policy,* June 1979.

14. University of Alabama, *A Survey of Photovoltaic Systems,* study prepared for the U.S. Department of Energy, August 1979.

15. For an excellent comprehensive account of planning and energy project management and financing constraints, see Owen T. Carroll, ed., *Energy Planning for the Developing Nations* (Stony Brook, N.Y.: The Institute for Energy Research, 1979).

16. For a review of the analysis of aggregate and sectoral energy and economic development models, see J. G. Leigh et al., *Energy and Development: Extended Analysis and Implications* (McLean, Va.: The Mitre Corporation, 1980). Regarding energy assessment strategies and methodologies, see Office of Country Energy Assessments, U.S. Department of Energy, "Cooperative Energy Assessments with Developing and Industrializing Countries" (October 1980).

17. See Herbert E. Hansen, "OPEC's Role in a Global Energy and Development Conference," *Journal of Energy and Development,* 5 (Spring 1980), pp. 182–93.

18. *New York Times,* 15 February 1981.

Chapter 10   Burdens of Debt and the New Protectionism

1. See Chapter 2.

2. See Table 2.8. The centrally planned economies are entered into the tables only with their net trade with the rest of the world. With OPEC demand for oil increasing by 6 percent per year, the total oil availability of non-OPEC, non-Comecon countries is projected to expand by .5 percent per year until 1990 under the Upper Bound scenario (and to decline thereafter), and to decline by 1.5 percent over the period until 2000 under the Lower Bound scenario. See Table 2.2.

3. Annual increase in energy supply: 2.6 percent and .8 percent, respectively. Plus annual increase in energy efficiency of 1½ to 2 percent. For the reasons mentioned by Stobaugh, it appears more reasonable to work with increments in energy efficiency rather than with alternative energy coefficients.

4. United Nations, *World Economic Survey, 1979–80* (New York: 1980), E/1980/38–ST/ESA/106, p. 14.

5. Under conditions of low output growth, the prices of raw materials are likely to be depressed, whereas the prices of more sophisticated manufactured products may still be rising.

6. World Bank, *World Development Report, 1981* (Washington: August 1981)

(hereafter cited as *World Development Report*), pp. 3ff, UN, *World Economic Survey*, pp. 69ff. The UN is projecting, for the period 1981–85, a GNP growth rate for the world of 3.5 percent per year, for developed market economies of 3.1 percent, and for developing market economies of 5.1 percent. Ibid., p. 71.

7. World Bank, *World Development Report* (Wahington: August 1980), p. 6.

8. According to the formula formerly under consideration, the relative (or "real") price of oil would be raised in line with real GNP growth in the OECD area.

9. UN, *World Economic Survey*, p. 14.

10. Twelve important OECD countries. See OECD, *Economic Outlook*, no. 27, July 1980, p. 21.

11. The seven most important OECD countries.

12. For an account of the European experience, see Roland Vaubel, "Monetary Divergencies and Exchange-Rate Changes in the European Community: The 1970s," paper prepared for the Conference on European Monetary Union, University of Salford, 15–17 September 1980.

13. The predictive value of the figures is impaired by a statistical discrepancy of $18 billion.

14. See GATT, *International Trade 1979/80* (Geneva: 1980), p. 5.

15. Ibid., p. 50.

16. Spain, Portugal, Greece, Yugoslavia, Brazil, Mexico, Hong Kong, Korea, Taiwan, and Singapore.

17. For an account of the performance of this group of countries, see OECD, *The Impact of the Newly Industrialising Countries on Production and Trade in Manufactures* (Paris: 1979), and Louis Turner, et al., *Living with the Newly Industrialising Countries* (London: The Royal Institute of International Affairs, 1980). The figures are quoted from the OECD source, pp. 18f.

18. Cf. OECD, *The Impact of the Newly Industrializing Countries,* p. 35; UN, *World Economic Survey*, p. 94.

19. It should again be emphasized that even the high case scenario allows only for an annual growth of per capita income of 1.8 percent in oil-importing low-income LDCs (see Table 10.1).

20. For instance, by the Cambridge Economic Policy Group, in *Cambridge Economic Policy Review,* University of Cambridge, Department of Applied Economics, nos. 1–5, February 1975–April 1979; F. Cripps and W. Godley, "Control of Imports as a Means to Full Employment and the Expansion of World Trade: The UK's Case," *Cambridge Journal of Economics,* September 1978.

21. See *World Development Report,* 1980, p. 7.

22. Cf. Morgan Guaranty Trust Company, *World Financial Markets,* December 1979, particularly the (most plausible) Scenarios II and III.

23. Of course, this need not be the end of the story; to the extent that foreign borrowing will be used to pay for *additional* goods and services, government and/or private corporate demand will fill the deflationary gap.

24. See Karl Otto Pöhl, "The Multiple-Currency Reserve System," *Euromoney,* October 1980, pp. 43ff. Also: International Monetary Fund, *Annual Report, 1980* (Washington: 1980), pp. 61ff. Foreign exchange reserves excluded gold, SDRs, and IMF reserve positions. The six reserve centers are the United States, Germany, France, the United Kingdom, Switzerland, and Japan.

25. Approaches to cooperative arrangements were already discussed by the Committee of Twenty in their deliberations on the reform of the Bretton Woods system. See IMF, *International Monetary Reform. Documents of the Committee of Twenty* (Washington: 1974).

26. The United States, Canada, Japan, Belgium, France, Germany, Italy, Nether-

lands, Sweden, and the United Kingdom. Switzerland is associated to the group.
27. The total does not add to 100 percent because of reserve accumulation and errors and omissions.
28. OECD, *External Debt Statistics for Developing Countries: Latest Trends* (Paris: 1980); quarterly banking statistics of the Bank for International Settlements.
29. World Bank, *Annual Report, 1980* (Washington: 1980), pp. 19f.
30. Informal rescheduling means the granting of new loans with the implicit purpose of refinancing expiring loans.
31. Cf. Rodney H. Mills, "US Banks are Losing Their Share of the Market," *Euromoney*, February 1980, pp. 50ff.
32. Cf. the remarks by Robert S. McNamara in his address to the Board of Governors, Washington, D.C., September 30, 1980, p. 14.
33. Mohammed M. Abudashi, in an interview with the German business magazine *Wirtschaftswoche*, no. 36, 3 September 1979.
34. Richard N. Cooper, U.S. Undersecretary of State for Economic Affairs, in a speech at Atlanta on 27 October 1980, as recorded in *Wireless Bulletin from Washington*, no. 203, 28 October 1980.
35. World Bank, *Annual Report, 1980*, pp. 68–70.
36. See *IMF Survey*, 13 October 1980, p. 310.
37. See *IMF Survey*, various issues in 1980 and early 1981.
38. See *IMF Survey*, 6 April 1981, pp. 97ff.
39. Communiqué of the Intergovernmental Group of 24 on International Affairs, 27 September 1980, reprinted in *IMF Survey*, 13 October 1980.
40. J. de Larosière, managing director of the IMF, in an address, "Recycling Needs and the Capital Markets," reproduced in *IMF Survey*, 10 November 1980, p. 351.
41. Estimates by Bankers Trust Company. See Susan Bluff, "OPEC's $350 Billion Balance Sheet," *Euromoney*, September 1980, and Bluff, "The OPEC Surplus," Bankers Trust.
42. For the following, cf. Ahmed Abdullatif, "A Strategy for Investing the OPEC Surplus," *Euromoney*, August 1980, pp. 23f.
43. Ibid.
44. Ibid., p. 24.
45. Ibid.
46. In view of the projected increases of the "real" oil price, this comparison tends to come out in favor of oil.
47. Abdullatif, "A Strategy for Investing," p. 24.
48. Ibid.
49. Of course, the need to undertake the "real" adjustments to the higher oil price is unaffected.
50. See Morgan Guaranty Trust Company, *World Financial Markets*, December 1979; Bluff, "OPEC Surplus."
51. For an ambitious proposal to harmonize these interests, see Armin Gutowski, "Reducing the Oil Hazards: A Proposal for a Contractual Solution," *Intereconomics*, September–October 1980, p. 213. For a discussion of the proposal, see the criticism by Otto Schlecht and the reply by Armin Gutowski in *Wirtschaftsdienst*, December 1980, pp. 602ff. "The limitations [on financial investments] imposed by the size of funds and the desire to avoid disturbing markets . . . could be considerably eased by the creation of the SDR Substitution Account." Abdul Aziz Alquraishi, governor of the Saudi Arabian Monetary Agency, in an interview in *International Investor*, August 1980. (Quoted from BIS *Press Review*, no. 170, 2 September 1980.)

52. The idea was first launched by Wilfried Guth, speaker of Deutsche Bank, at the International Monetary Conference held in New Orleans in the first week of June 1980. It was taken up in early 1981 by Rinaldo Ossola, the president of Banco di Napoli and former deputy governor of the Central Bank of Italy, in a modified version. Ossola, in an article in the *Banca del Lavoro Quarterly Review* (September 1980, pp. 291ff.), is proposing swapping arrangements between IMF and BIS on the one side and between the twenty most important international banks on the other.

Chapter 11   Cohesion and Disruption in the Western Alliance

1. *Le Figaro* (Paris), 31 May 1979. Also *Le Matin,* 1 June 1979.
2. On the role of "miracles," see Chapter 8 in Robert Stobaugh and Daniel Yergin, eds., *Energy Future: Report of the Energy Project at the Harvard Business School* (New York: Ballantine, 1980).
3. Robert J. Lieber, *Oil and the Middle East War: Europe in the Energy Crisis* (Cambridge, Mass.: Harvard Center for International Affairs, 1976), pp. 13ff.
4. See Robert Lieber, "America and Europe in the World Energy Crisis," *International Affairs* (London), October 1979, pp. 533–34.
5. Calculated from figures in *OECD Economic Outlook,* July 1979, p. 57.
6. My percentage calculations, based on OECD data, particularly from *OECD Economic Outlook,* July 1979, Table 32, p. 63, and Table 61, p. 140. Figures exclude marine bunkers.
7. This information is based on interviews by the author at the International Energy Agency and OECD, Paris, 19–21 December 1979, and 3–7 July 1980. Also see IEA. *Energy Policies and Programmes of IEA Countries: 1979 Review,* p. 12.
8. For a more detailed discussion of the incident, see Robert J. Lieber, "Energy, Economics and Security in Alliance Perspective," *International Security* (Spring 1980), pp. 139–63.
9. IEA Communiqué, Meeting of Governing Board at Ministerial Level, IEA/Press (79) 28 (Paris), 10 December 1979.
10. Communication to the author, 10 November 1980.
11. The reactor order was signed in November 1975. In April 1979, just before its completion and export from France, the reactor core was sabotaged — possibly by Israelis, but not inconceivably by (or with the collusion of) the French secret services. The ensuing delay of nearly two years created an opportunity for the French to provide the reactor with a newly developed 7 percent enriched "caramel" fuel, with far less proliferation potential. Iraq refused this change and the French government delivered the reactor on the basis of the original agreement. An understanding that deliveries of Iraqi petroleum to France would increase from 20 million tons in 1978 to 30 million tons in 1980 cannot have been absent from French calculations. French officials, however, stressed the safeguards under which the enriched uranium was supplied. On 7 June 1981, Israeli jets bombed and destroyed the facility.
12. Jean Klein, "France and the Arms Trade," in *The Gun Merchants: Politics and Policies of the Major Arms Suppliers,* ed. C. Cannizzo (Elmsford, N.Y.: Pergamon Press, 1980).
13. In 1974–75, there was a dispute within the IEA over a U.S. proposed floor price for oil (to ensure the viability of investments in new energy sources) at the — now quaint — price of $13 per barrel. This was followed by a desultory debate within the European Community on Britain's proposals for a $7 floor price (in current,

not constant, dollars) to protect its investments in costly North Sea oil development.

14. Øystein Noreng, "The European Energy Exporters," in *The European Energy Transition Away from Oil,* ed. Gordon Goodman et al. (New York: Academic Press, 1981), pp. 255–70.

15. *Washington Post,* 28 March 1979.

16. Japanese crude oil stocks (including natural gas liquids and feedstocks) rose from 38.268 million metric tons at the end of 1978 to 42.756 a year later. In the same period, German stocks rose from 17.014 to 19.069. OECD, IEA. *Quarterly Oil Statistics: Fourth Quarter 1979* (Paris: 1980), no. 1, pp. 14, 98.

17. The criticism has focused on Henry Kissinger. Statements to this effect by George Ball, William Simon, and James Akins, former U.S. ambassador to Saudi Arabia, were televised on "Sixty Minutes," on 4 May 1980. It has also been a recurrent theme among the French. See Lieber, *Oil and the Middle East War,* p. 29 and fn. 88.

18. Quoted in *Europe* (Brussels: Agence Internationale d'Information pour la Press, no. 2649, 29 March 1979), and statement at IEA ministerial meeting, *Europe,* no. 2683, 21–22 May 1979.

19. *OECD Economic Outlook,* no. 26, December 1979, p. 12.

20. Benjamin Cohen treats this problem lucidly, with an emphasis on monetary policy, in "Europe's Money, America's Problem," *Foreign Policy,* no. 35, Summer 1979, p. 41.

21. IEA Communiqué. Meeting of Governing Board at Ministerial Level, 9 December 1980; IEA/Press (80) 20. Paris, 9 December 1980.

22. In 1979, Europe obtained approximately 63 percent of its oil from the Arabian/Persian Gulf area.

23. From Energy balances of the OECD countries, as compiled by the IEA, 4 December 1980. The calculation is based on tons of oil equivalent per $1000 of real U.S. GNP at 1975 prices and exchange rates.

24. For French responses to the energy problem, see Robert J. Lieber, "Energy Policies of the Fifth Republic: Autonomy versus Constraint," in *The Impact of the Fifth Republic on France,* eds. William G. Andrews and Stanley Hoffmann (New York: SUNY Press, 1981).

25. Figures from Alan Marsh, "Environmental Issues in Contemporary European Politics," in Goodman, *The European Energy Transition,* p. 144.

26. Based on revised proposals of the Ministry of Industry, presented to the French Council of Ministers. See *Le Monde,* 3–4 April 1980.

27. Calculated from *International Energy Statistical Review,* National Foreign Assessment Center, U.S., CIA, GI IESR 82-001, 26 January 1982, p. 20.

27. Figures for oil import percentages are from an address by Michel Pecqueur, chairman of the French Atomic Energy Commission, Washington, 17 November 1980. French Embassy Press and Information Service, 80/99. The figure of 67 percent import dependence is from *Les chiffres cles de l'energie* (Paris: Ministere de l'Industrie, 1979), p. 31. The 1990 projection is in the 1980 edition, p. 31.

29. France brought 24.9 million tons of crude oil from Iraq in the year ending 29 February 1980. This represented 20.2 percent of total French petroleum imports, compared to 13.8 percent in 1973. By contrast, Saudi Arabia supplied 34.5 percent (22.4 percent in 1973). Iraq also ordered 8 billion francs worth of arms from France in 1977 and 1978. From *Les chiffres cles de l'energie,* 1979, pp. 60–61, and *Le Monde,* 30 April 1980.

30. *OECD Economic Outlook,* December 1981, table 51, p. 115.

31. *The Financial Times* (London), 4 February 1982. The Soviet gas price was believed to be $4.65 per hcm.

32. Proposed in January 1980 by President Carter, it announced: "Let our position be absolutely clear: An attempt by any outside force to gain control of the Persian Gulf region will be regarded as an assault on the vital interests of the United States of America, and such an assault will be repelled by any means necessary, including military force." U.S. Department of State, Bureau of Public Affairs, *President Carter, State of the Union Address,* Current Policy no. 132, 23 January 1980.

33. For planning for responses to energy emergencies, see David A. Deese and Joseph S. Nye, eds., *Energy and Security* (Cambridge, Mass.: Ballinger, 1981).

34. The Independent Commission on International Development Issues under the Chairmanship of Willy Brandt, *North-South, A Program for Survival* (Cambridge, Mass.: MIT Press, 1980).

Chapter 12   Energy and the Power of Nations

1. The myth of homogeneity is also attached to the idea of power. A distinction between putative and actualized power is made in Klaus Knorr, *The Power of Nations: The Political Economy of International Relations* (New York: Basic Books, 1975), pp. 9–15.

2. Frederick Soddy, *Matter and Energy* (1912), quoted in Gerald Foley, *The Energy Question* (London: 1976), p. 9.

3. A case is that of France, which has conserved its domestic uranium resources while drawing on uranium from former colonies in Africa, in contrast to, say, Argentina, which has based its nuclear power program on domestic uranium alone. Similarly, Britain (in contrast to the Netherlands) has hitherto chosen to import a certain amount of liquefied natural gas as an alternative to depleting its indigenous gas resources at a higher rate.

4. For the particular relation between oil import dependence and supply security in the United States, see Joseph S. Nye, "Energy and Security" in David A. Deese and Joseph S. Nye, eds., *Energy and Security* (Cambridge, Mass.: Ballinger, 1981), pp. 12–13.

5. The effect is named for Peter Forsyth and John Kay, who first formulated this thesis; see *The Sunday Times* (London), 4 January 1981.

6. A somewhat similar view is in Joseph S. Nye, "Energy Nightmares," *Foreign Policy,* 40 (Fall 1980), pp. 132–54.

7. A summary discussion of the main conditions for exerting "producer power" effectively, as they apply to fuels and other natural resources, is in Ian Smart, "Uniqueness and Generality," in *The Oil Crisis,* ed. Raymond Vernon (New York: Norton, 1976), pp. 264–75.

8. The role of these stocks is explored in Edward M. Krapels, *Oil Crisis Management: Strategic Stockpiling for International Security* (Baltimore: Johns Hopkins University Press, 1980).

9. A comparable point about imports in general is made in Sam H. Schurr et al., *Energy in America's Future: The Choices before Us* (Baltimore: Johns Hopkins University Press, 1979), p. 428.

10. *Energy Conservation in Industry in IEA Countries* (Paris: IEA, 1979), p. 45, Table 22. No similarly comparable statistics seem to have been reported recently.

11. See, for example, Dorothy Nelkin, *Technological Decision and Democracy* (London: 1977), and the report by the OECD Committee for Scientific and Tech-

nological Policy, *Technology on Trial* (Paris: OECD, 1979). For international variation of public attitudes to nuclear energy policy, see also Mans Lonnroth and William Walker, *The Viability of the Civil Nuclear Industry,* International Consultative Group on Nuclear Energy (New York: 1979).

12. The earlier trend was not entirely unnoticed. One 1965 assessment in which I had some part, basing itself on changes only from 1959, suggested that American oil imports could reach 265 mt in 1980 (as against the actual figure of about 340 mt); see *Sources of Conflict in the Middle East* (London: Institute for Strategic Studies, 1966), p. 5, Adelphi Paper no. 26.

13. The parallel point, that American import dependence may tend in favor of stability to the extent that exporters are reluctant to antagonize a major power, is made in Schurr, *Energy in America's Future*, p. 429.

14. These figures are based on data from the *1980 World Bank Atlas* (Washington: 1981), and the *BP Statistical Review of the World Oil Industry, 1980* (London: 1981). In the case of Soviet GNP, however, the World Bank estimate, reflecting a conversion to U.S. dollars at the official exchange rate, has been modified to take account of relative purchasing power by applying the ratio adopted in the U.S. Central Intelligence Agency's *Handbook of Economic Statistics 1979* (Washington: 1979), Table 7.

15. For a balanced view of those difficulties, see Jeremy Russell, *Energy as a Factor in Soviet Foreign Policy* (London: 1976), usefully brought up to date by Marshall I. Goldman, "The Role of Communist Countries," in Deese and Nye, *Energy and Security,* pp. 111-30.

16. *1980 World Bank Atlas* (Washington: 1981); *BP Statistical Review of the World Oil Industry, 1980* (London: 1981).

Index

Contributors

DANIEL YERGIN is chairman of the International Energy Security Seminar at Harvard University and a lecturer at the Energy and Environmental Policy Center at Harvard's Kennedy School of Government. He is an associate of the Atlantic Institute for International Affairs and a former lecturer at the Harvard Business School. He is also a Fellow of the German Marshall Fund of the United States. He has written *Shattered Peace: The Origins of the Cold War,* and is a co-author of *Energy Future: Report of the Energy Project at the Harvard Business School.*

MARTIN HILLENBRAND is director general of the Atlantic Institute for International Affairs. From 1939 to 1976 he served in the American Foreign Service. He was formerly the U.S. ambassador to the Federal Republic of Germany and to Hungary as well as assistant secretary of state for European affairs. He is the co-author and editor of *The Future of Berlin* and has written *Power and Morals* and *Zwischen Politik und Ethik.*

* * *

ALTHEA DUERSTEN is a senior economist at the World Bank, specializing in the financing of oil and gas projects in Africa and Latin America.

ROBERT DOHNER is an assistant professor of economics at the Fletcher School of Law and Diplomacy of Tufts University. A former staff economist of the President's Council of Economic Advisers, he specializes in international trade and in the macroeconomic aspects of energy price changes.

ROBERT J. LIEBER is a professor of government at Georgetown University, specializing in energy security and U.S.-European relations. Among his works are *British Politics and European Unity* and *Oil and the Middle East War: Europe in the Energy Crisis.* He is also the co-editor of both *Eagle Entangled: U.S. Foreign Policy in a Complex World* and *Eagle Re-*

surgent: The Foreign Policy of the Reagan Administration. He is completing a book about the effects of the energy problem on the Western alliance.

TERUYASU MURAKAMI is a senior consultant in the Energy Studies Department of Nomura Research Institute in Japan. The co-author of *Business in Japan,* he is currently working on a project concerning the maritime security of Japanese oil supplies, with particular emphasis on the Strait of Hormuz and the Strait of Malacca.

G. F. RAY is a senior research fellow at the National Institute of Economic and Social Research in London and, since 1975, a professor at the University of Surrey. He is president of the Association d'Instituts Européens du Conjoncture Economique. Among the many books he has written or co-authored are *The Diffusion of New Industrial Processes: An International Comparison, Western Europe and the Energy Crisis,* and *The European Energy Market: Outlook to 1990.*

JEAN SAINT-GEOURS is the chairman of the consulting firm SEMA-METRA and an adviser to the prime minister of France. Among many government positions, he was head of the Economic and Finance Studies Department and the Economic Forecast and Planning Department of the French Ministry of Finance. Subsequently, he was managing director of Credit Lyonnais. Among his books are *La Politique économique (des principaux pays industriels de l'Occident), Vive la société de consommation,* and *Pouvoir et finance.*

HANS-ECKART SCHARRER is the director of the Department of International Finance and Industrial Countries at the HWWA-Institute für Wirtschaftsforschung in Hamburg. He is the president of the Société Universitaire Européene de Recherches Financières and a member of the board of the Institut für Europäische Politik. He has edited or co-edited *Europäische Wirtschaftspolitik, Die Europäische Gemeinschaft in der Krise,* and *Die neue Weltwirtschaftsordnung.* His current work focuses on the international monetary system and German and Japanese foreign trade.

IAN SMART is an adviser on international energy affairs and chairman of Ian Smart Ltd. Formerly assistant director of the International Institute for Strategic Studies and director of studies at the Royal Institute of International Affairs (Chatham House), he is the author of *Arrangements for the Nuclear Fuel Cycle* and *Multinational Arrangements for the Nuclear Fuel Cycle.* He co-edited *The Political Implications of North Sea Oil and Gas* and is currently writing a book on the politics of the world energy trade.

ROBERT STOBAUGH is a professor of business administration and the director of the Energy Research Project at the Harvard Business School. He spent eighteen years in the oil and petrochemical industries in the United States and abroad. A former president of the Academy of International Business, he is a co-author of *Energy Future: Report of the Energy Project at the Harvard Business School, Money in the Multinational Enterprise: A Study in Financial Policy, Energy: The Next Twenty Years,* and of the forthcoming *International Technology Flows: Choice, Transfer, and Management.*

ARPAD VON LAZAR is a professor of international development and energy at the Fletcher School of Law and Diplomacy at Tufts University. He has written *Latin American Politics* and *Reform and Revolution* and is currently completing a book on the international oil market.

JOJI WATANUKI is a professor of sociology at the Institute of International Relations of Sophia University in Tokyo. He is a co-author of *The Crisis of Democracy* and *Politics in Postwar Japanese Society.* His current projects include a study of political management of the economy in Japan and West Germany.